T0351963

Behavioral and Distributional Effects of Environmental Policy

**A National Bureau
of Economic Research
Conference Report**

Behavioral and Distributional Effects of Environmental Policy

Edited by **Carlo Carraro and Gilbert E. Metcalf**

The University of Chicago Press

Chicago and London

CARLO CARRARO is professor of environmental economics at the
University of Venice and research director of the Fondazione Eni
Enrico Mattei. GILBERT E. METCALF is professor of economics at
Tufts University and a research associate of the National Bureau of
Economic Research.

The University of Chicago Press, Chicago 60637
The University of Chicago Press, Ltd., London
© 2001 by the National Bureau of Economic Research
All rights reserved. Published 2001
Printed in the United States of America
10 09 08 07 06 05 04 03 02 01   1 2 3 4 5
ISBN: 0-226-09481-2 (cloth)

Library of Congress Cataloging-in-Publication Data

Behavioral and distributional effects of environmental policy / edited
by Carlo Carraro and Gilbert E. Metcalf.
    p.   cm.—(A National bureau of economic research conference
report)
    ISBN 0-226-09481-2 (cloth : alk. paper)
    1. Environmental policy—Economic aspects—Congresses.
    I. Carraro, Carlo.   II. Metcalf, Gilbert E.   III. Series.

    GE170 .B44 2001
    363.7—dc21
                                                        00-041799

Contents

Acknowledgments

This volume includes 10 papers that were prepared as part of a joint research project on environmental policy carried out by the Fondazione Eni Enrico Mattei (FEEM) and the National Bureau of Economic Research (NBER). The papers were presented at a conference hosted by FEEM at their headquarters in Milan, Italy, on 11–12 June 1999. We are grateful to Martin Feldstein, president of the NBER, and James Poterba, program director for Public Economics at the NBER, for their constant support and advice over the course of this project. We also thank Domenico Siniscalco, director of the FEEM, for his support. Finally, we acknowledge financial support from the NBER, FEEM, and a grant by the National Science Foundation to the Yale/NBER Program in International Environmental Economics.

Organizing a transoceanic conference requires tremendous logistical support and we were well supported on both sides of the Atlantic. The staff of the NBER Conference Department provided outstanding support in preparing for the conference. Rob Shannon, Brett Maranjian, and Kirsten Foss Davis were always ready with a calm demeanor and ready answer for the most complex question or problem. In Milan, Rita Murelli made our stay very enjoyable and handled numerous logistical issues on short notice with great efficiency. The entire conference went off without a hitch in large part due to the expert assistance from these individuals at NBER and FEEM.

After the conference, Helena Fitz-Patrick of the NBER Publications Department provided consistent support to the authors and editors as we moved these papers and comments through the editorial process and into print.

Introduction

Carlo Carraro and Gilbert E. Metcalf

In recent years, environmental policymakers have adopted a set of instruments quite different from those usually prescribed in environmental policy textbooks. Economists have traditionally encouraged the use of incentive-based instruments in place of command and control regulation. The starting point for a discussion of efficient environmental policy has been the Pigouvian prescription: to set taxes on pollution equal to marginal environmental damages. In recent years, however, economists have come to recognize that the standard Pigouvian prescription needs to be modified in the face of other important economic and political considerations.

The reasons for this modification of the standard Pigouvian prescription can be found in the nature of the environmental problems to be managed. These problems are often characterized by a transnational dimension, by links to other economic issues, and by an interrelationship with several types of economic externalities. All this implies that environmental policy has to be redesigned in order to be effective even in a world where the policymakers may have multiple interrelated targets and an incomplete set of policy instruments.

Some examples may clarify this point. If markets are not perfectly competitive, and if the regulator is unable to restore perfect competitiveness—sometimes because a single regulator does not exist, as in the case of multinationals—then an environmental policy designed to correct the environmental externalities produced in the imperfectly competitive industry cannot be the usual "optimal" tax scheme where the tax rate is equal

Carlo Carraro is professor of environmental economics at the University of Venice and research director of the Fondazione Eni Enrico Mattei. Gilbert E. Metcalf is professor of economics at Tufts University and a research associate of the National Bureau of Economic Research.

to the social marginal damage produced by the industry. In general, the optimal tax is lower than the social marginal damage, because the market externality induced by imperfect competition must be taken into account.[1]

As another example, if the regulator is concerned about unemployment, capital flows and the location of firms, or distributional considerations across different sectors in the economy, then he may set a tax and mandate the use of the resulting tax revenue in such a way as to identify the optimal trade-off among multiple objectives (environment, employment, capital stock, economic structure and growth, etc.).[2]

If the environmental problem is transnational or global, optimal policy design is even more complicated. Trade-offs may be necessary to ensure the adoption of environmental policy tools by all or most countries involved in the management of the environmental problem. Again the standard economic principle that implies the equalization of marginal abatement costs and benefits across countries may not be the "optimal" one.[3]

It is often argued that environmental policy may induce firms to innovate, either by carrying out more and newer research and development (R&D) or by adopting existing environmentally benign technology. Even in this case, environmental policy needs to take into account external effects induced by innovation, market distortions in the R&D markets, free-riding incentives induced by R&D spillovers, possible delays in innovation, and benefits and costs of R&D cooperation.[4]

Finally, actual policymakers often attach considerably more importance to the distributional effects of the policy measures that they adopt than they do to issues of efficiency. As a consequence, environmental policy, given the incompleteness of the policy tools that governments have at their disposal and the presence of multiple externalities, faces the well-known trade-off between efficiency and equity.[5]

The departure of actual environmental policy from "optimal" textbook

1. The design of environmental policy under imperfect competition is analyzed in several recent theoretical contributions. See Carraro, Katsoulacos, and Xepapadeas (1996) and the survey by Carraro (1999b). See also an analysis of monopolies and optimal environmental taxation in Fullerton and Metcalf (1997).

2. The EU proposal of a 50-50 carbon/energy tax to reduce greenhouse gases (GHGs) emissions is an example of a policy instrument that balances environmental effectiveness, distributional burden across sectors, and cross-country equitable burden sharing. Another example is provided by the green tax reforms and by their attempts to provide both an environmental and an economic benefit (double dividend). An interesting discussion of this problem is in Bovenberg (1997) and Fullerton and Metcalf (1998). A theoretical analysis of environmental policy in the presence of multiple targets is provided by Anastasios Xepapadeas (chap. 9 in this volume) and Michael Rauscher (chap. 6 in this volume).

3. The design of policy instruments in a world where economic issues are often global, but decisions are taken by (many) sovereign states, is the core of the analysis carried out in Carraro and Siniscalco (forthcoming).

4. These problems are more fully analyzed in Carraro (1999a). New results are provided by Katsoulacos, Ulph, and Ulph (chap. 10 in this volume).

5. See chapter 2, by Bovenberg and Goulder, and chapter 6, by Rauscher, in this volume.

recommendations may also be induced by political constraints. For example, the failure to use environmental taxation in Europe can be explained by objections to new taxes in countries where the tax burden is already quite high. This has to do with the acceptability of policy measures on which a consensus needs to be found. This is also why voluntary environmental agreements seem to have become a widely adopted policy tool in Europe. Even though this tool may not produce the optimal management of a given environmental problem, it may be the only way to achieve some satisfactory results on emission control in some industries or countries.[6] In other cases, environmental policy is not optimal because it is simpler or less costly in terms of administrative costs to adopt a policy that is built on existing policies or that is implemented through an existing set of procedures or administrative structures.[7]

These remarks suggest some common themes. First, environmental policy has to manage new environmental problems in a world in which there is an insufficient number of appropriate policy tools because of inertia, lack of consensus, or an excessive number of policy targets. Hence, policy analysis inevitably deals with second-best worlds and yields second-best policy outcomes.

In addition, most environmental problems are closely linked to industrial, trade, and financial problems. Hence, any analysis of these problems should not be carried out by specialists in environmental economics only, but by economists who can integrate environmental economics with public finance, industrial economics, and trade theory, among other fields.

These conclusions led us to design a book in which chapters are written by leading economists who are able to integrate environmental analysis with the analysis of other intertwined economic issues. Hence, we invited experts from public finance, industrial organization, and trade theory in order to increase "leakages" and "transplants" across fields. The chapters also analyze economic and policy problems in which distributional issues and multiple market failures inevitably lead to second-best policy analyses.

Two other novel features of this book deserve some comment. First, most chapters adopt a microeconomic, behavioral approach to environmental problems. This is not to say that a macroeconomic analysis of environmental issues, as is presented in most textbooks, is inappropriate. We believe, however, that greater insight into the mechanisms through which environmental policy works to correct externalities can be obtained by looking at the behavioral responses of economic agents to environmental policy initiatives. It is thus possible to single out the economic incentives

6. For example, the Danish and Italian governments have recently implemented a policy scheme to achieve the Kyoto targets in which a modest environmental tax is linked to widespread adoption of industry-specific voluntary agreements.

7. See chapter 2 by Fullerton, Hong, and Metcalf, and chapter 3 by Smulders and Vollebergh in this volume.

provided by environmental policy, their effects on a firm's strategy, and their distribution across sectors.

A second important feature is the focus on empirical microeconomic analysis. Most chapters not only provide important theoretical advances, but also attempt to validate the theoretical analysis through careful empirical research, which is often carried out on newly constructed data sets.[8] Hence, this book can be distinguished from other books because of its ability to integrate theoretical and empirical research, environmental economics and environmental policy, behavioral and distributional concerns, and environmental issues in a wide set of economic analytical tools.

The book is the outcome of a fruitful partnership between the National Bureau of Economic Research (NBER) and the Fondazione Eni Enrico Mattei (FEEM). In order to prepare this book, a conference was held in Milan, Italy, in June 1999, where papers were presented and discussed. Discussants were asked to prepare written discussions, which are also published in this book. These discussions outline directions for future research and can be quite useful to those who might extend and generalize the results presented in the chapters of this book.

The papers published in this volume fall into four broad groups. The first group is concerned with issues that arise in the design and implementation of tax or other market-based instruments. The second group of papers addresses compliance cost issues in environmental policy. The third group of papers addresses environmental policy design when trade and development issues are considered. Finally, the last group of papers takes up the issues of incentives, information, and R&D as they affect optimal policy design. As the categories presented suggest, the papers in this volume span a wide range of topics in the area of environmental policy, in keeping with the overall design we had in mind when planning the volume.

Taxes and Other Economic Instruments

As noted, environmental policy is rarely applied in a first-best world in which pollution is the only externality to be corrected and in which policymakers have all the information necessary to implement standard Pigouvian tax schemes. Moreover, distributional concerns and administrative costs may also affect the decision to adopt a given environmental policy instrument. These issues are discussed in depth in the first part of this volume.

Chapter 1, by Don Fullerton, Inkee Hong, and Gilbert E. Metcalf, entitled "A Tax on Output of the Polluting Industry Is Not a Tax on Pollution: The Importance of Hitting the Target," considers the efficiency implications

---

8. Three good examples are chapter 5 by Becker and Henderson, chapter 4 by Levinson, and chapter 8 by Siniscalco et al. in this volume.

of imprecisely targeted instruments. While the standard undergraduate environmental economics textbook prescribes the use of Pigouvian taxes on emissions, real-world environmental taxes generally tax inputs or outputs that are imperfectly correlated with pollution. Given the practical difficulties in monitoring pollution, it is often administratively less costly to tax inputs or outputs associated with pollution. These goods may already be taxed and are traded in markets, making the imposition and collection of a tax relatively straightforward. The authors calculate optimal tax rates on goods associated with pollution, as well as tax rates on pollution itself under different assumptions about the availability and use of various tax instruments. Among other findings, the authors confirm the result that the optimal tax rate on emissions in the presence of preexisting taxes on other goods or factors is less than the social marginal damages of pollution, the level prescribed by Pigou (1932) in his classic analysis of pollution taxes. The authors also consider the relative gains from changes in either pollution or output taxes and find that the welfare gain from increasing output taxes is roughly half of the welfare gain from a comparable increase in a tax on pollution itself.

Distributional concerns and their impact on the optimal environmental policy are at the heart of chapter 2, by A. Lans Bovenberg and Lawrence H. Goulder, entitled "Neutralizing the Adverse Industry Impacts of CO_2 Abatement Policies: What Does It Cost?" One of the significant obstacles to any major tax reform or new tax initiative is the existence of windfall gains and losses that accrue to various industry sectors. This concern has manifested itself in the construction of a market for tradable permits for SO_2 emissions in the 1990 Clean Air Act Amendments in the United States. During the debate leading up to this new law, an important question was whether firms should be given the permits (grandfathered) or should be sold the permits. Bovenberg and Goulder draw attention to the possibility of an intermediate position in which some of the permits are given and the remainder sold. Given the large potential of any CO_2 policy to generate industry rents (windfall profits), the authors find that only a small fraction of emissions permits need be grandfathered to preserve profits and equity values for a firm. The computable general equilibrium (CGE) simulations that the authors run suggest that the most affected industries (coal and oil and gas) need have no more than 15 percent of their permits grandfathered to preserve profits and equity values.

In the context of an emissions tax, only a small fraction of the emissions need be exempted from taxation. The authors also point out the large difference between preserving a firm's profits and preserving its tax payments. Since a large fraction of the carbon tax burden is shifted forward to consumers, compensating firms for the taxes they owe (the statutory burden) will overcompensate them (relative to the economic burden of the tax). Bovenberg and Goulder carry out a variety of simulations in a CGE model to consider a full range of possible policy plans with their corresponding incidence effects across industries. The key finding—that potential rents are large relative to profits in the absence of regulation and so grandfathering a small fraction of permits suffices to compensate firms for their losses—holds generally in their model.

A frequent obstacle to the introduction of efficient environmental policies is the presence of relevant administrative costs. Chapter 3, by Sjak Smulders and Herman Vollebergh, is entitled "Green Taxes and Administrative Costs: The Case of Carbon Taxation." Smulders and Vollebergh explore the trade-off between efficiency and administrative costs in the design of environmental tax instruments with a particular focus on carbon taxes in Europe. As emphasized by Fullerton, Hong, and Metcalf, emissions taxes are more efficient than output or input taxes, which only indirectly tax emissions. Smulders and Vollebergh focus more sharply in this chapter on the specific administrative costs that arise in the construction and operation of a pollution tax system. In addition, they allow for a continuum of policy choices between the use of emissions taxes and product or input taxes.

Smulders and Vollebergh note that where emissions are closely linked to inputs (or outputs), abatement possibilities are few, and if administrative costs of emissions taxes are high, then emissions taxes should not be implemented. Those conditions precisely apply to carbon emissions and the current tax treatment of carbon in several European countries. The carbon content of fuels is relatively fixed for each fuel type and carbon scrubbing is very costly with current technologies. Moreover, carbon taxes can build on an extensive system of energy taxation and so do not require that a large new collection system be developed. This piggybacking can lead to considerable administrative cost savings. The authors then go on to examine current carbon taxation in various European countries to investigate the extent to which the taxes have been optimally implemented. They find that there is considerable scope for broadening the tax base for the carbon tax at little additional administrative cost.

Compliance Costs

Environmental policy often gives rise to substantial compliance costs. This is an argument raised by companies and industrial lobbies to oppose environmental regulation and policy, who often threaten to relocate their industrial plants in countries where compliance costs are lower. To assess the validity of this argument, it is crucial to define a correct methodology to quantify compliance costs. Only then is it possible to analyze the impact that compliance costs have on the investment decisions of domestic and foreign firms. This is done by the second group of papers in this volume.

In chapter 4, entitled "An Industry-Adjusted Index of State Environmental Compliance Costs," Arik Levinson constructs an index of state-level environmental regulatory stringency from 1977 to 1994 across the 50 states of the United States. The index has a number of advantages over existing indexes of environmental regulation. Previous measures were often subjective and simple cross-sectional measures that made analysis of trends across time and states impossible to carry out. Moreover, Levinson notes that the indexes ignore differences in industrial composition across states. Ignoring industrial composition means that states with a heavy concentration of pollution-intensive industry will show up as a highly regulated state (as measured by costs of compliance), regardless of the state's regulatory structure. Levinson's measure explicitly controls for industrial composition. After constructing the measure, Levinson applies it to an analysis of foreign direct investment (FDI) in the United States. He finds that the index performs better than other measures of regulatory stringency in measuring the impact of environmental policy on FDI.

Chapter 5, by Randy Becker and J. Vernon Henderson, is entitled "Costs of Air Quality Regulation" and investigates the role of air quality regulations in the United States in capital investment as well as the cost structure of firms. For the past 20 years, each county in the United States has been designated as in or out of attainment with the National Ambient Air Quality Standards (NAAQS) for ground-level ozone (O_3). Firms in nonattainment areas are subject to stricter scrutiny and regulation than are firms in attainment areas, and the contrast between investment behavior in nonattainment versus attainment areas can serve to identify the impact of environmental policy on firm investment. The identification process is confounded, however, by the fact that firms can self-select into or out of nonattainment areas, and Becker and Henderson pay careful attention to the endogeneity of firm behavior. Focusing on the industrial organic chemical and miscellaneous plastic parts industries—major emitters of volatile organic compounds and nitrogen oxides—the authors find that the stricter regulation in nonattainment areas leads to greater amounts of up-front investment in those areas but lower overall size in mature firms. The large up-front fixed costs associated with the permitting process for construction of new plants makes it efficient for firms to increase their initial plant investment. Moreover, as regulations tighten over time, these plants are grandfathered under the old rules and so avoid increased production costs. Becker and Henderson then quantify the regulatory costs by estimating cost functions and comparing cost differences between attainment and nonattainment areas. The cost differences are substantial. Not surprisingly, these cost differences are considerably larger than the costs of pollution abatement estimates from data sets that directly measure pollution abatement investment and costs. Among other things, the latter data sets only measure costs of activities directly and substantially related to pollution abatement and so ignore many of the costs that firms bear in response to regulation.

International Trade and Development

Environmental policy often has an international dimension. This is because environmental problems are often international or even global. But the international dimension of environmental policy is also related to its distributional and behavioral impacts. Indeed, policy decisions taken in one country may affect economic variables in other countries through their effects on trade, investment, energy prices, and other economic variables. This may modify both the growth and the distribution of world income (this point should be obvious if, for example, one thinks of the effects of past oil shocks). Some of these aspects are taken into account by the papers in the third part of this volume. For instance, chapter 6 extends the analysis of the double dividend question to the case in which international factor movements are explicitly considered in the model,[9] whereas chapter 7 focuses on the interdependence between environmental policy and income distribution at the world level.

Chapter 6, by Michael Rauscher, is entitled "International Factor Movements, Environmental Policy, and Double Dividends." Rauscher investigates how the design of environmental policies affects unemployment and welfare in a model with unemployment due to sticky wages. While a fixed-wage model ignores the complexities that arise when unemployment is due to a wage-bargaining-type model, it allows Rauscher to focus on the interactions between unemployment and environmental tax reforms. As in Bovenberg and Goulder's chapter, environmental policy can lead to scarcity rents that accrue to those who employ capital as a factor of production. A tightening of environmental policy raises the scarcity rents, which in turn makes capital more desirable (since the rents are associated with the use of capital). As the demand for capital rises, labor is more productive and firms respond by hiring more workers, thus lowering unemployment. Rauscher considers a variety of environmental tax reforms and measures the impact of this rents effect on unemployment. He finds that this rents-unemployment connection persists in general equilibrium and that it is possible that tighter environmental policy could in fact attract mobile capital from outside the country and thus lower unemployment. A corollary finding is that a shift from command and control regulation that creates scarcity rents to a tax regime that allows the government to appro-

---

9. The double dividend literature explores the implications of nonenvironmental benefits in addition to environmental benefits when taxes on pollution are implemented. For a recent survey on this issue, see Bovenberg (1999).

priate those rents could lead to capital flight. How large this effect is remains an empirical question to be answered in future research.

Chapter 7, by Raghbendra Jha and John Whalley, is entitled "The Environmental Regime in Developing Countries." Again the chapter is concerned with distributional issues, but at the international level. Jha and Whalley make a number of points relating to differences in environmental problems and solutions between developing and developed countries. First, they note that environmental problems in developing countries are more typically problems of degradation rather than of pollution. Soil erosion, congestion, and common property resources, for example, create externalities that may dwarf the externalities arising from traditional pollutants. Jha and Whalley argue that to contrast environmental policies in developed and developing countries without taking into account the very different environmental issues is to paint a very misleading picture of the problems and possible solutions. They next point out that measures of the social costs arising from environmental problems are highly misleading if the measures focus on traditional pollutants that are the focus of policy in developed countries. Third, the relation among economic growth, policy reform, and environmental quality may differ considerably between developed and developing countries, suggesting that grafting policy responses from developed countries onto developing countries without considering the differences between the two regions may be misguided. The authors conclude by arguing that an index of the severity of environmental problems across countries that measures the welfare gain from moving to full internalization of the externalities would work best to transcend the large differences in institutional and economic structure, as well as environmental problems across developed and developing countries.

Information, Incentives, and Environmental R&D

The last group of papers in this volume focuses more deeply on the behavioral effects of environmental policy. In order to assess the effectiveness of different sets of policy instruments, it is important to analyze how these instruments modify firms' strategic incentives to promote their products, to innovate, to invest in R&D, to locate their plants in different countries or regions. The chapters contained in the fourth part of the volume extend previous results by introducing uncertainty and irreversibilities, by allowing for endogenous cooperative environmental R&D among firms in the same industry, and by accounting for the possibility that firms adopt appropriate incentive schemes to increase the environmental performance of their own management.

In chapter 8, entitled "Environmental Information and Company Behavior," Domenico Siniscalco, Stefania Borghini, Marcella Fantini, and Federica Ranghieri present some preliminary results from a fascinating new data set on voluntary environmental initiatives at the firm level. These voluntary initiatives range from company environmental reports to environmental audit and management schemes, such as the recent International Standards Organization (ISO) 14000 standards. Critics of these policies argue that they are simply public relations schemes to make the companies look environmentally benign. It may be, however, that these types of policies can have real effects, given the complex structure of large corporations. Agency theory can help explain why these policies might have real effects. Shareholders care about the value of the company, which can be adversely affected by liabilities arising from harmful environmental activities in which the firm may engage. In addition to the liability risk, firms can lose valuable goodwill and reputation, which in turn can diminish the value of the firm. While shareholders may have a stake in their company engaging in environmentally clean activities, managers may not share this stake. Voluntary information-based policies can play a role in aligning the two groups' interests. Siniscalco and his coauthors have a unique data set that may allow formal tests of some of the theoretical underpinnings of agency theory associated with these voluntary environmental policies.

Since 1992, FEEM has been collecting data from firms that produce corporate environmental reports. In 1998, approximately 150 firms had produced these reports. The authors first provide some background information on the firms and the types of reports they prepare and then turn to a data analysis of a number of large firms based in 16 countries in the petrochemical, oil and gas, and electric power-generation industries. It is important that the researchers have gathered additional information on the companies' environmental compensation and award schemes for employees as well as economic data for the firms. Their preliminary data analysis suggests that the quality of information provided in these corporate environmental reports has important explanatory power for corporate environmental management. While tentative, their results are compatible with some of the emerging agency theories on voluntary information-based environmental policies. Their empirical analysis suggests that firms that implement some form of environmental reporting perform better from both an economic and an environmental point of view. Moreover, firms that implement incentive schemes to induce managers to do careful environmental reporting achieve better results than firms that do not. While these results are quite preliminary, the chapter provides valuable information about the range of information-based activities taken at a large number of important international firms and suggests fruitful research opportunities as this data set grows over time.

Chapter 9, by Anastasios Xepapadeas, is entitled "Environmental Policy and Firm Behavior: Abatement Investment and Location Decisions under Uncertainty and Irreversibility." Xepapadeas allows for uncertainty over output prices, environmental policy, and technological parameters as firms consider the optimal level of investment in pollution abatement capital as well as the location of new plants. Investment and relocation costs are in many cases irreversible and optimal decision making takes into account the value of options that are killed by making irreversible investments. Xepapadeas constructs a framework in which policymakers can construct optimal emissions taxes and pollution abatement subsidies under a variety of models of uncertainty.

Yannis Katsoulacos, Alistair Ulph, and David Ulph take up the topic of environmental research joint ventures (RJVs) in chapter 10. Their chapter, entitled "The Effects of Environmental Policy on the Performance of Environmental Research Joint Ventures," investigates the optimal amount of investment in R&D related to pollution abatement technologies when firms can form voluntary research joint ventures. From the point of view of firms, RJVs help firms avoid costly duplication of R&D effort. This is beneficial from society's point of view, although there is the countervailing loss from the possibly slower development of new technologies that could significantly reduce pollution due to the lack of competition. In the context of this model, policymakers have two instruments at their disposal. They can levy environmental taxes that will encourage R&D activity. They can also prohibit the formation of RJVs if they feel that the costs resulting from a slowdown in pollution abatement innovations from RJVs outweigh the benefits of avoiding duplication of investment costs. The authors show that the magnitude of environmental damages that can be averted through innovations resulting from R&D activity is a critical parameter for understanding whether government should allow or prohibit RJVs.

Conclusion

There are a few lessons that can be learned from the set of papers published in this book. First, in a world in which multiple externalities and market imperfections interact, and in which implementation costs are relevant—mainly because of information asymmetries—environmental policy needs to be careful designed and is inevitably more complex than the first-best policies usually studied in textbooks. In particular, a policy mix formed by environmental and nonenvironmental policy instruments is often adopted as part of the price paid for political or administrative viability. Second, innovation in policy instrument design is also important for managing new and complex environmental problems. Third, the social costs of environmental policy cannot be neglected. There are various strategies through which a regulator can reduce the costs paid by firms and consumers. For example, when designing the environmental policy mix, the regulator can use the possible revenue of environmental taxation to provide incentives to spur economic growth. But the regulator can also adopt low-cost economic instruments, such as information-based policies and incentive schemes, that improve firms' environmental performance while minimizing impacts on firm profitability. Alternatively, the regulator can adopt policies to encourage environmental technological cooperation, or policies to discourage firm relocation or capital flight.

The chapters presented in this book suggest that further efforts should be made to analyze the costs and benefits of policy mixes specifically designed to manage new and complex environmental problems. Empirical analyses can be a great help in this matter. Another lesson that can be derived from the chapters of this book is that new empirical information can be very helpful for assessing the performance and cost of different policy mixes. However, empirical analysis is still at a preliminary stage, particularly in Europe, mainly because reliable and broad data sets on environmental issues are still relatively rare. It is our hope that the research presented in this book will play a part in spurring further empirical work—both in Europe and North America—that will extend our understanding of environmental policy design in a second-best world.

References

Bovenberg, A. Lans. 1997. Environmental policy, distortionary labor taxation, and employment: Pollution taxes and the double dividend. In *New directions in the economic theory of the environment,* ed. Carlo Carraro and Domenico Siniscalco, 69–104. Cambridge: Cambridge University Press.

———. 1999. Green tax reforms and the double dividend: An updated reader's guide. *International Tax and Public Finance* 6 (3): 421–43.

Carraro, Carlo. 1999a. Environmental technological innovation and diffusion. In *Frontiers of environmental economics,* ed. H. Folmer, L. Gabel, S. Gerking, and A. Rose, 343–71. Cheltenham, U.K.: Edward Elgar.

———. 1999b. Imperfect markets and environmental policy instruments. In *Handbook of environmental and resource economics,* ed. J. Van den Bergh, chap. 15. Cheltenham, U.K.: Edward Elgar.

Carraro, Carlo, Yannis Katsoulacos, and Anastasios Xepapadeas, eds. 1996. *Environmental policy and market structure.* Dordrecht: Kluwer Academic.

Carraro, Carlo, and Domenico Siniscalco, eds. Forthcoming. *Global issues, sovereign states.* Cambridge: Cambridge University Press.

Fullerton, Don, and Gilbert E. Metcalf. 1997. Environmental controls, scarcity rents, and pre-existing distortions. NBER Working Paper no. 6091. Cambridge, Mass.: National Bureau of Economic Research.

———. 1998. Environmental taxes and the double dividend hypothesis: Did you really expect something for nothing? *Chicago-Kent Law Review* 73 (1): 221–56.

Pigou, Arthur C. 1932. *The economics of welfare.* 4th ed. London: Macmillan.

1

A Tax on Output of the Polluting Industry Is Not a Tax on Pollution
The Importance of Hitting the Target

Don Fullerton, Inkee Hong, and Gilbert E. Metcalf

1.1 Introduction

A tax per unit of pollution can induce all the cheapest and most efficient forms of pollution abatement (Pigou 1932). To reduce its tax liability, the firm can switch to a less-polluting fuel, add a scrubber, change disposal methods, or otherwise adjust its production process. These methods of substitution in production reduce the pollution per unit of output. In addition, the tax raises the overall cost of production, so the higher equilibrium output price chokes off demand for the output. Thus the tax has a substitution effect that reduces pollution per unit, and an output effect that reduces the number of units.

Yet few actual taxes are targeted directly on pollution (Barthold 1994). Taxes on gasoline are prevalent around the world, and the use of gasoline is indeed correlated with vehicle emissions. This gas tax might provide some incentive to reduce emissions by driving less, but it provides no incentive to reduce emissions per gallon (such as by adding pollution-control equipment). The United States taxes chemical feedstocks associated with contaminated Superfund sites, and this tax may help reduce pollution, but

Don Fullerton is professor of economics at the University of Texas at Austin and a research associate of the National Bureau of Economic Research. Inkee Hong is a doctoral candidate in the Department of Economics at the University of Texas at Austin. Gilbert E. Metcalf is professor of economics at Tufts University and a research associate of the National Bureau of Economic Research.

For funding, the authors thank the National Bureau of Economic Research and the National Science Foundation (SBR-9811324). For helpful suggestions, the authors thank Eliana de Bernardez Clark, Larry Goulder, Gilbert H. A. van Hagen, Rob Williams, and conference participants. This paper is part of the NBER's Public Economics research program. The views expressed in this paper are those of the authors and do not reflect those of the National Science Foundation or the National Bureau of Economic Research.

it provides no incentive to use a cleaner production process, to avoid spills, or to use any other method of reducing pollution per unit of chemical input (Fullerton 1996). In Europe, some industrial effluent taxes are calculated using an assumed industrywide rate of effluent per unit output, so the firm cannot reduce its tax by reducing its own effluent per unit (Hahn 1989). These taxes miss the substitution effect.

In this paper, we measure the welfare effect of improperly targeted instruments. We build a simple analytical general equilibrium model with substitution in production and demand by consumers, and we derive second-best optimal tax rates on emissions or on output. These rates are based on preference parameters, technological parameters, and preexisting tax rates. We discuss these optimal tax rates, and then we choose plausible values of the parameters to calculate the effects of a small change in each tax rate. For alternative initial conditions, we use the model to calculate the cost of missing the target: the welfare gain from a targeted tax on emissions minus the gain from an imperfectly targeted tax on output of the polluting industry.

Actual taxes may miss the target for several reasons. First, actual policy may not fully appreciate the importance of hitting the target. Policymakers may have been concerned primarily with equity considerations, trying to ensure that polluting industries are made to pay for pollution—without realizing that the form of these taxes affects incentives to reduce pollution. Second, actual emissions may be difficult or impossible to measure. In these cases, the best available tax may apply to a measurable activity that is closely correlated with emissions. To reduce vehicle emissions, for example, the gasoline tax may be the best available instrument. Third, the technology of emissions measurement is improving over time. Policymakers may be slow to adjust the tax base to reflect the newly reduced cost of measuring a particular pollutant.

We do not measure or model the costs of targeting the tax on pollution, that is, the costs of measurement, monitoring, and enforcement. We only measure the benefits of properly targeting the tax. Thus our results can be taken as a measure of the importance of developing new measurement or enforcement technologies and of reforming the law to take advantage of those technologies. That is, we calculate the improvement over an output tax that can be obtained by a targeted tax on pollution that can capture the substitution effect as well as the output effect.

The next section reviews actual environmental taxes around the world and describes the extent to which they miss the target. Section 1.3 reviews existing economic literature on this subject. Most early economic models ignored the substitution effect, assuming that pollution was associated only with output. More recently, others model substitution in production, but assume that the emissions tax is fully available. Schmutzler and Goulder (1997) provide a partial equilibrium model of the difference between an output tax and an emissions tax. Our paper contributes to this literature by providing a general equilibrium model to compare the welfare effects of these taxes.

If emissions cannot be monitored at reasonable cost and policy is limited to a tax on the output of the polluting industry, then how should that tax rate be set? One might think that the imperfection of this blunt instrument would reduce the optimal rate of tax. In our results section, we show that is not the case: The second-best output tax should be set to capture exactly the same output effect that would have been captured by the emissions tax. If the unavailable emissions tax would have raised output price by 12 percent, for example, then the output tax should be set to 12 percent. We also solve for the optimal emissions tax in a second-best world with some fixed preexisting output tax.

Finally, we use plausible parameters to calculate the incremental effects on welfare of slight increases in any preexisting output tax or emissions tax, and we show the welfare gap. We find that the welfare gain from an initial emissions tax is more than twice the gain from an initial output tax. This cost of missing the target does not depend on the size of the preexisting output tax or on the size of the elasticity of substitution in utility, but it does depend on the elasticity of substitution in production. A larger ability to substitute between emissions and other inputs in production substantially raises the importance of hitting the target.

1.2 Environmental Taxes around the World

While the economics literature has long championed the use of market-based instruments (e.g., environmental taxes and tradable permits), most countries have long relied on a system of regulations, including command and control regulations. In the past 10 years, however, countries have begun to shift to the use of environmental taxes of some sort. In this section, we review the types of taxes that are typically used and consider to what extent these taxes "hit the target."[1]

As we noted previously, the problem of targeting environmental taxes accurately in most cases follows from a difficulty in monitoring emissions. This has led Eskeland and Devarajan (1996) to distinguish between direct and indirect instruments to control pollution. Direct instruments require knowledge of actual emissions, while indirect instruments do not. A Pigovian tax, as developed in textbooks, is a tax on emissions themselves. The difficulty with direct taxes is that monitoring emissions is technologically difficult and administratively complex. Thus, most actual policies fall back

---

1. Roughly speaking, market-based instruments may be either price-based or quantity-based instruments. Taxes are a form of price-based instrument, while tradable permits are a quantity-based instrument. This paper compares various kinds of taxes, while Fullerton and Metcalf (1997) use a similar model to consider quantity instruments.

on indirect approaches to reduce emissions; the problem of hitting the target can be reframed as a problem of the administrative need to use indirect instruments.

1.2.1   Air Pollution

A variety of taxes are employed around the world to combat air pollution. Sweden applies a charge on actual nitrous oxide (NO_x) emissions of large heat and power producers (final sale only) at a rate of roughly 40 Swedish crowns per kilogram of NO_x ($7.17 per kg) (OECD 1994). For companies without emissions measurement equipment, standard emissions rates (in grams of NO_x per joule) exceed typical average actual emissions. The higher assumed emissions rate provides an incentive for companies to install measurement equipment. Tax collections are rebated to firms on the basis of final energy production. Thus the combination is revenue neutral because it provides a subsidy to low-emitting firms and a tax on high-emitting firms. The Swedish experience suggests that technological limitations on the use of directly targeted taxes may fall with technological progress. Moreover, this tax provides an interesting example of allowing firms to choose whether to be subject to a direct or an indirect tax. For firms that do not adopt monitoring equipment, the tax becomes a tax on fuel consumption; the actual NO_x emissions are irrelevant.

Japan levies a charge on SO_2 emissions, with the rate varying across regions. The tax is based partly on historic emissions (1982–86) and partly on emissions from the previous year. The tax rate in 1992 was 124 yen per cubic nanometer for historic emissions, and it was between 95 and 860 yen per cubic nanometer depending on the geographic region in which emissions occurred (OECD 1994). Allowing the rate to vary across geographic region provides the possibility of linking the rate more closely to marginal environmental damages. Whether Japan does in fact link the rates closely to marginal environmental damages is a question beyond the scope of this paper.

Taxes on coal illustrate how technological differences can significantly affect the ability to target emissions directly. A number of European countries levy a tax on the sulfur content of various fuels. Norway levies a charge on the sulfur content of oil. Sweden levies a charge on the sulfur content of oil, coal, and peat. A strict sulfur-content tax is an indirect tax in that it does not require any monitoring of emissions. It also does not provide any incentives to use scrubbers or otherwise reduce sulfur emissions (other than by shifting from high- to low-sulfur-content fuel). Sweden rebates the tax to firms that can demonstrate significant reductions in SO_2 emissions from the use of technologies such as flue gas cleaning. As of 1993, Finland levied a tax differential between standard and sulfur-free oil.

A tax on carbon content can be viewed as a direct tax on CO_2 emissions in the sense that it is economically infeasible to alter the ratio of carbon emissions to carbon content of the fuel in the industrial process.[2] Thus, whatever carbon is embodied in a fuel will be released to the atmosphere upon burning. As of 1992, six Organization for Economic Cooperation and Development (OECD) countries had either explicit or implicit carbon taxes (Denmark, Finland, Italy, Netherlands, Norway, and Sweden). Most of the countries tax different sectors at different rates, with some sectors exempted altogether. These taxes, to our knowledge, do not provide any incentive for carbon scrubbing.

The Montreal Protocol of 1989 required the eventual phasing out of halons and chlorofluorocarbons (CFCs). In the United States, Congress imposed taxes on these ozone-depleting chemicals at the same time that it implemented quantity regulations. The tax rate depends to some extent on the degree of ozone depletion. Merrill and Rousso (1991) note that the purpose of this tax was to capture monopoly rents arising from quantity restrictions. While the tax rate is not explicitly set equal to social marginal damage, it is a direct tax in that these chemicals have a direct relationship to the ozone-depletion damage stemming from their use. They are indirect and thus imprecisely targeted, however, in the sense that CFC emissions to the atmosphere are assumed rather than measured. No distinction is made in the use of CFCs regarding circumstances in which release to the atmosphere is more or less likely.

1.2.2   Water Pollution

Taxes related to water pollution are of two general types: user charges for sewage treatment and wastewater effluent charges. The latter is of more concern to us than the former. Better-targeted taxes would be based on "load," a measure of the pollutants contained in the wastewater. It can be measured on an instantaneous basis (so many parts per million) or on a flow basis (so many grams per hour or day). Sewage-treatment charges for households are based on water consumption rather than load on the treatment plant, and so they serve as an indirect charge. Industry is more likely to be metered with charges based on load. In 1992, for example, Denmark, Finland, Norway, and Sweden all levied charges on firms based on pollution loads exceeding some minimum amount.

In 1976, Germany implemented a water effluent charge for firms (to go into effect in 1981), with different rates for different pollutants (e.g., chemical oxygen demand [COD] and heavy metals). Firms are taxed on the basis of "damage units," defined approximately as the amount of pollution generated by one individual. Damage units are defined in terms of the amount

---

2. Processes for carbon scrubbing are described in Astarita, Savage, and Bisio (1983), but these technologies generate such a vast amount of solid waste (e.g., carbonic acid or forms of carbonates) that disposal costs become prohibitive. Sulfur scrubbing, on the other hand, involves a small amount of sulfur per ton of coal (or barrel of oil) and so does not lead to such significant sulfur-disposal problems.

of discharge of various pollutants.[3] While the tax is a tax on emissions, certain features of the system make it more an enforcement mechanism for technology standards. In particular, firms get a 75 percent reduction in rates if they can demonstrate compliance with specific technology standards. Thus, the tax might be viewed as a tax on old technology rather than on pollution. To the extent that the tax induces a shift to new, less polluting technology, however, these standards may improve targeting relative to standards that do not induce technology improvement.

1.2.3 Solid and Hazardous Waste

The OECD distinguishes between municipal-waste user charges and waste-disposal taxes. Municipal-waste user charges may be collected as a flat rate, but they are increasingly based on actual waste, and they are used to finance the cost of collection and disposal. In contrast, waste-disposal taxes generate revenues that either go into the general budget or are earmarked for environmental expenditures (e.g., subsidies for recycling). Since no effort is made to monitor the contents of waste, any pollution to groundwater from solid waste in a landfill does not affect the charge to households or to firms producing the waste. In other words, these taxes are indirect taxes with no incentive for shifting the composition of the waste stream.

As of 1992, five OECD countries had some form of tax that they characterize as hazardous-waste taxes. As the U.S. experience makes clear, these taxes may be described only very loosely as taxes on hazardous waste. The United States levies a number of Superfund taxes detailed in Fullerton (1996). These are taxes on petroleum ($0.097 per barrel) as well as on 42 organic and inorganic chemical feedstocks—with rates ranging from $0.22 to $4.87 per ton in 1992. The chemicals to be taxed were chosen to some extent on the basis of their presence in hazardous-waste sites to be cleaned up under the Superfund law. In particular, the tax rates are set to raise a specified sum necessary to clean up Superfund sites, where required collections on oil and chemicals are based on their relative importance in waste sites. These are indirect taxes at best, and, like many of the taxes discussed here, they do not provide incentives for emissions reduction. They might reduce the purchase of the petroleum or chemical products, but they do not influence their handling, their use, or the amount that becomes waste.

For hazardous waste, the form of disposal affects the marginal environmental damages quite dramatically. This fact suggests that a tax on the disposal of hazardous waste could reduce welfare if it shifts the mode of disposal from safe, monitored disposal sites to illegal dumping in unmonitored, unsecured sites. The welfare impact of a tax on hazardous-waste disposal will depend in an important way on the cost of monitoring dis-

---

3. The following amounts of effluents (in addition to others not listed) each add up to 1 damage unit: 50 kg of organic matter (COD), 3 kg of phosphorus, or 1 kg of copper and copper compounds (Anderson and Lohof 1997).

posal activities as well as the cost of enforcement and illegal disposal activities. The more costly it is to monitor disposal activities and enforce rules for proper disposal, the more likely it is that a tax on hazardous-waste disposal will reduce welfare.

As of 1992, several countries levied taxes on the disposal of automobile batteries (Canada, Denmark, Portugal, and Sweden) with differing rates based on the type of battery. Some countries levy charges on waste-oil disposal (Finland, France, Italy, Norway, and the United States). Numerous countries levy charges on packaging. Other taxes are levied on disposable diapers (Canada), car tires (Canada and the United States), and plastic shopping bags (Italy).

1.2.4   Taxes on Products Associated with Pollution

Product taxes are indirect taxes by definition. As of 1992, 10 OECD countries levied some form of one-time sales tax differential on cars based on weight (Canada), degree of compliance with emissions standards (Belgium, Greece, Japan, Netherlands, and Sweden), fuel efficiency (Canada, Japan, and the United States), or the lack of a catalytic converter (Finland, Germany, and Norway).

In 1992 some OECD countries levied higher annual taxes on vehicles that lack a catalytic converter (Austria and Denmark) and on average emissions for major pollutants for each class of car (Germany). All OECD countries levy excise taxes on gasoline. In addition, many OECD countries levy a higher tax on leaded than unleaded fuel. For example, Denmark, Finland, and Norway levied a surtax on leaded fuel of $0.11 per liter in 1992.

1.2.5   Summary

This brief survey of environmental taxes suggests that few taxes anywhere are precisely targeted taxes on emissions. The failure to target emissions precisely may follow from significant costs associated with measuring emissions, from costs associated with monitoring point- and non-point-source emissions at reasonable cost, and—as a consequence—difficulties with preventing tax evasion and illegal disposal activities.[4] To some extent, however, the imprecise targeting may result when policymakers do not fully appreciate the costs of missing the target.

1.3   Prior Literature

The literature on environmental taxes is extensive. Most papers, however, do not focus on the distinction between taxes on emissions and taxes on inputs or outputs that are imperfectly correlated with emissions. In an early example that is typical of this literature, Sandmo (1975) carries out

---

4. Siniscalco et al. (chap. 8 in this volume) discuss the possibilities of voluntary compliance with pollution control rules.

an optimal tax exercise in the presence of externalities. One of the consumption goods enters the utility function directly as a negative externality. Because of the one-for-one relation between the good itself and pollution, a tax on the good corresponds exactly to a tax on pollution. If the good itself is associated with pollution, instruments to discourage pollution can only operate through an output effect (as discussed previously). Actually, the tax on output can still perfectly correct for pollution associated with an input—if output must be produced using a fixed amount of pollution per unit.[5] With substitution in production, however, the output tax is no longer equivalent to a tax on pollution.

A recent paper by Cremer and Gahvari (1999) extends the Sandmo analysis to allow pollution to be associated with one of several inputs in the production of a "dirty good." In a standard optimal tax analysis, Cremer and Gahvari show first that emissions taxes and output taxes are not equivalent and, second, that both emissions and output taxes may be needed to achieve optimality in a second-best world. In effect, the emissions tax corrects externalities, while the output tax rates handle tax collections for general revenue needs in an optimal fashion. While it is an important extension of the original Sandmo analysis, the Cremer and Gahvari paper does not consider the loss from using an output tax instead of an emissions tax. That is, it still assumes that taxes on emissions are feasible.

A paper by Schmutzler and Goulder (1997) directly examines the trade-off between the use of emissions taxes and output taxes in the presence of imperfect monitoring of emissions. They note that previous authors (e.g., Cropper and Oates 1992) have recognized that output taxes may be preferable to emissions taxes if emissions are difficult to monitor, and they attempt to make more precise what it means to be "difficult to monitor." They enumerate four factors that affect the choice between emissions and output taxes: (1) monitoring costs, (2) technological factors, (3) the regulator's information structure, and (4) social preferences for consumption goods versus environmental quality. As the costs of monitoring emissions rise, the advantage of precisely targeted emissions taxes falls. This effect relates to evasion possibilities, as discussed in the previous section about hazardous-waste-disposal taxes. Technological factors come into play by determining the scope of substitution in production away from pollution. If emissions are a fixed proportion of output, then an output tax would be equivalent to an emissions tax without the need to measure emissions di-

---

5. See, for example, the recent paper by Sandmo and Wildasin (1999). This statement is true as long as emissions per unit of output are constant across firms. Nichols (1984) notes that the cost of using an output tax rather than an emissions tax rises with variation across firms in the emissions-to-output ratio (even if this ratio is fixed for each firm). Nichols also notes that targeting emissions per se is not precisely correct because the tax should distinguish among emissions according to their marginal damages. This point justifies, for example, time-varying emissions taxes where marginal damages vary during the day.

rectly. The regulator's information structure determines what it can monitor. Regulators face difficulty monitoring emissions, but they may face even more difficulty trying to tax certain inputs or output (thereby affecting the relevant target). Finally, the loss from poorly targeted instruments is a loss in the value of output, while the loss from the high cost of monitoring may be a loss in environmental quality, so the trade-off in utility between consumption and the environment may also affect the choice of instruments. Smulders and Vollebergh (chap. 3 in this volume) explore many similar issues.

Policy can miss the target in another important sense that we note here, but do not pursue in this paper. In the presence of multiple pollutants, targeting one pollutant may cause the substitution of other pollutants for that pollutant. Devlin and Grafton (1994) explore this topic in the context of determining the optimal number of tradable permits for a pollutant when multiple pollutants coexist.[6]

All of these papers ignore general equilibrium considerations. A large literature starting with Bovenberg and de Mooij (1994) explores the welfare consequences of environmental taxes and other instruments in a general equilibrium context with preexisting taxes.[7] These papers have typically focused on the interactions among taxes rather than on the issue of emissions taxes versus output taxes (or otherwise imperfectly targeted taxes). In the model that we present here, we allow for general equilibrium considerations as well as the existence of other distorting taxes.[8] We turn now to that model.

1.4   A General Equilibrium Model of Production and Consumption

The review of environmental taxes in section 1.3 indicates a slow movement away from output taxes and toward emissions taxes. However, the predominant existing environmental taxes still miss the target in that they tax a purchased input to production or an output sold, but not emissions per se. In this section, we carry out a general equilibrium analysis of the costs and effects of using mistargeted environmental instruments. The model allows us to investigate the welfare effects of a commodity or emissions tax in a second-best world with a preexisting labor distortion. We allow for choices in production and consumption, using the same notation

6. A similar idea is analyzed by Metcalf, Dudek, and Willis (1984), who consider the effect of controlling one form of disposal medium for a pollutant in a situation with multiple disposal media.

7. A very partial list includes Bovenberg and Goulder (1996), Parry (1995), Goulder (1995), Goulder, Parry, and Burtraw (1997), and Fullerton and Metcalf (1997). This literature is surveyed in Fullerton and Metcalf (1998).

8. The paper by Schmutzler and Goulder (1997) is closest in spirit to our paper. Their analysis is explicitly partial equilibrium, however, and they do not consider other preexisting tax distortions.

Table 1.1   **Policy Experiments**

1. Preexisting tax on Y only
   A. Increase tax on Y
   B. New tax on Z
2. Preexisting tax on Z only
   A. New tax on Y
   B. Increase tax on Z

as in Fullerton and Metcalf (1997). The model has a homogeneous population (of size N) and the possibility of substituting inputs in production. We assume perfect competition, complete information, and perfect factor mobility. The model has a clean good (X) and a dirty good (Y) that is produced using labor (L_y) and emissions (Z).

A number of policies can be analyzed with this model. We can solve for the optimal second-best tax rate (either on emissions, Z, or on output, Y) as a function of preference and production parameters as well as preexisting tax rates. In addition, we can consider various incremental tax reforms. With respect to the latter, we consider the four possible scenarios listed in table 1.1.

Most actual taxes fall into category 1, as noted in our review, and the relevant policy reform is either an increase in one of these taxes or the introduction of a new, more targeted tax. Taxes on gasoline are an output tax, for example, so proposals in the United States to increase the gasoline tax are an example of scenario 1A. On the other hand, proposals for a new carbon tax in the context of current taxes on gasoline are an example of scenario 1B. As an example of a preexisting tax on emissions, a carbon tax was implemented in the Scandinavian countries in the early 1990s. Policy reforms to implement new taxes in those countries on goods associated with pollution are examples of scenario 2A, while proposals to increase the carbon taxes are examples of scenario 2B. We begin by developing the model in the case with preexisting taxes on either emissions or output, but we first consider only an incremental tax on output (t_y).

1.4.1 Production

The clean good is produced in a constant-returns-to-scale technology using only labor (L_x) as an input:[9]

(1) $$X = L_x.$$

---

9. Our model assumes one factor of production, for simplicity called "labor," but under some circumstances this factor can be taken to represent a homogeneous composite of all clean resources used in production.

For convenience, the numeraire is taken to be labor (or, equivalently, the clean good). The dirty good (Y) is produced in a constant-returns-to-scale production function using labor (L_Y) and emissions (Z):

(2) $$Y = F(L_Y, Z).$$

Also, emissions entail some private cost in terms of resources (labor), and we can define a unit of emissions as the amount that requires one unit of resources:[10]

(3) $$Z = L_Z.$$

Aggregate emissions adversely affect environmental quality:

(4) $$E = e(NZ) \quad e' < 0.$$

Finally, a public good is produced using labor:

(5) $$G = NL_G.$$

The amount of this public good is held constant in revenue-neutral reforms later.

1.4.2   Consumption

In this model, the N identical households derive utility from the two private goods (X and Y), leisure (L_H), the public good (G), and environmental quality (E):

(6) $$U = U(X, Y, L_H; G, E).$$

The household budget constraint is given by

(7) $$X + (p_Y + t_Y)Y = (1 - t_L)L,$$

where

(8) $$L = L_X + L_Y + L_Z + L_G.$$

Government finances the public good with a preexisting tax on labor income (t_L) and possibly a tax on output (t_Y). The nominal net wage is $1 - t_L$. A fixed amount of time ($\overline{L}$) can be allocated between work (L) and leisure (L_H).

---

10. Note that emissions are positively related to the use of these resources: L_Z is not to clean up or reduce emissions, but just to cart it away. Abatement is undertaken by substituting L_Y for Z. This overall production function is still constant returns to scale, since Z is a linear function of L_Z. The private cost for emissions helps justify our assumption of an internal solution with a finite choice for Z, even without corrective government policy.

1.4.3 Comparative Statics

Later we consider the effect of changing various prices through the use of taxes. We employ a log-linearization technique for the analysis, an approach that is appropriate when considering small changes. This technique allows us to capture important behavioral attributes of producers and consumers with a few key parameters. It also makes for a tractable analysis by allowing us to solve a system of linear equations. The goal in this section is to develop the various equations that trace through the impacts of a tax change on prices, quantities, and welfare. We begin by noting how any changes affect utility:[11]

$$(9) \qquad \frac{dU}{\lambda L} = t_L \hat{L} + t_Y \left(\frac{Y}{L}\right) \hat{Y} + (t_Z - \mu)\left(\frac{Z}{L}\right)\hat{Z},$$

where dU is the change in a representative agent's utility and λ is the private marginal utility of income. The term μ equals $-NU_E e'/\lambda$ and is the marginal social damage from pollution. A hat over a variable indicates a percentage change (e.g., $\hat{Z} = dZ/Z$). The left-hand side of this expression is the change in welfare (in dollars) as a fraction of the total resource in the economy. The right-hand side is composed of three parts. The first two parts are the welfare effect of the environmental policy through its impact on labor supply and the amount of the dirty good. Since labor is already discouraged as a result of the tax on wage income, any policy that further discourages labor supply will reduce welfare. A similar effect holds for any preexisting tax on the dirty good. If either t_L or t_Y is 0, the corresponding welfare effect on labor or the dirty good disappears from the equation. The third term is the welfare impact resulting from the change in pollution.

In order to find tractable solutions to the welfare equation (9), we make some simplifying assumptions about consumer preferences. In particular, we assume that environmental quality and the public good are separable from the consumption goods and that the consumption goods enter utility in a homothetic subutility function:[12]

$$(10) \qquad U(X,Y,L_H,E,G) = U(V(Q(X,Y),L_H),E,G),$$

where V and Q are both homothetic. For later use, define p_Q as a price index on $Q(X, Y)$ such that

---

11. This equation follows from totally differentiating the utility function and substituting in the consumer's first-order conditions. Details are available from the authors.

12. The assumption of separability is standard in this second-best tax literature because it is tractable and because it is a central case with neither complements nor differential substitutes. We only have two private goods (X and Y). With more disaggregation, particular private goods would undoubtedly be complements to leisure or to the environment and receive unique tax treatments for those reasons.

(11) $$p_Q Q = X + (p_Y + t_Y)Y,$$

and let w be the real net wage,

(12) $$w = (1 - t_L)/p_Q.$$

Thus the change in the real net wage ($\hat{w} \equiv dw/w$) will be related to the change in the labor tax ($\hat{t}_L$, defined as $dt_L/(1 - t_L)$) and the change in $p_Q(\hat{p}_Q \equiv dp_Q/p_Q)$. From equation (11), the change in p_Q depends on the change in the producer price p_Y and the change in the output tax ($\hat{t}_Y \equiv dt_Y/(1 + t_Y)$). Finally, let c_Y be the consumer price for Y,

(13) $$c_Y = p_Y + t_Y.$$

Our assumptions about consumer preferences allow us to characterize the general equilibrium response to a change in the tax on Y with four equations:

(14) $$\hat{Y} - \hat{X} = \sigma_Q(\hat{p}_X - \hat{c}_Y) = -\sigma_Q \hat{c}_Y,$$

(15) $$\hat{L} = \varepsilon \hat{w},$$

(16) $$\hat{w} = -\hat{t}_L - \phi \hat{t}_Y,$$

(17) $$(1 - \phi)\hat{X} = \hat{L} - \hat{t}_L - \phi(\hat{Y} + \hat{t}_Y).$$

In these equations, σ_Q is the elasticity of substitution in consumption between X and Y, ε is the uncompensated labor supply elasticity, and ϕ is the share of the consumer's after-tax income spent on Y. Equations (14) and (15) follow directly from our definition of σ_Q and our assumptions about consumer preferences. Equation (16) follows from totally differentiating equation (12) and using equation (11), while equation (17) follows from differentiating the consumer's budget constraint.

We totally differentiate the government budget constraint, hold G fixed, and assume that the revenue from the change in t_Y is offset by a change in t_L. These assumptions provide the fifth equation in our system:

(18) $$\hat{t}_L = -\left(\frac{t_L}{1 - t_L}\right)\hat{L} - \left(\frac{t_Z Z}{(1 - t_L)L}\right)\hat{Z} - \phi\left[\hat{t}_Y + \left(\frac{t_Y}{1 + t_Y}\right)\hat{Y}\right].$$

This is the change in t_L necessary for government to balance the budget when changing t_Y. Next, we turn to the equations implied by production. As yet, we do not allow for a change in the tax on emissions. Thus any change in inputs or output comes entirely from an output effect. Because of constant returns to scale in production, we have

(19) $$\hat{Y} = \hat{Z} = \hat{L}_Y.$$

Also, the producer price is fixed,[13] and so

(20) $$\hat{c}_Y = \hat{t}_Y.$$

Equations (14)–(20) represent eight linear equations that can be solved for the eight variables ($\hat{Y}$, $\hat{X}$, $\hat{w}$, $\hat{t}_L$, $\hat{c}_Y$, $\hat{L}$, $\hat{L}_Y$, and $\hat{Z}$), all as functions of the exogenous $\hat{t}_Y$.

After we solve for changes in Y, L, and Z as functions of the change in t_Y, we substitute these expressions into equation (9) and express the welfare change as a function of the incremental tax reform:[14]

(21) $$\frac{dU}{\lambda L} =$$

$$-\left\{\frac{\sigma_Q(1-\phi)\left\{(1-t_L)\left[t_Y\left(\frac{Y}{L}\right) + (t_Z - \mu)\left(\frac{Z}{L}\right)\right] + \varepsilon t_L\mu\left(\frac{Z}{L}\right)\right\}}{(1-t_L-\varepsilon t_L) - (1+\varepsilon)\left[t_Y\left(\frac{Y}{L}\right) + t_Z\left(\frac{Z}{L}\right)\right]}\right\}\hat{t}_Y.$$

Despite the complexity of this equation, we can make some general observations about the welfare impact of an incremental tax reform. First, note that the welfare impact does not depend on the ability to substitute other inputs for emissions in production (σ_Y). Since we have limited our instrument to a tax on the dirty output, the only welfare gain comes about from an equilibrium output effect (arising from substitution in consumption). The change in the output tax provides no substitution effect in production.

Second, the first term in the denominator must be positive to ensure that the government is on the upward-sloping side of the Laffer curve for wage taxation. We will assume that this is always the case. A condition for the entire denominator to be positive is for $\varepsilon < (NL - G)/G$, or that ε be bounded above by the ratio of private output to government output.[15] We will also assume that this condition holds throughout.

Third, note that the formula simplifies considerably with no preexisting taxes:

---

13. We normalize the initial producer price of Y to be 1, for any given emissions tax. In this section, where we do not allow for the emissions tax to change, the producer price will be unaffected by changes in the output tax.

14. Details are available from the authors upon request.

15. This follows from the fact that government spending is financed entirely by taxes: $G/N = t_L + t_Z Z + t_Y Y.$

$$(21') \qquad \frac{dU}{\lambda L} = \sigma_Q(1 - \phi)\mu\left(\frac{Z}{L}\right)\hat{t}_Y.$$

Welfare is unambiguously increased by an initial output tax, as long as consumers can substitute X for Y.[16]

Next we turn to a model for the case where the policy shock is a change in the tax on emissions rather than on output. The relevant equations (14) and (15) are unchanged, and other equations change, as noted by primes:

$$(14) \qquad \hat{Y} - \hat{X} = \sigma_Q(\hat{p}_X - \hat{c}_Y) = -\sigma_Q\hat{c}_Y,$$

$$(15) \qquad \hat{L} = \varepsilon\hat{w},$$

$$(16') \qquad \hat{w} = -\hat{t}_L - \left(\frac{(1 + t_Z)Z}{(1 - t_L)L}\right)\hat{t}_Z,$$

$$(17') \qquad (1 - \phi)\hat{X} = \hat{L} - \hat{t}_L - \phi\left[\hat{Y} + \left(\frac{(1 + t_Z)Z}{Y}\right)\hat{t}_Z\right],$$

$$(18') \qquad \hat{t}_L = -\left(\frac{t_L}{1 - t_L}\right)\hat{L} - \phi\left(\frac{t_Y}{1 + t_Y}\right)\hat{Y} - \frac{(1 + t_Z)Z}{(1 - t_L)L}\left[\hat{t}_Z + \left(\frac{t_Z}{1 + t_Z}\right)\hat{Z}\right],$$

$$(19.1') \qquad \hat{L}_Y = \hat{Z} + \sigma_Y\hat{t}_Z,$$

$$(19.2') \qquad \hat{Y} = \left(\frac{L_Y}{Y}\right)\hat{L}_Y + \left(\frac{(1 + t_Z)}{Y}\right)\hat{Z},$$

$$(20') \qquad \hat{c}_Y = \left(\frac{(1 + t_Z)Z}{(1 + t_Y)Y}\right)\hat{t}_Z.$$

Equations (19.1') and (19.2') require a bit of explanation. The first equation is the behavioral relationship in production given by the elasticity of substitution in production (σ_Y). Then the second equation follows from the first-order conditions in production.

Combining these equations and using the zero-profits condition, we can solve for the welfare impact of a change in the tax on emissions:

---

16. The output effect disappears if σ_Q equals 0, because then consumers do not substitute X for Y when the latter's price rises. In addition, if ϕ equals 0 or 1, then the consumer is at a corner and again does not substitute X for Y (note that $\phi = 0$ implies $Z = 0$).

(22) $\dfrac{dU}{\lambda L} =$

$$\left\{ \begin{array}{c} \sigma_Q(1-\phi)(1+t_Z)\left(\dfrac{Z}{Y}\right)\left\{(1-t_L)\left[t_Y\left(\dfrac{Y}{L}\right)\right.\right. \\[2mm] \left. + (t_Z - \mu)\left(\dfrac{Z}{L}\right)\right] + \varepsilon t_L \mu\left(\dfrac{Z}{L}\right)\right\} + \sigma_Y\left(\dfrac{L_Y}{Y}\right)(1+t_Y) \\[4mm] -\dfrac{\left\{(1-t_L)(t_Z-\mu)\left(\dfrac{Z}{L}\right) + \left[\varepsilon t_L + (1+\varepsilon)t_Y\left(\dfrac{Y}{L}\right)\right]\mu\left(\dfrac{Z}{L}\right)\right\}}{(1+t_Y)\left\{(1-t_L-\varepsilon t_L) - (1+\varepsilon)\left[t_Y\left(\dfrac{Y}{L}\right) + t_Z\left(\dfrac{Z}{L}\right)\right]\right\}} \end{array} \right\} \hat{t}_Z.$$

The first term in the numerator of equation (22) is very similar to the whole numerator in equation (21) and reflects substitution in consumption (i.e., the effect on output). The second term in the numerator reflects the substitution effect in production because we now have the possibility of changing relative input prices. With no preexisting taxes of any kind, the expression simplifies to

(22') $\qquad \dfrac{dU}{\lambda L} = \left[\sigma_Q(1-\phi)\left(\dfrac{Z}{Y}\right) + \sigma_Y\left(\dfrac{L_Y}{Y}\right)\right]\left(\dfrac{\mu Z}{L}\right)\hat{t}_Z.$

The first term in equation (22') corresponds to equation (21'), adjusted for the fact that the tax is on emissions rather than output. It represents the output effect from the emissions tax. In addition, a substitution effect is captured by the second term.

1.5 Model Analysis

1.5.1 Optimal Tax Rates

We begin the analysis by considering the optimal tax in the various scenarios described. First consider the optimal emissions tax in the case with a preexisting tax on labor, but no tax on output. This is, we ask what is the tax t_Z in equation (22), where $t_Y = 0$, such that no further change $\hat{t}_Z$ can affect welfare ($dU = 0$). We set equation (22) to 0 and solve for the tax rate on emissions:

(23) $\qquad t_Z^* = \mu\left[1 - \left(\dfrac{t_L}{1-t_L}\right)\varepsilon\right].$

Unless the tax rate t_L or the uncompensated labor supply elasticity is 0, the term in brackets is less than 1, and the optimal emissions tax is less than the social marginal damages ($t_Z^* < \mu$). This result is consistent with Bovenberg and de Mooij (1994).

To see how our expression (23) relates to other results in the literature, let ψ be the partial-equilibrium marginal cost of public funds for the labor tax. Goulder and Williams (1999) show that

$$(24) \qquad \psi = 1 + \frac{t_L(\partial L_H/\partial t_L)}{L - t_L(\partial L_H/\partial t_L)}.$$

Then some simple manipulation of this formula provides

$$(25) \qquad \psi = \left[1 - \left(\frac{t_L}{1 - t_L}\right)\varepsilon\right]^{-1}.$$

With a positive tax rate and positive labor supply elasticity ε, the marginal cost of funds is $\psi > 1$. Thus equation (23) can be rewritten as

$$(23') \qquad t_Z^* = \frac{\mu}{\psi},$$

as noted by Sandmo (1975) and Bovenberg and van der Ploeg (1994).[17]

Analogously, if the emissions tax is unavailable ($t_Z = 0$), we can solve for the second-best tax on output as the t_Y in equation (21) such that a change $\hat{t}_Y$ does not raise welfare ($dU = 0$). We set the numerator of equation (21) to 0 and find

$$(26) \qquad t_Y^* = \left(\frac{\mu}{\psi}\right)\left(\frac{Z}{Y}\right).$$

A striking result is that the optimal tax on output is very similar to the optimal tax on emissions.[18] This tax is second-best in two respects. First, this t_Y is reduced when divided by $\psi > 1$, to account for the preexisting tax on labor. Second, one might think that it should be reduced even more, to account for missing the target. The output tax is a blunt instrument for dealing with pollution. On the other hand, perhaps t_Y should be increased to get more of an output effect, since it misses the substitution effect. Yet equation (26) shows that the output tax should be set to generate exactly the same output effect as the ideal emissions tax. To see this, note that the

17. This result is also consistent with Cremer and Gahvari (1999). They solve for optimal second-best tax rates on emissions and on outputs in the general case without separability, but in our case with separability, their emissions tax would be μ/ψ and their output tax would be 0 (using t_L for revenue).

18. For the special case where $Y = Z$, equation (26) collapses to (23').

second-best emissions tax ($t_Z^* = \mu/\psi$) would raise production costs by $(\mu/\psi)Z$. Divide this amount by Y to get the extra cost per unit of output, which is exactly the amount that t_Y^* would raise the price of output.[19]

In other words, the fact that the output tax cannot achieve the desired substitution effect should not deter policymakers from its use to achieve the desired output effect. The optimal t_Y^* is the damage per unit of output—calculated as the desired tax per unit of emissions (t_Z^*) times emissions per unit output (Z/Y).

If the ideal t_Z^* were unavailable, could the authorities set t_Y^* and enforce it? If firms differ, one might think that authorities would need to know each firm's Z (or equivalently Z/Y) to set that firm's output tax rate using equation (26). Yet, if authorities knew Z, it seems they could employ an emissions tax directly. However, authorities only need to measure (or estimate) Z/Y once to set the output tax rate. The tax can then be enforced simply by counting units of output. In contrast, the emissions tax requires continuous measurement of Z, especially after firms change their Z/Y ratio in response to the tax. Moreover, if firms are similar, authorities need only the average Z/Y to set the output tax rate. Even if firms are similar, the emissions tax requires authorities to measure (or at least threaten to measure) each firm's emissions.

Next, consider the possibility of preexisting taxes either on emissions or on output. Equation (26) generalizes readily in the presence of a preexisting tax on Z:

$$(27) \qquad t_Y^* = \left(\frac{\mu}{\psi}\right)\left(\frac{Z}{Y}\right) - t_Z\left(\frac{Z}{Y}\right).$$

The first term in equation (27) is the output effect from the optimal output tax if t_Z is 0. The second term adjusts the tax rate to account for the output effect already obtained from taxing emissions. If the emissions tax is fixed suboptimally, then the additional required output tax is simply the additional desired output effect to account for the undertaxation of emissions. If emissions are taxed optimally ($t_Z = \mu/\psi$), then (27) shows that the optimal tax on output is 0.

To find the optimal tax on emissions (t_Z^*) in the case of a preexisting tax on output, we find the tax rate on emissions that cannot raise utility. That is, we find t_Z in equation (22) such that $dU = 0$. The solution to this equation is more complicated:

---

19. Actually, the optimal t_Y^* in equation (26) uses the Z/Y without any t_Z, without any substitution effect, so that Z/Y is higher than the optimal Z/Y. The rule in equation (26) gives the same output effect, but the level of t_Y^* in equation (26) is higher than the output effect of t_Z^* (at optimal Z/Y).

(28) $\sigma_Q(1 - \phi)(1 + t_Z^*)Z\{(1 - t_L)[t_Y Y + (t_Z^* - \mu)Z] + \varepsilon t_L \mu Z\}$

$$+ \sigma_Y L_Y(1 + t_Y)\left\{(1 - t_L)(t_Z^* - \mu)Z + \left[\varepsilon t_L + (1 + \varepsilon)t_Y\left(\frac{Y}{L}\right)\right]\mu Z\right\} = 0.$$

While solving for t_Z^* is not possible, we can rewrite equation (28) to make a basic point:

(28′) $t_Z - \mu =$

$$-\left\{\begin{array}{c}(1 - t_L)\sigma_Q(1 - \phi)(1 + t_Z)t_Y Y \\ + \left\{[\sigma_Q(1 - \phi)(1 + t_Z)Z]\varepsilon t_L + \left[\sigma_Y L_Y(1 + t_Y)\varepsilon t_L + (1 + \varepsilon)t_Y\left(\frac{Y}{L}\right)\right]\right\}\mu \\ \hline (1 - t_L)[\sigma_Q(1 - \phi)(1 + t_Z)Z + \sigma_Y L_Y(1 + t_Y)]\end{array}\right\}.$$

While we have not explicitly solved for the optimal emissions tax (since t_Z appears on both sides), we can show that the right-hand side is less than 0. Thus the optimal emissions tax rate is less than the social marginal damages ($t_Z^* < \mu$), with preexisting t_Y and t_L. A sufficient condition for the emissions tax rate to equal the social marginal damages is that ε and t_Y both equal 0.[20] Note that if either t_L or ε is 0 (so $\psi = 1$), then a nonzero t_Y still means that the optimal tax on emissions is less than μ.

For the special case where $Y = Z$ and $\sigma_Y = 0$, equation (28) collapses to

(28″) $$t_Y + t_Z = \frac{\mu}{\psi}$$

In this case, we need not distinguish between taxes on emissions or output, since the production function is such that output itself is polluting. Once again, the optimal tax is marginal social damages divided by the marginal cost of public funds.

1.5.2   Incremental Tax Reforms

We now turn to a numerical analysis of tax reforms. We measure the impact on welfare of a small change in either t_Z or t_Y. In order to carry out these calculations, we need values for a number of key parameters. Table 1.2 presents the assumed values for our base-case calculations; justification for these selections appear in Fullerton and Metcalf (1997).

Little evidence exists on some of these parameters, especially σ_Q and σ_Y, and so we present a sensitivity analysis in subsection 1.5.3. Also, marginal environmental damages (μ) could be considerably higher for some pollut-

---

20. Alternatively, the optimal tax rate equals μ if t_Y and t_L are 0 (i.e., a first-best world).

Table 1.2   Parameter Assumptions

Parameter	Value
μ	0.3
ε	0.3
σ_Q	1.0
Y/L	0.3
σ_Y	1.0
Z/L	0.15
t_L	0.4

ants (and lower for others). Tax rate results are proportional to μ, however, so it is easy to see how results change with that parameter.

Consider scenario 1, with no preexisting tax on emissions, but with a preexisting tax on labor and (perhaps) on output. With these values, the first-best Pigouvian tax would be $\mu = 0.3$, but the marginal cost of funds is $\psi = 1.25$, so the second-best tax on emissions is 0.24, from equation (23'). Then, since emissions constitute half of output, equation (26) says that the second-best tax on output is 12 percent. For our measure of welfare, we use $dU/\lambda L$, the monetary value of the change in utility as a fraction of total income.

Figure 1.1 depicts the general welfare effects of a small change in either the output tax or the emissions tax, assuming no preexisting emissions tax, for alternative values of a preexisting output tax. The horizontal axis indicates the level of the output tax prior to the reform, and the vertical axis shows the net change in welfare as a proportion of national income. The absolute welfare change may be in billions of dollars, but dividing by GDP makes the relative gains look small on the vertical axis. Consider the lower line, which indicates the change in welfare for a change in t_Y. First note that it crosses the horizontal axis at $t_Y = 0.12$. Since the optimal output tax with this configuration of parameters is 0.12, welfare does not change when the tax rate is altered from this level. At tax rates below this optimum, welfare rises when the tax on Y is increased a small amount. The maximum gain occurs with no preexisting tax on output.[21] The line falls below the horizontal axis in the region where t_Y exceeds 12 percent, indicating that a further increase in the tax rate would reduce welfare.

The upper line shows the welfare gain from introducing a small emissions tax. First, note that this line is everywhere above the line for raising

---

21. It is tempting to integrate under this curve to measure the welfare impact of a large change in t_Y, say from 0 to 12 percent. This would be a legitimate exercise if the private marginal utility of income were constant across this interval. In general, λ is not constant, however, so the increments to welfare are measured in different units of income and are not additive.

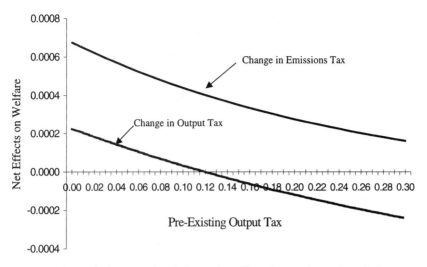

Fig. 1.1   Scenario 1: proportional change in welfare from a change in emissions tax or output tax (with preexisting labor and output taxes)

the output tax. For any preexisting tax on output, welfare is raised more by introducing a tax on emissions than by increasing the tax on output. Recall that the major distinguishing difference is that the emissions tax provides both output and substitution effects while the output tax only provides an output effect. As an approximation, then, the substitution effect is the gap between these two lines. With no initial taxes, the welfare gain from a small output tax is less than half that of a small emissions tax: More than half of the gain from an emissions tax comes from the shift in production processes as emissions become more expensive. This decomposition depends directly on σ_Y because the crucial distinction between an output tax and emissions tax is the ability of the latter to operate through the substitution effect. Later we provide a sensitivity analysis of this parameter.

Second, the welfare gain from introducing an emissions tax is everywhere positive. At high preexisting output tax rates, the additional output effect reduces welfare, but the initial substitution effect from the first introduction of t_Z is sufficiently strong to overwhelm any negative output effect.

Figure 1.2 corresponds to scenario 2 (no preexisting output tax). The horizontal axis now indicates the level of a preexisting emissions tax. The optimal emissions tax for this set of parameter assumptions is 24 percent, so the line measuring the incremental gains from an incremental emissions tax crosses the horizontal axis at 0.24 (where welfare cannot be raised by any change in t_Z). Interestingly, the output tax curve also crosses the horizontal axis at 24 percent. In other words, if the preexisting emissions tax is already at the second-best optimal rate of 0.24, then the initial intro-

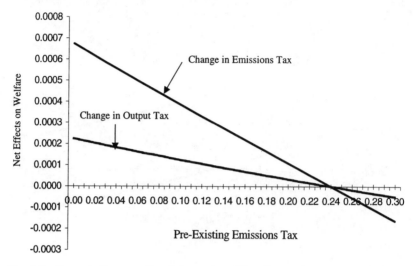

Fig. 1.2   Scenario 2: proportional change in welfare from a change in emissions tax or output tax (with preexisting labor and emissions taxes)

duction of t_Y has no first-order effect on welfare. Back at the vertical axis, where the initial t_Y and t_Z are both 0, the introduction of an initial t_Z dominates the introduction of an initial t_Y (since t_Z has both substitution and output effects). In other words, the emissions-tax curve starts out higher than the output-tax curve, and they must both cross the horizontal axis at 0.24 (the second-best optimum). If the emissions tax rate exceeds the second-best optimum, a further increase in this tax is more welfare reducing than an increase in the output effect, since the increase in the emissions tax has both an unwanted substitution effect and an unwanted output effect.

1.5.3   Sensitivity Analysis

Figures 1.1 and 1.2 are drawn using one set of preference parameters and technological parameters. Clearly, however, the size of the substitution effect depends on the elasticity of substitution in production (σ_Y), and the size of the output effect depends on consumer demand (the elasticity of substitution in utility, σ_Q). We vary those parameters in figures 1.3 and 1.4, but we still show the effect of missing the target—the welfare gap—defined as the gain from adding a tax on emissions minus the gain from adding to the tax on output.

Figure 1.3 shows this welfare gap (on the vertical axis) for different values of the elasticity of substitution in utility (on the horizontal axis). The three curves in the figure correspond to three initial values of t_Y (0.0, 0.12, and 0.24). Since σ_Q most directly affects the output effect, and not the substitution effect, it does not much affect the cost of missing the target.

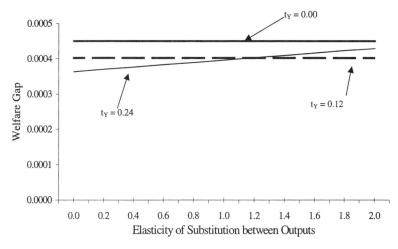

Fig. 1.3   How the welfare gap depends on substitution between outputs: the proportional gain from adding a tax on emissions minus the gain from adding to the tax on output

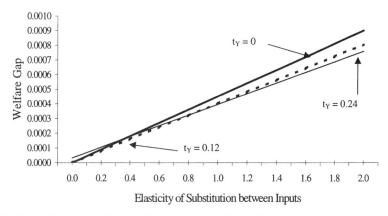

Fig. 1.4   How the welfare gap depends on substitution between inputs: the proportional gain from adding a tax on emissions minus the gain from adding to the tax on output

When the initial tax rate is 12 percent or lower, the welfare gain from the emissions tax exceeds the gain from the output tax by a relatively constant amount. At higher levels of preexisting t_Y, a higher σ_Q raises this amount.

Figure 1.4 shows the welfare gap for different values of the elasticity of substitution in production (again for initial t_Y equal to 0.0, 0.12, or 0.24). The assumed value of σ_Y clearly affects the size of the welfare gap. For any initial tax on output, the ability to substitute in production dramatically increases the importance of hitting the target.

1.6 Conclusion

A tax on pollution has been suggested by Pigou (1932) and thoroughly analyzed in the economics literature ever since, but a true Pigouvian tax is essentially never employed by actual policy. Most actual environmental taxes apply to the output of a polluting industry or to an input that is correlated with emissions, rather than directly to emissions. Perhaps policymakers think that the "polluter pays" principle is satisfied, since the polluters bear the burden of the output tax, but this paper shows the loss in welfare from missing the target in this fashion. Using plausible parameters, the introduction of a tax on emissions raises welfare by more than twice as much as a tax on the output of the polluting industry. We find that the ability of producers to substitute away from emissions directly affects the cost of missing the target, but it does not affect the second-best optimal tax on output. In the case in which emissions cannot be taxed, perhaps for technological reasons, we find that the second-best output tax should still be set to obtain the same effect on output price as would occur with the desired but unavailable emissions tax.

Other research directions are not explored in this paper but represent important avenues for further study. First of all, we ignore the administrative cost of trying to monitor emissions. If the ability to measure and tax emissions is a matter of degree, then we would expect a trade-off at the margin between the falling marginal benefits of hitting closer to the target and the rising marginal costs of doing so. The optimum might then involve some optimal degree of effort to measure and tax emissions.

Second, our model considers a tax on the output of the polluting industry for comparison with the ideal emissions tax, but some of the actual environmental taxes apply to an input to production that is correlated with pollution. To analyze such a tax, our model would have to be modified such that the polluting industry uses three inputs to production: labor, emissions, and some other input that is correlated to emissions.

Third, our model is rather stylized, with one clean output, one dirty output, and one very general technology of switching from emissions to the other input in production. Our results are valuable for a conceptual understanding of the importance of hitting the target, but specific policy problems should be analyzed for particular industries with carefully specified technologies of pollution abatement.

Fourth, as indicated in our review of actual taxes, some programs may allow the firm to choose between paying an output tax or purchasing abatement and monitoring equipment to pay a lower emissions tax. In addition, waste taxes may be earmarked for public spending on abatement. Hazardous-waste taxes may increase illegal, unmonitored activities.

Finally, we note that our model relies on many other standard simplifying assumptions and thus could be extended to consider the effects of uncertainty, imperfect competition, heterogeneity among firms, distributional effects among consumers, traded goods, transboundry pollution, and many other interesting problems.

References

Anderson, Robert, and Andrew Lohof. 1997. The United States experience with economic incentives in environmental pollution control policy. Washington, D.C.: U.S. Environmental Protection Agency, Office of Policy, Planning, and Evaluation.

Astarita, Giovanni, D. Savage, and A. Bisio. 1983. *Gas treating with chemical solvents.* New York: Wiley.

Barthold, Thomas A. 1994. Issues in the design of environmental excise taxes. *Journal of Economic Perspectives* 8:133–51.

Bovenberg, A. Lans, and Ruud de Mooij. 1994. Environmental levies and distortionary taxation. *American Economic Review* 84:1085–89.

Bovenberg, A. Lans, and Lawrence H. Goulder. 1996. Optimal environmental taxation in the presence of other taxes: General equilibrium analyses. *American Economic Review* 86:985–1000.

Bovenberg, A. Lans, and Frederic van der Ploeg. 1994. Environmental policy, public finance, and the labor market in a second-best world. *Journal of Public Economics* 55:349–90.

Cremer, Helmuth, and Firouz Gahvari. 1999. What to tax: Emissions or polluting goods? University of Illinois. Mimeo.

Cropper, Maureen, and Wallace Oates. 1992. Environmental economics: A survey. *Journal of Economic Literature* 30:675–740.

Devlin, R. A., and R. Q. Grafton. 1994. Tradable permits, missing markets, and technology. *Environmental and Resource Economics* 4:171–86.

Eskeland, Gunnar S., and Shanta Devarajan. 1996. *Taxing bads by taxing goods: Pollution control with presumptive charges.* Washington, D.C.: World Bank.

Fullerton, Don. 1996. Why have separate environmental taxes? *Tax Policy and the Economy* 10:33–70.

Fullerton, Don, and Gilbert E. Metcalf. 1997. Environmental controls, scarcity rents, and pre-existing distortions. NBER Working Paper no. 6091. Cambridge, Mass.: National Bureau of Economic Research.

———. 1998. Environmental taxes and the double dividend hypothesis: Did you really expect something for nothing? *Chicago-Kent Law Review* 73 (1): 221–56.

Goulder, Lawrence H. 1995. Environmental taxation and the "double dividend": A reader's guide. *International Tax and Public Finance* 2:157–83.

Goulder, Lawrence H., Ian Parry, and Dallas Burtraw. 1997. Revenue-raising vs. other approaches to environmental protection: The critical significance of preexisting tax distortions. *RAND Journal of Economics* 28:708–31.

Goulder, Lawrence H., and Roberton Williams. 1999. The usual excess-burden approximation usually doesn't come close. Stanford University. Mimeo.

Hahn, Robert. 1989. Economic prescriptions for environmental problems: How the patient followed the doctor's orders. *Journal of Economic Perspectives* 3: 95–114.

Merrill, Peter R., and Ada S. Rousso. 1991. Federal environmental taxation. In

Proceedings of the eighty-third annual conference of the National Tax Association, ed. Frederick D. Stocker, 191–98. Columbus, Ohio: National Tax Association, Tax Institute of America.

Metcalf, Gilbert E., Daniel Dudek, and Cleve Willis. 1984. Cross-media transfers of hazardous wastes. *Northeast Journal of Agricultural and Resource Economics* 13:203–9.

Nichols, Albert. 1984. *Targeting economic incentives for environmental protection.* Cambridge, Mass.: MIT Press.

OECD. 1994. *Managing the environment: The role of economic instruments.* Paris: Organization for Economic Cooperation and Development.

Parry, Ian. 1995. Pollution taxes and revenue recycling. *Journal of Environmental Economics and Management* 29:S64–S77.

Pigou, Arthur C. 1932. *The economics of welfare.* 4th ed. London: Macmillan.

Sandmo, Agnar. 1975. Optimal taxation in the presence of externalities. *Swedish Journal of Economics* 77:86–98.

Sandmo, Agnar, and David E. Wildasin. 1999. Taxation, migration, and pollution. *International Tax and Public Finance* 6:39–59.

Schmutzler, Armin, and Lawrence H. Goulder. 1997. The choice between emission taxes and output taxes under imperfect monitoring. *Journal of Environmental Economics and Management* 32:51–64.

Comment   Gilbert H. A. van Hagen

Introduction

In their contribution to this conference volume, Don Fullerton, Inkee Hong, and Gilbert Metcalf explore the difference between a tax on the output of a polluting industry and a direct tax on emissions. According to their numerical estimates, the welfare gain from the introduction of an emissions tax may be twice the welfare gain from an (imperfectly targeted) output tax. Thus, they emphasize the importance of looking for direct taxes on emissions that can replace the currently employed taxes on the output of polluting industries and on inputs that are imperfectly correlated with emissions.

In addition, the analysis of Fullerton, Hong, and Metcalf provides an important guideline for the design of taxes on the output of the polluting industry in the case that an emissions tax is unavailable. They indicate that the second-best output tax would still be set to obtain the same effect on output price as would occur with the desired but unavailable emissions tax. In this comment, I shall focus on this conclusion. In particular, I will provide some additional insight into the relationship between the second-

Gilbert H. A. van Hagen is an economist at the CPB Netherlands Bureau for Economic Policy Analysis in The Hague, and a research associate of the OCFEB Research Centre for Economic Policy at the Erasmus University in Rotterdam.

best tax on the output of a polluting industry and the second-best emissions tax.

Restrictions on the Availability of Tax Instruments

The central idea behind the analysis of second-best taxation is that policymakers are faced with various restrictions on the availability of tax instruments. In particular, let us assume that the objectives of the tax system are threefold: (1) to finance an exogenous level of public expenditure; (2) to redistribute income from high-wage to low-wage households; and (3) to internalize the external effects from production and consumption activities on the level of pollution and, thereby, on environmental quality. Ideally, the government should use a set of personalized lump-sum taxes and transfers to achieve the first two goals and an emissions tax to achieve the third objective. However, both a set of personalized lump-sum taxes and a precisely targeted emissions tax are generally unavailable to the policymaker.

First, households differ in terms of their earnings capacity and, thereby, their abilities to pay taxes. Fairness considerations prescribe that individuals with relatively low earning capacity ought to pay less taxes than people with higher levels of ability. The government faces great difficulties, however, in determining the precise ability level (e.g., the hourly wage rate) of each person. Moreover, private agents lack an incentive to truthfully disclose this information if they can reduce their tax bill by lying; that is, by claiming to possess a lower earnings capacity than their true hourly wage rate. As a result, a set of lump-sum, nondistortionary ability taxes is generally unavailable to the policymaker.

In contrast, observations of annual earnings levels (rather than implicit hourly wage rates) are relatively straightforward to obtain. Moreover, differences in earnings levels across households can be expected to feature a strong correlation with underlying differences in earning capacities. Consequently, policymakers typically rely on distortionary income taxes (and transfers) to finance public outlays and redistribute income across households.

Second, environmental taxes often cannot be targeted directly on emissions, as evidenced by the comprehensive review of Fullerton, Hong, and Metcalf of actual environmental taxes employed by various governments around the world. Instead, taxes are usually imposed on commodities that are imperfectly correlated with the level of emissions. Such levies raise the price of, and thereby lower the demand for, dirty inputs and outputs. As a result, their introduction will generally succeed in reducing the total level of emissions. These output and input levies fail, however, to provide an incentive for the polluting firms to reduce the ratio of the level of emissions to the level of the taxed input or output.

A Simple Rule for the Second-Best Tax on Output of a Polluting Industry

How should the tax system be designed if we take into account (1) the restriction that income taxes rather than lump-sum taxes are used to finance public expenditures, and (2) the constraint that an imperfectly targeted tax on the output of the polluting industry is used rather than an emissions tax? Fullerton, Hong, and Metcalf provide a preliminary answer to this question by analyzing the welfare effects of a tax on output of the polluting industry in an illustrative general equilibrium model with a single representative firm and household, and a preexisting distortionary tax on labor income. In particular, they find that when an emissions tax is unavailable, the constrained efficient solution requires setting the tax on output of the polluting industry equal to the social marginal damages from pollution divided by the marginal cost of public funds and multiplied by the emissions-output ratio.

This is a particularly simple and useful tax rule, as a numerical implementation by Fullerton, Hong, and Metcalf illustrates. Environmental economists have developed a number of methods, such as contingent valuation analysis, to provide an empirical estimate of the social marginal damages from (different sorts of) pollution. Similarly, public finance economists have developed methods to estimate the value of the marginal cost of public funds, which measures the marginal excess burden from the preexisting tax on labor income. Finally, an estimate of the average emissions-output ratio in the particular industry under consideration is required.

The Second-Best Emissions Tax versus the Second-Best Output Tax

The Fullerton-Hong-Metcalf (FHM) rule for the second-best output tax implies that the output tax should still be set to obtain the same effect on the output price as would occur with the desired but unavailable emissions tax, *for given values of the social marginal damages from pollution and of the marginal cost of public funds.* Fullerton, Hong, and Metcalf emphasize this result, but without the proper qualification that I have added in italics. The values of the social marginal damages from pollution and of the marginal cost of public funds are not given, however.

Let us start from the situation in which no tax on emissions or on output of the polluting sector has been imposed, while a proportional income tax is used to finance an exogenous level of public expenditures. Now let us compare the effects from the introduction of a tax on emissions versus the introduction of an (imperfectly targeted) output tax, where both have the same effect on the output price.

The introduction of a tax on emissions will induce a larger increase in the level of environmental quality than the introduction of an output tax. A tax on output reduces the level of pollution only through an increase in the price of, and an associated reduction in demand for, the output of the polluting industry. In addition to the effect on the demand for output of the polluting industry, an emissions tax generates a further reduction in pollution by encouraging firms to switch to a cleaner production technology and a lower emissions-output ratio. Hence, environmental quality will be greater after the introduction of an emissions tax than after the introduction of an output tax that has the same effect on the output price as the emissions tax.

A higher level of environmental quality implies a lower value for the social marginal damages from pollution (measured in terms of private income). As environmental quality expands, people start to care less for the environment relative to income. Since an emissions tax is more successful in reducing the level of pollution than an output tax, the introduction of a tax on emissions will lead to a larger decline in the value of the social marginal damages from pollution than the introduction of an output tax with the same effect on the output price as the emissions tax.

The FHM rule for the optimal second-best tax states that the emissions tax or, if unavailable, the output tax should be raised until its rate corresponds to the social marginal damages from pollution divided by the marginal cost of public funds (multiplied, in the case of an output tax, by the emissions-output ratio). Ignore the value of the marginal cost of public funds for the moment. We have seen that an increase in the emissions tax will lower the value of the social marginal damages from pollution more rapidly than an equivalent increase in the output tax. Hence, the FHM rule must be satisfied at a lower rate of the emissions tax than of the output tax (multiplied by the emissions-output ratio).

That is, even though the second-best tax *rules* for the emissions and output taxes correspond, their second-best tax *rates* differ. In particular, the second-best output tax will have a greater effect on the output price than the desired but unavailable second-best emissions tax, while the optimal level of environmental quality will be lower in the case of an output tax. If the government lacks access to an emissions tax, the efficient level of environmental quality is lower because a tax on output of the polluting industry is less efficient than a tax on emissions in achieving the same reduction in the level of pollution. Yet the output price should be raised more strongly in the case of the second-best output tax in order to partially compensate for missing the target (which results in a higher level of pollution and therefore in a larger value of the social marginal damages from pollution).

Interaction with the Preexisting Income Tax

The FHM rule for the optimal second-best environmental tax reveals that due to the presence of a distortionary income tax, the social marginal damages from pollution must be divided by the marginal cost of public funds in the calculation of the second-best tax rate. This result holds both in the case of an emissions tax and in the case of an output tax. This result appears to suggest that the quantitative adjustment to the environmental tax in order to account for the presence of a preexisting, distortionary income tax is the same in both cases. In other words, the unavailability of lump-sum taxation does not seem to interact with the unavailability of a direct tax on emissions.

In the previous section, however, we arrived at the conclusion that the second-best output tax will have a greater effect on the output price than the desired but unavailable second-best emissions tax. This implies that the tax revenue from the second-best output tax exceeds the revenue from the second-best emissions tax. These surplus tax revenues could then be returned to households through an additional cut in the rate of the proportional income tax, at a given level of public expenditure. Then the income tax rate, and hence the marginal cost of public funds, is smaller in the case of a second-best output tax than in the case of a second-best emissions tax.

Intuitively, the output price should be raised more strongly in the case of an output tax in order to partially compensate for missing the target. As a result, the total revenues from environmental taxation expand, and the income tax and the size of the marginal cost of public funds (MCPF) can be reduced. In other words, the unavailability of a direct tax on emissions appears to somewhat alleviate the welfare losses from the unavailability of lump-sum taxation, as reflected in a lower value of the MCPF (although, of course, an emissions tax, if available, is still preferable to a tax on output of the polluting sector).

Conclusion

In my comments on the contribution of Fullerton, Hong, and Metcalf, I have explored some of the implications of the tax rule that they derive for the optimal second-best tax on the output of a polluting sector. In particular, the government's inability to tax emissions implies a lower level of environmental quality because a tax on output of the polluting industry is less efficient than a tax on emissions in achieving the same reduction in the level of pollution. However, the efficient output price would be raised more strongly in the case of the (second-best) output tax, in order to partially compensate for missing the target. As a result, the total revenues from environmental taxation expand, so that the rate of the income tax and consequently the size of the MCPF can be reduced relative to the case of a second-best tax on emissions.

These implications show that relevant lessons can be learned from a study of the optimal design of a second-best tax system, in which we take into account (1) the constraint that income taxes rather than lump-sum taxes are used to finance public outlays, and (2) the restriction that an imperfectly targeted tax on the output of the polluting industry is used rather than an emissions tax. Fullerton, Hong, and Metcalf provide an important first contribution to this subject. Nevertheless, it should be noted that their analysis rests on a number of simplifying, unrealistic assumptions (as they acknowledge in the conclusion of their paper). For example, their model abstracts from the heterogeneity of firms (regarding the emissions-output ratio) and from the heterogeneity of households (in terms of wage rates). Hence, ample room is still available for extensions and improvements to the study of the optimal design of a second-best tax on output (and inputs) of a polluting sector in the presence of a distortionary income tax.

Neutralizing the Adverse Industry Impacts of CO_2 Abatement Policies
What Does It Cost?

A. Lans Bovenberg and Lawrence H. Goulder

2.1 Introduction

Most studies of U.S. CO_2 abatement policies have focused on the aggregate costs and benefits of these initiatives. Yet the desirability and political feasibility of these policies tend to hinge critically on their distributional impacts. A full assessment of CO_2 abatement options therefore requires attention to distributional impacts.

Some studies—including Poterba (1991), Bull, Hassett, and Metcalf (1994), Schillo et al. (1996), and Metcalf (1998)—have focused on the distribution of impacts of CO_2 abatement policies across household income groups. In these papers, the instrument for CO_2 abatement is usually a carbon tax. This tax is generally found to produce a regressive impact, although this impact is fairly small, especially when one ranks households by measures of lifetime income (such as expenditure) rather than by annual income (which includes transitory shocks and lifetime variations that make annual income a bad indicator of permanent income). Moreover, as indicated by Schillo et al. (1996) and Metcalf (1998), the government can reduce the regressive effect by lowering personal income tax rates at the bottom end of the income scale and by raising public transfers.

A second important distributional dimension is the variation in impacts

A. Lans Bovenberg is professor of economics at the CentER for Economic Research, Tilburg University, and a research fellow of the Centre for Economic Policy Research, London. Lawrence H. Goulder is associate professor of economics at Stanford University, a research associate of the National Bureau of Economic Research, and a university fellow of Resources for the Future.

The authors are grateful to Gilbert Metcalf, Ruud de Mooij, Peter Orszag, Ian Parry, Jack Pezzey, Robert Stavins, Roberton Williams III, and conference participants for helpful comments, and to Derek Gurney and Rudolf Schusteritsch for excellent research assistance.

across industries. CO_2 abatement policies such as carbon taxes or carbon quotas can reduce net output prices in the fossil fuel (carbon supplying) industries and raise costs in industries that intensively employ fossil fuels as inputs. These price and cost impacts have the potential to seriously harm profits, employment, and equity values. The distribution of impacts along these dimensions crucially influences political feasibility, since representatives of fossil fuel producers carry significant weight in the political process.[1] CO_2 abatement policies that pose serious burdens on these industries may stand little chance of political survival.[2]

This paper explores the distributional impacts of various U.S. CO_2 abatement policies in terms of their impacts on profits and equity values for the industries supplying fossil fuels (the coal industry and crude petroleum and natural gas industry) and the industries that rely heavily on fossil fuels as intermediate inputs (e.g., petroleum refining and electric utilities). We examine a range of abatement policies, including policies designed to avoid adverse consequences for the regulated industries. As discussed later, some of the adverse consequences can be avoided through industry-specific corporate tax cuts, direct transfers, and the government's free provision (or "grandfathering")[3] of emissions permits to firms. A main purpose of our paper is to assess the efficiency cost of avoiding adverse impacts through such policies.

To perform this investigation we employ an intertemporal general equilibrium model of the United States.[4] The general equilibrium framework

1. Indeed, the industry-distribution impacts may be more important politically than the household-distribution impacts, since the stakes for each firm from these policies are high, while the impacts of abatement policies on households, although important in the aggregate, are fairly small for each individual household. Under these circumstances, affected firms may be more willing to incur the costs of political mobilization than affected households are. This discussion invokes the notion of political mobilization bias, an idea originated by Olson (1965). For a discussion of this bias and other political transactions cost issues, see Williamson (1996). For an analysis of the implications of political mobilization bias for legislators' choice of environmental policy instruments, see Keohane, Revesz, and Stavins (1998). For a general discussion of political resistance to market-based environmental policy instruments, see Pezzey (1988).

2. This will be the case irrespective of the efficiency properties of abatement policies. In the real world, winners often cannot compensate losers through costless, lump-sum transfers. Hence the most efficient policies—those with the largest net benefits in the aggregate—may not yield actual Pareto improvements. They may only be potentially Pareto-improving. In such a world, political feasibility may require designing policies that avoid serious negative impacts on key stakeholders. This may involve a sacrifice of some of the efficiency gains from the most efficient policy.

3. Grandfathering is a special case of free provision. It is a legal rule whereby "old" entities (e.g., firms subject to previous environmental rules) are exempt from new regulatory requirements and remain bound only to the earlier (and perhaps more lax) regulatory provisions. Under grandfathering, the free provision of permits is linked to current production factors. Newly entering firms are not eligible for a free provision, and investments in new capital are not rewarded.

4. This is the same model as that in Goulder (1995a) and Bovenberg and Goulder (1996, 1997), with some extensions to allow for attention to the industry-specific revenue-recycling and tradable-permits provisions described below.

is especially useful for assessing the incidence of carbon policies. Nearly all of the studies of distributional impacts of carbon policies have employed a partial equilibrium framework that ignores behavioral responses to environmental taxes.[5] These studies impose exogenous incidence assumptions and cannot analyze how behavioral responses affect pollution, efficiency, and distribution. An applied general equilibrium analysis, in contrast, derives tax incidence endogenously from the model-generated behavioral responses. An important distributional consideration is the impact of CO_2 abatement policies on returns to labor and capital. A general equilibrium framework is appropriate for this purpose, since it captures important links between energy markets and factor markets. The model used here is especially useful in this regard because it incorporates forward-looking investment behavior and the adjustment costs associated with the installation or removal of physical capital. These features enable us to consider the capitalization effects of unanticipated policies.[6] Most other general equilibrium models treat capital as perfectly mobile, and thus they cannot successfully examine impacts on profits or equity values. In such models, the impacts on industries are measured largely in terms of the effects on outputs. The results from this paper show that output effects are unreliable indicators of distributional effects.

Earlier analyses of CO_2 abatement policies reveal a tension between promoting economic efficiency, on the one hand, and avoiding serious adverse distributional consequences to key industries, on the other. Goulder, Parry, and Burtraw (1997), Farrow (1999), Fullerton and Metcalf (1998), and Parry, Williams, and Goulder (1999) show that policies that raise revenues and use these revenues to finance cuts in preexisting distortionary taxes have lower costs than policies that do not generate and recycle revenues in this way. The differences in the costs of the two types of policies can be large enough to determine whether the overall efficiency impact—environment-related benefits minus economy-wide costs of abatement—is positive or negative.[7] Yet the latter, less efficient policies impose a smaller financial burden on the regulated industries because they do not charge firms for every unit of their pollution.[8] These considerations suggest a conflict be-

5. An exception is Jorgenson, Slesnick, and Wilcoxen (1992).

6. Our model does not incorporate adjustment costs for industry-specific labor; indeed, labor is perfectly mobile across industries. To the extent that labor faces adjustment costs, one should explore capitalization effects on labor along lines similar to this paper's exploration of such effects on capital.

7. Parry, Williams, and Goulder (1999) show that while a carbon tax with revenues recycled through cuts in marginal income tax rates produces efficiency gains, reducing CO_2 emissions through a system of freely provided (grandfathered) CO_2 permits may be an efficiency-reducing proposition: For any level of emissions reduction, the environmental benefits will fall short of society's costs of abatement!

8. Two revenue-raising policies are an emissions tax and a system of auctioned emissions permits, where the permits are initially auctioned. Under these policies, firms endure costs of emissions abatement and, for whatever emissions they continue to produce, must either pay a tax or purchase permits. In contrast, under a system of freely provided emissions

tween efficiency and political feasibility: The efficient policy—the carbon tax—appears less politically acceptable because it puts too much of a burden on politically mobilized, fossil fuel industries, while the politically more acceptable policy of grandfathered carbon (or CO_2) permits involves serious inefficiencies.

The present study finds that the choice between efficiency and insulating profits of key industrial stakeholders (to enhance political feasibility) may be less problematic than previously thought. We find that desirable distributional outcomes at the industry level can be achieved at relatively low cost in terms of efficiency; without substantial added cost to the overall economy, the government can implement carbon abatement policies that protect profits and equity values in fossil fuel industries. The key to this conclusion is that CO_2 abatement policies have the potential to generate rents that are very large in relation to the potential loss of profit. Under a standard carbon tax policy, these potential rents do not materialize; instead they become revenues collected by the government. In contrast, under a policy involving freely allocated emissions permits, or a policy in which some (inframarginal) emissions are exempted from a carbon tax, firms realize some of the potential rents.[9] Because the potential rents are very large in relation to potential lost profit, the government can protect firms' profits and equity values in fossil fuel industries by enabling firms to retain only a very small fraction of the potential rents. Thus, the government needs to freely allocate (as opposed to auction) only a small percentage of CO_2 emissions permits or, similarly, must exempt only a small fraction of emissions from the base of a carbon tax.[10] Each of these gov-

---

permits requiring equivalent emissions reductions, firms endure the same costs of abatement but do not pay for remaining emissions. Hence, the burden on polluting firms is smaller.

9. Buchanan and Tullock (1975) pointed out that environmental policies can generate significant rents to firms to the extent that such policies cause output to be restricted. They showed that because of such rents, regulated firms can experience higher profits than in the absence of regulation. Our findings in this paper are consistent with Buchanan and Tullock's analysis. Fullerton and Metcalf (1998) emphasize the importance of rents to the overall efficiency costs of policies to reduce pollution. They indicate that efficiency costs are substantially higher under policies producing rents that are not taxed away, in comparison with policies that do not produce rents that are left in private hands. A parallel line of investigation was conducted by Goulder, Parry, and Burtraw (1997), who show that policies that fail to tax away rents are at a disadvantage in terms of efficiency because they fail to generate revenues that can be used to reduce preexisting distortionary taxes. Such policies thus fail to exploit an efficiency-enhancing revenue-recycling effect. In the present study, we examine the extent to which policy-generated rents affect the impacts of CO_2 abatement policies on the profitability and equity values of regulated firms.

10. Correspondingly, if the government were to freely allocate 100 percent of the emissions permits, or exempt 100 percent of inframarginal emissions from the base of a carbon tax, it would generate substantial windfalls to firms: The rents produced and enjoyed by producers would be many times larger than the income losses otherwise attributable to the policy. The government does not need to be nearly this generous in order to safeguard firms' profits and equity values. Our focus on the use of inframarginal exemptions to accomplish distributional objectives is in the spirit of Farrow (1999), who employs a model along the lines of Bovenberg

ernment policies involves only a small sacrifice of potential government revenue. Such revenue has an efficiency value because it can be used to finance cuts in preexisting distortionary taxes. Since these policies give up little of this potential revenue, they involve only a small sacrifice in terms of efficiency. In suggesting that the revenue sacrifice is small relative to potential revenues, our findings complement those obtained in a paper by Vollebergh, de Vries, and Koutstaal (1997), who employ a partial equilibrium model to compare the potential tax revenues and abatement costs that could stem from a carbon tax in the European Union.[11]

Because the potential rents are quite large, it is also possible to devise policies that, with relatively little loss of efficiency, protect not only the fossil fuel industries, but also certain industries (such as petroleum refining and electric utilities) that intensively use fossil energy. These industries would suffer significant profit losses under a standard carbon tax.

We also find that there is a very large difference between preserving firms' profits and preserving their tax payments. Allowing firms to enjoy a dollar-for-dollar offset to their payments of carbon taxes (e.g., through industry-specific cuts in corporate tax rates) substantially overcompensates firms, raising profits and equity values significantly relative to the unregulated situation. This reflects the fact that producers can shift onto consumers most of the burden from a carbon tax. The efficiency costs of such policies are far greater than the costs of policies that do not overcompensate firms. To maintain firms' profits, the government needs to offer tax relief representing only a small fraction of carbon tax payments.[12]

The remainder of the paper is organized as follows. Section 2.2 provides a brief description of the numerical general equilibrium model employed to evaluate the various policy alternatives. Section 2.3 then indicates the links in the model between the various policy alternatives and firms' profits and equity values. Section 2.4 briefly describes the model's data and parameters. Section 2.5 indicates the policies under consideration, and section 2.6 conveys and discusses the results from numerical simulations. The final section offers conclusions.

---

and de Mooij (1994) with one factor of production (labor). Our analysis differs from Farrow's in its consideration of imperfectly mobile capital and its attention to the implications of pollution-abatement policies for firms' profits and equity values.

11. Vollebergh, de Vries, and Koutstaal (1997) calculate the tax revenues and abatement costs that would stem from a carbon tax sufficient to reduce CO_2 emissions by 13 percent in the countries of the European Union. Their results indicate that the revenues from this tax would be many times the policy-generated abatement costs. This suggests that exempting a small share of inframarginal emissions from the carbon tax (or grandfathering a small share of the permits under a permits policy) would be sufficient to compensate the fossil fuel suppliers involved.

12. Felder and Schleiniger (1999) consider the efficiency costs of meeting the constraint that there be no "monetary transfers" (i.e., no change in overall tax payments) as a result of a carbon tax policy. Using a numerical general equilibrium model of Switzerland, they meet this constraint through industry-specific output subsidies or labor subsidies.

2.2 The Model

This section outlines the structure of the intertemporal general equilibrium model employed in this study. The model is an intertemporal general equilibrium model of the U.S. economy with international trade. It generates paths of equilibrium prices, outputs, and incomes for the U.S. economy and the rest of the world under specified policy scenarios. All variables are calculated at yearly intervals beginning in the benchmark year 2000 and usually extending to the year 2075.

The model combines a fairly realistic treatment of the U.S. tax system and a detailed representation of energy production and demand. It incorporates specific tax instruments and addresses effects of taxation along a number of important dimensions. These include firms' investment incentives, equity values, and profits;[13] and household consumption, saving, and labor-supply decisions. The specification of energy supply incorporates the nonrenewable nature of crude petroleum and natural gas and the transitions from conventional to synthetic fuels.

U.S. production divides into the 13 industries indicated in table 2.1. The energy industries consist of (1) coal mining, (2) crude petroleum and natural gas extraction, (3) petroleum refining, (4) synthetic fuels, (5) electric utilities, and (6) gas utilities. The model also distinguishes the 17 consumer goods shown in the table.

2.2.1   Producer Behavior

General Specifications

In each industry, a nested production structure accounts for substitution between different forms of energy as well as between energy and other inputs. Each industry produces a distinct output (X), which is a function of the inputs of labor (L), capital (K), an energy composite (E), a materials composite (M), and the current level of investment (I):

(1) $$X = f[g(L,K), h(E,M)] - \phi(I/K) \cdot I.$$

The energy composite is made up of the outputs of the six energy industries, while the materials composite consists of the outputs of the other industries:

(2) $$E = E(\bar{x}_2, \bar{x}_3 + \bar{x}_4, \bar{x}_5, \bar{x}_6, \bar{x}_7),$$

(3) $$M = M(\bar{x}_1, \bar{x}_8, \ldots, \bar{x}_{13}),$$

---

13. Here the model applies the asset-price approach to investment developed in Summers (1981).

Table 2.1   **Industry and Consumer Goods**

	Gross Output, Year 2000[a]	
	Level	Percent of Total
Industries		
1. Agriculture and noncoal mining	993.6	6.2
2. Coal mining	50.5	0.3
3. Crude petroleum and natural gas	193.7	1.2
4. Synthetic fuels	0.0	0.0
5. Petroleum refining	324.6	2.0
6. Electric utilities	234.6	1.5
7. Gas utilities	211.5	1.3
8. Construction	1508.8	9.5
9. Metals and machinery	799.1	5.0
10. Motor vehicles	541.2	3.4
11. Miscellaneous manufacturing	3365.2	21.3
12. Services (except housing)	5183.6	32.8
13. Housing services	2420.8	15.3
Consumer goods		
1. Food		
2. Alcohol		
3. Tobacco		
4. Utilities		
5. Housing services		
6. Furnishings		
7. Appliances		
8. Clothing and jewelry		
9. Transportation		
10. Motor vehicles		
11. Services (except financial)		
12. Financial services		
13. Recreation, reading, and miscellaneous		
14. Nondurable, nonfood household expenditure		
15. Gasoline and other fuels		
16. Education		
17. Health		

[a] Billions of year 2000 dollars.

where $\bar{x}_i$ is a composite of domestically produced and foreign made input i.[14] Industry indices correspond to those in table 2.1.

Managers of firms choose input quantities and investment levels to max-

14. The functions f, g, and h, and the aggregation functions for the composites E, M, and $\bar{x}_i$, feature a constant elasticity of substitution (CES) and exhibit constant returns to scale. Consumer goods are produced by combining outputs from the 13 industries in fixed proportions.

imize the value of the firm. The investment decision takes account of the adjustment (or installation) costs represented by $\phi(I/K) \cdot I$ in equation (1), where ϕ is a convex function of the rate of investment I/K.[15]

Special Features of the Oil and Gas and Synthetic Fuels Industries

The production structure in the oil and gas industry is somewhat more complex than in other industries to account for the nonrenewable nature of oil and gas stocks. The production specification is

(4) $$X = \gamma(Z) \cdot f[g(L,K), h(E,M)] - \phi(I/K) \cdot I,$$

where γ is a decreasing function of Z, the cumulative extraction of oil and gas up to the beginning of the current period. This captures the idea that as Z rises (or, equivalently, as reserves are depleted), it becomes increasingly difficult to extract oil and gas resources, so that greater quantities of K, L, E, and M are required to achieve any given level of extraction (output). Each oil and gas producer perfectly recognizes the impact of its current production decisions on future extraction costs.[16] Increasing production costs ultimately induce oil and gas producers to remove their capital from this industry.

The model incorporates a synthetic fuel—shale oil—as a backstop resource, a perfect substitute for oil and gas.[17] The technology for producing synthetic fuels on a commercial scale is assumed to become known in 2020. Thus, capital formation in the synthetic fuels industry cannot begin until that year.

All domestic prices in the model are endogenous, except for the domestic price of oil and gas. The path of oil and gas prices follows the assumptions of the Stanford University Energy Modeling Forum.[18] The supply of imported oil and gas is taken to be perfectly elastic at the world price. So long as imports are the marginal source of supply to the domestic economy, domestic producers of oil and gas receive the world price (adjusted for tariffs or taxes) for their own output. However, rising oil and gas prices stimulate investment in synthetic fuels. Eventually, synthetic fuel production plus domestic oil and gas supply together satisfy all of the domestic demand. Synthetic fuels then become the marginal source of supply, so that the cost of synthetic fuel production rather than the world oil price dictates the domestic price of fuels.[19]

---

15. The function ϕ represents adjustment costs per unit of investment. This function expresses the notion that installing new capital necessitates a loss of current output because existing inputs (K, L, E, and M) are diverted to install new capital.

16. We assume representative oil and gas firms: initial resource stocks, profit-maximizing extraction levels, and resource-stock effects are identical across producers.

17. Thus, inputs 3 (oil and gas) and 4 (synthetic fuels) enter additively in the energy aggregation function shown in equation (2).

18. The world price is $19 per barrel in 2000 and rises in real terms by $5.00 per decade. See Gaskins and Weyant (1996).

19. For details, see Goulder (1994, 1995a).

2.2.2   Household Behavior

Consumption, labor supply, and saving result from the decisions of a representative household maximizing its intertemporal utility, defined on leisure and overall consumption in each period. The utility function is homothetic, and leisure and consumption are weakly separable (see the appendixes). The household faces an intertemporal budget constraint requiring that the present value of consumption not exceed potential total wealth (nonhuman wealth plus the present value of labor and transfer income). In each period, overall consumption of goods and services is allocated across the 17 specific categories of consumption goods or services shown in table 2.1. Each of the 17 consumption goods or services is a composite of a domestically and foreign-produced consumption good (or service) of that type. Households substitute between domestic and foreign goods to minimize the cost of obtaining a given composite.

2.2.3   The Government Sector

The government collects taxes, distributes transfers, and purchases goods and services (outputs of the 13 industries). The tax instruments include energy taxes, output taxes, the corporate income tax, property taxes, sales taxes, and taxes on individual labor and capital income. In the benchmark year, 2000, the government deficit amounts to approximately 2 percent of GDP. In the reference case (or status quo) simulation, the real deficit grows at the steady-state growth rate given by the growth of potential labor services. In the policy-change cases, we require that real government spending and the real deficit follow the same paths as in the reference case. To make the policy changes revenue-neutral, we accompany the tax-rate increases that define the various policies with reductions in other taxes, either on a lump-sum basis (increased exogenous transfers) or through reductions in marginal tax rates.

2.2.4   Foreign Trade

Except for oil and gas imports, imported intermediate and consumer goods are imperfect substitutes for their domestic counterparts.[20] Import prices are exogenous in foreign currency, but the domestic-currency price changes with variations in the exchange rate. Export demands are modeled as functions of the foreign price of U.S. exports and the level of foreign income (in foreign currency). The exchange rate adjusts to balance trade in every period.

2.2.5   Equilibrium and Growth

The solution of the model is a general equilibrium in which supplies and demands balance in all markets at each period of time. The requirements

20. Thus, we adopt the assumption of Armington (1969).

of the general equilibrium are that supply equal demand for labor inputs and for all produced goods, that firms' demands for loanable funds match the aggregate supply by households, and that the government's tax revenues equal its spending less the current deficit. These conditions are met through adjustments in output prices, in the market interest rate, and in lump-sum taxes or marginal tax rates.[21]

Economic growth reflects the growth of capital stocks and of potential labor resources. The growth of capital stocks stems from endogenous saving and investment behavior. Potential labor resources are specified as increasing at an exogenous rate.

2.3   Relationships between Carbon Abatement Policies, Profits, and Equity Values

An important component of this study is the impact of CO_2 abatement policies on the profitability of firms that supply fossil fuels. The first part of this section describes in fairly general terms the model's treatment of firms' profits and equity values, while the second part focuses on how abatement policies affect these elements. In all of this section, we concentrate on the fossil fuel industries: coal, and oil and gas.

2.3.1   Profits, Dividends, and Equity Values

Let α denote the (fixed) ratio of carbon emissions to units of fuel (output) in the industry in question, and let τ_c denote the carbon tax rate per unit of emissions. Then the carbon tax τ_c requires a payment of $\tau_c \alpha$ per unit of output X. The equity value of the firm can be expressed in terms of dividends and new share issues, which in turn depend on profits in each period. The firm's profits during a given period are given by

$$(5) \qquad \pi = (1 - \tau_a)[(p - \tau_c\alpha)X - w(1 + \tau_L)L - EMCOST$$
$$- iDEBT - TPROP + LS] + \tau_a(DEPL + DEPR),$$

where τ_a is the corporate tax rate (or tax rate on profits), p is the output price, w is the wage rate net of indirect labor taxes, τ_L is rate of the indirect tax on labor, $EMCOST$ is the cost to the firm of energy and materials inputs, i is the gross-of-tax interest rate paid by the firm, $DEBT$ is the firm's current debt, $TPROP$ is property tax payments, LS is a lump-sum receipt (if applicable) by the firm, $DEPL$ is the current gross depletion allowance, and $DEPR$ is the current gross depreciation allowance. $TPROP$ equals $\tau_p p_{K,s-1} K_s$, where τ_p is the property tax rate, p_K is the purchase price of a unit of new capital, and s is the time period. Current depletion allow-

---

21. Since agents are forward-looking, equilibrium in each period depends not only on current prices and taxes but on future magnitudes as well.

ances, $DEPL$, are a constant fraction β of the value of current extraction: $DEPL = \beta pX$. Current depreciation allowances, $DEPR$, can be expressed as $\delta^T K^T$, where K^T is the depreciable capital stock basis and δ^T is the depreciation rate applied for tax purposes.[22]

The firm's sources and uses of revenues are linked through the cash-flow identity

(6) $$\pi + BN + VN = DIV + IEXP.$$

The left-hand side represents the firm's sources of revenues: profits, new debt issue (BN), and new share issues (VN). The uses of revenues on the right-hand side are investment expenditure ($IEXP$) and dividend payments (DIV). Negative share issues are equivalent to share repurchases and represent a use rather than a source of revenue.

Firms pay dividends equal to a constant fraction, a, of profits gross of capital gains on the existing capital stock and net of economic depreciation. They also maintain debt equal to a constant fraction, b, of the value of the existing capital stock. Thus,

(7) $$DIV_s = a[\pi_s + (p_{K,s} - p_{k,s-1})K_s - \delta p_{K,s}K_s],$$

(8) $$BN_s \equiv DEBT_{s+1} - DEBT_s = b(p_{K,s}K_{s+1} - p_{K,s-1}K_s).$$

Investment expenditure is expressed by

(9) $$IEXP_s = (1 - \tau_K)p_{K,s}I_s,$$

where τ_K is the investment tax credit rate. Of the elements in equation (6), new share issues, VN, are the residual, making up the difference between $\pi + BN$ and $DIV + IEXP$.[23]

Arbitrage possibilities compel the firm to offer its stockholders a rate of return comparable to the rate of interest on alternative assets:

(10) $$(1 - \tau_e)DIV_s + (1 - \tau_v)(V_{s+1} - V_s - VN_s) = (1 - \tau_b)i_s V_s.$$

The parameters τ_e, τ_v, and τ_b are the personal tax rates on dividend income (equity), capital gains, and interest income (bonds), respectively. The return to stockholders consists of the current after-tax dividend plus the after-tax capital gain (accrued or realized) on the equity value (V) of the firm net of the value of new share issues. This return must be comparable to the after-tax return from an investment of the same value at the market rate of interest, i.

Recursively applying equation (10) subject to the usual transversality

---

22. For convenience, we assume that the accelerated depreciation schedule can be approximated by a schedule involving a constant rate of exponential tax depreciation.

23. This treatment is consistent with the "old view" of dividend behavior. For an examination of this and alternative specifications, see Poterba and Summers (1985).

condition ruling out eternal speculative bubbles yields the following expression for the equity value of the firm:

(11)
$$V_t = \sum_{s=t}^{\infty} \left[\frac{1 - \tau_e}{1 - \tau_v} DIV_s - VN_s \right] \mu_t(s),$$

where

$$\mu_t(s) \equiv \prod_{u=t}^{s} \left[1 + \frac{r_u}{1 - \tau_v} \right]^{-1}.$$

Equation (11) indicates that the equity value of the firm is the discounted sum of after-tax dividends net of new share issues.

2.3.2 Abatement Policies and Equity Values

A Standard Carbon Tax

Abatement policies affect equity values by altering firms' profits and the stream of dividends paid by firms. A carbon tax, in particular, will tend to lower the profits of firms in the industries on which the tax is imposed. Figure 2.1 heuristically indicates the carbon tax's implications for the coal industry.[24]

The line labeled S_0 in the figure is the supply curve for coal in the absence of a tax. This diagram accounts for the quasi-fixed nature of capital resulting from capital-adjustment costs. The supply curve S_0 should be regarded as an average of an infinite number of supply curves, beginning with the curve depicting the marginal cost of changes in supply in the first instant and culminating with the marginal cost of changing supply over the very long term, when all factors are mobile. This curve therefore indicates the average of the discounted marginal costs of expanding production, given the size of the initial capital stock. We draw the supply curve as upward sloping, in keeping with the fact that in all time frames except the very long run capital is not fully mobile and production exhibits decreasing returns in the variable factors—labor and intermediate inputs.[25]

The supply curve represents the marginal costs associated with increments in the use of variable factors to increase supply. Capital is the fixed factor underlying the upward-sloping supply curve. The return to this factor is the producer surplus in the diagram. With an upward-sloping supply curve, this producer surplus is positive. The existence of producer surplus does not necessarily imply supernormal profits. Indeed, in an initial long-

---

24. Pezzey and Park (1998, fig. 1) offer a somewhat similar analysis for policies involving tradable emissions permits.

25. In the long run, in contrast, capital is fully mobile, production exhibits constant returns to scale, and the supply curve is infinitely elastic.

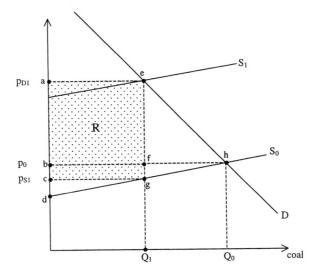

Fig. 2.1   CO$_2$ abatement and profits

run equilibrium, the producer surplus is just large enough to yield a normal return on the capital stock. To illustrate, at the initial equilibrium with a market price p_0 and aggregate quantity supplied Q_0, the producer surplus amounts to the triangular area *bhd*. On a balanced-growth path, this producer surplus yields a normal (market) return on the initial capital stock so that the value of the initial capital stock equals the price of investment (and thus Tobin's q is unity).

Now consider the impact of an unanticipated carbon (coal) tax. The introduction of this tax shifts the supply curve upward to S_1. As a direct consequence, the output price paid by coal consumers increases from p_0 to p_{D1}. However, since supply is not infinitely elastic, the suppliers of coal are not able to shift the entire burden of the tax onto demanders. Indeed, the producer price of coal declines to p_{S1}. This causes producer surplus to shrink to the area *cgd*. Since this triangle is smaller than the initial producer surplus, the return on the initial capital stock (valued at the price of investment goods) falls short of the market rate of return. Hence, to satisfy the arbitrage condition, Tobin's q falls below unity and the owners of the capital stock suffer a capital loss.

This analysis is complicated by the fact that the carbon tax can finance reductions in other taxes, which may imply reductions in costs to firms. This will tend to offset the carbon-tax-induced losses in profits and the associated reductions in equity values. To the extent that the carbon tax revenues finance general (economy-wide) reductions in personal or corporate income taxes, the reductions in tax rates will be small and thus will exert only a small impact on costs to the fossil fuel industries. If the revenues are recycled through tax cuts targeted for the fossil fuel industries, however, the changes in marginal rates can be significant and the beneficial offsetting impact on profits and equity values may be more pronounced.

Effects of Rent-Generating Policies

In figure 2.1, the shaded rectangle R (with area $aegc$) represents the firms' payments of the carbon tax. If the government forgoes some of the carbon tax revenue, and allows producers to retain this potential revenue as a rent, the impact on profits, dividends, and equity values can be fundamentally different. Consider, for example, the case in which the government restricts CO_2 emissions through a system of carbon permits. Since such emissions are proportional to coal combustion, the government can accomplish a given percentage reduction in emissions from coal by restricting coal output by that same percentage through the sale of a limited number of coal-supply permits. For comparability, suppose that the number of permits restricts supply to the level Q_1 in the figure. If the permits are auctioned competitively, then the government (ideally) collects the revenue R from sale of the permits and the effects on firms are the same as under the carbon tax. In contrast, if the permits are given out free (or grandfathered), then the area R represents a rent to firms. The government-mandated restriction in output causes prices to rise, but there is no increase in costs of production (indeed, marginal production costs are lower).

As suggested by figure 2.1, this rent can be quite large and, indeed, can imply substantial increases in profits and equity values to the regulated industries. In the figure, the postregulation profits enjoyed by the firm are given by the sum of areas R and cgd. Here postregulation profits are many times higher than the profit prior to regulation (bhd). Owners of industry-specific capital enjoy a capital gain as Tobin's q jumps above unity. Intuitively, by restricting output, government policy allows producers as a group to exploit their market power and reap part of the original consumer surplus.

Using comparable diagrams, it is straightforward to verify that the magnitude of the profit increase under a system of grandfathered emissions permits depends on

1. The extent of abatement (or number of permits issued relative to "business-as-usual" emissions). The regulation-induced increase in profit is represented by the difference between the areas of the rectangle $aefb$ and the triangle fhg. For incremental restrictions in supply, the former will be larger than the latter (if demand is less than infinitely elastic). Thus, producers must gain. However, this is not necessarily the case as the magnitude of the required reduction in supply gets larger. If the demand curve has a "choke price" (a price above which demand is strictly 0), then the potential rent will shrink to 0 as the extent of abatement approaches 100 percent.

2. The elasticity of supply. The potential to enjoy significant additional profits from restrictions on output is larger, the higher the elasticity of supply. In this case, most of the burden associated with reductions in output is borne by consumers, and a large share of the rents rectangle R represents an increase in producer surplus—that is, most of R will lie above p_0. In contrast, if supply is inelastic (as in the case where adjustment costs are substantial), very little of the rents rectangle R represents an increase in producer surplus because much of it extends below the initial output price p_0. In this case, restrictions in output do not enable producers to expropriate much of the consumer surplus. Thus, the rectangle $aefb$ will be smaller than the triangle fhg, and profits will fall.

A small income share of the fixed factor (i.e., capital in our model) contributes to a large supply elasticity. A large supply elasticity (or flat supply curve) implies that the producer surplus bhd (i.e., the income to the fixed factor) is small while most of the area R will lie above p_0. Hence the additional profits will be large compared to the initial producer surplus.

3. The elasticity of demand. If demand is highly elastic, the policy-induced reduction in output gives firms relatively little market power—only a small part of R will lie above p_0. In contrast, if demand is inelastic, the abatement policy enables firms to exercise substantial market power. In this case, much of R will lie above p_0, and firms will be able to expropriate a considerable amount of the consumer surplus. The aggregate elasticity of demand for a given fossil fuel will reflect the elasticities of substitution inherent in the production functions of domestic users of coal. In addition, the response of demand will reflect the extent to which the government insulates domestic fossil fuel producers from foreign competition. In particular, the elasticity of demand will be smaller, and the potential to enjoy large rents larger, to the extent that the government accompanies taxes on domestic production with levies on imports on fossil fuels and subsidies to exports of such fuels. Carbon taxes or auctioned emissions permits applicable to imported fuels cause the imported fuel prices to rise in tandem with the prices of domestic fuels, thus preventing domestic consumers from shifting demand to imported fuels. Export subsidies ensure that the prices of exported fuels do not rise relative to foreign fuel prices, and thus they help to sustain foreign demand for domestically produced fuels.

Under rent-generating policies, the rectangle R corresponds, in a dynamic context, to

$$(12) \qquad \sum_{s=t}^{\infty} (1 - v)(1 - \tau_a)(p_{D1,t} - p_{S1,t})Q_{1,t}\mu_t(s),$$

where the factors $1 - v$ and $1 - \tau_a$ address the fact that the rents are subject to the personal and corporate income taxes, respectively. Here $Q_{1,t}$ represents gross output (under the policy change) at time t.

A system of grandfathered permits is not the only form of regulation that would enable firms to capture much of R. Firms could capture some of R under a carbon tax policy in which inframarginal emissions (emissions below some trigger level) are exempt from the tax, while all emissions beyond that level face the tax.[26]

In sum, the impact on firms' profits and equity values can be fundamentally different, depending on how much of the area R is retained by firms rather than collected by the government. It also depends on how much of the area R lies above the initial equilibrium price. This, in turn, will depend on the extent of abatement and on elasticities of supply and demand. We return to these issues in the discussion of policy results in section 2.6.

2.4   Data and Parameters

Data documentation is in Cruz and Goulder (1992), which is available on request. Industry input and output flows (used to establish share parameters for production functions) were obtained from the 1988 input-output table from the Bureau of Economic Analysis of the U.S. Department of Commerce. This table is also the source for consumption, investment, government spending, import, and export values by industry. Data on industry capital stocks derive from Bureau of Economic Analysis (1994). Employment by industry was obtained from the October 1990 Survey of Current Business. To form the benchmark data set, these data are projected to the year 2000, based on the average growth of real GDP from the relevant historical period to 1998. Data on the carbon content of fossil fuels were obtained from the 1998 U.S. Department of Energy *Annual Energy Outlook.*

Elasticities of substitution determine the industry and household price elasticities of demand. We derive the production function elasticities by transforming parameters of translog production functions estimated by Dale Jorgenson and Peter Wilcoxen (personal communication). The capital-adjustment cost parameters are based on Summers (1981).

Other important parameters apply to the household side of the model. The elasticity of substitution in consumption between goods and leisure, v, is set to yield a compensated elasticity of labor supply of 0.4.[27] The intertemporal elasticity of substitution in consumption, σ, equals 0.5.[28] The intensity parameter α_c is set to generate a ratio of labor time to the total time endowment equal to 0.44. These parameters imply a value of 0.19 for the interest elasticity of savings between the current period and the next.

---

26. Farrow (1999) describes and evaluates a policy of this sort. See also Pezzey (1992).

27. This lies midway in the range of estimates displayed in the recent survey by Russek (1996).

28. This value falls between the lower estimates from time-series analyses (e.g., Hall 1988) and the higher ones from cross-sectional studies (e.g., Lawrance 1991).

2.5   Abatement Policies Investigated

In nearly all simulations, the tool for abatement is the carbon tax (although we also consider CO_2 quotas or tradable permits, as discussed later). All policies are unanticipated and phased in smoothly (with equal increments to the carbon tax) over a 3-year period beginning in the base year, 2000. The carbon tax is levied "upstream"; that is, the tax is imposed on suppliers of fossil fuels (producers of coal and of oil and gas). To prevent an adverse impact on the international competitive position of fossil-fuel-producing industries, exports of fossil fuels are exempt from the carbon tax, while imports of these fuels are subject to the carbon tax. Nearly all proposals for a U.S. carbon tax include export and import elements of this type.

The policies differ in two main ways: (1) the way the gross revenues from the carbon tax are recycled to the private sector, and (2) the extent to which the policies create and leave rents for the regulated firms. We normalize the carbon tax so that discounted carbon emissions are the same across the policies.[29] We do not allow for public debt policy. Hence, all gross revenues from the carbon tax are immediately returned to the private sector.

2.5.1   Starting Point: Policies without Distributional Adjustments

The first set of policies involves broad-based revenue recycling and thus does not attend to distributional concerns. These policies involve three alternative ways to recycle the revenues: higher lump-sum transfers to households, lower personal income tax rates, and lower corporate income tax rates. We implement these recycling options by using the recycling instrument to endogenously balance the government budget.

The other policies involve additional elements to address important distributional considerations. Thus these policies involve not only environmental neutrality (the reductions in emissions are normalized across policies) and revenue neutrality (all gross revenues are recycled), but also some form of distributional neutrality. The attention to distributional neutrality is motivated by concerns about equity and political feasibility.

2.5.2   Imposing the Requirement of Equity-Value Neutrality

The first group of policies to attend to distributional neutrality adds the constraint that the real[30] value of equity of the principally affected industries must not be changed (i.e., reduced) at the time the abatement policy

---

29. In the simulations, we have approximated environmental neutrality by scaling the results of a uniform carbon tax of $25 per ton carbon by discounted emissions reductions. We find that efficiency outcomes from the model are close to linear within the small range of variation in emissions reductions, so that this type of scaling does not significantly affect the interpretation of results.

30. To express the equity values in real terms, we adopt the ideal price index that is associated with the utility function of the representative household.

is announced and implemented. We call this the requirement of *equity-value neutrality*. The most vulnerable industries are fossil-fuel-supplying industries (coal, and oil and gas), the petroleum refining industry, and the electric utilities industry.[31] The constraint on the value of equity can be interpreted as the requirement that industry-specific production factors not be hurt by the carbon tax. In the model, labor is perfectly mobile across industries, while capital is subject to adjustment costs. Since capital is the only the industry-specific production factor, the effect on the value of capital represents the impact of the carbon tax on industry-specific production factors. Unanticipated policies yield instantaneous changes in the value of industry-specific wealth, as measured by changes in the equity values of various industries.

We consider three mechanisms for achieving equity-value neutrality: (1) industry-specific cuts in corporate tax rates, (2) lump-sum transfers to capital employed in particular industries, and (3) inframarginal exemptions to the carbon tax. Our model abstracts from uncertainty and from heterogeneity across firms in a given industry. In such a model, a policy involving emissions permits, in which a certain fraction of the permits is given out free (rather than auctioned), is equivalent to a carbon tax policy in which the same fraction of (inframarginal) emissions is exempt from the carbon tax.[32] Thus, the third policy can be interpreted as one in which the government controls emissions through emissions permits and freely allocates, or grandfathers, some of these permits. We simulate this policy by imposing a $25 per ton carbon tax and rebating to the firm a share of its tax payment, with the share corresponding to the percentage of emissions that are exempt from the tax. The rebate is lump-sum from the firm's point of view. Under this simulation, output and emissions from the coal and the oil and gas industries rise through time. Hence, this corresponds to an emissions permit policy in which the number of permits in circulation increases through time.

In our simulations, the policies with inframarginal exemptions have the potential to produce dramatic impacts on profits and equity values. For this reason, we perform additional policy experiments involving inframarginal exemptions of various magnitudes. In these experiments, we do not aim to achieve equity-value neutrality, but instead focus on how profits, equity values, and other important variables are affected by the magnitude of the exemptions. The policies introduced under this heading are an emissions permit system in which 100 percent of the permits are grandfathered,

---

31. Thus we focus on four of the six energy industries identified in the model. We give less attention to the natural gas-delivery industry, which experiences considerably smaller impacts from the abatement policies, and the synthetic fuels industry, which does not emerge significantly until 2025.

32. They are equivalent under appropriate scaling of the two policies: The limit on emissions under the permits program must be the same as the level of emissions that occurs under the carbon tax.

and a carbon tax with inframarginal exemptions equal to 50 or 90 percent of first-period emissions under the unregulated status quo. The rents generated by each of these policies face the same taxes as other producer income and thus are subject to the corporate income tax.

These policies involve three instruments to achieve three targets. The carbon tax rate assures environmental neutrality (the same emissions reductions), the adjustment to the personal income tax rate yields revenue neutrality (all additional revenues from the carbon tax must be recycled), and the industry-specific corporate income tax cuts, lump-sum payments, or inframarginal exemptions bring about equity-value neutrality.

2.5.3   Imposing the Requirement of Tax-Payment Neutrality

In the political arena, a popular indicator of distributional neutrality is the tax payment of a particular industry.[33] According to this alternative notion of distributional neutrality, a given industry's overall tax payments from carbon taxes, corporate taxes, property taxes, and indirect labor taxes should remain constant. We term this *tax-payment neutrality*. As instruments for this type of neutrality we consider industry-specific corporate tax cuts and explicit lump-sum transfers to sector-specific capital. As with the simulations involving equity-value neutrality, the tax-payment neutrality simulations involve policy packages in which three instruments achieve three targets.

2.6   Simulation Results

2.6.1   Policies without Distributional Adjustments

Policy 1: Lump-Sum Recycling

We begin by examining the effects of the $25 per ton carbon tax with lump-sum recycling. Results are displayed in column (1) of table 2.2. The table shows the impacts on prices, output, and after-tax profits for years 2002 (2 years after implementation) and 2025.

The coal industry experiences the largest impact on prices and output. In this industry, prices rise by approximately 54 percent by the time the policy is fully implemented (year 2002) and the price increase is sustained at slightly above that level. The price increase implies a reduction in output of about 24 percent in the long run. The other major impacts on prices and output are in the oil and gas, petroleum refining, and electric utilities industries. Although the carbon tax is imposed on the oil and gas industry, the price increase is considerably smaller than in the coal industry, reflecting the lower carbon content (per dollar of fuel) of oil and gas as

---

33. Indeed, in several countries, such as Denmark and the Netherlands, additional environmental taxes raised from energy-intensive industries are earmarked for technology subsidies to this sector.

Table 2.2 Industry Impacts of CO$_2$ Abatement Policies (percentage changes from reference case)

| | No Distributional Adjustments Revenue-Recycling via | | | Equity-Value Neutrality Imposed via | | |
	Lump-Sum Transfer (1)	PIT Rate Reduction (2)	CIT Rate Reduction (3)	Industry-Specific CIT Rate Cut (4)	Industry-Specific Lump-Sum Payment (5)	Partial Grandfathering of Emissions Permits (6)
Gross of tax output price (2002, 2025)						
Coal mining	54.5, 57.0	54.5, 57.0	54.5, 57.1	54.3, 55.9	54.5, 57.0	54.5, 57.0
Oil and gas	13.2, 8.3	13.2, 8.3	13.2, 8.3	13.2, 8.3	13.2, 8.3	13.2, 8.3
Petroleum refining	6.4, 5.1	6.4, 5.1	6.4, 5.1	6.3, 4.7	6.4, 5.1	6.4, 5.1
Electric utilities	2.5, 5.6	2.5, 5.5	2.5, 5.7	2.5, 5.1	2.5, 5.5	2.5, 5.5
Average for other industries	−0.6, −0.6	−0.6, −0.7	−0.6, −0.7	−0.6, −0.6	−0.6, −0.7	−0.6, −0.7
Output (2002, 2025)						
Coal mining	−19.2, −23.6	−19.1, −23.3	−19.2, −23.5	−18.9, −21.9	−19.1, −23.3	−19.1, −23.3
Oil and gas	−2.0, −3.9	−2.1, −4.4	−1.3, −2.5	1.5, −0.4	−2.1, −4.4	−2.1, −4.3
Petroleum refining	−7.9, −5.6	−7.8, −5.3	−7.9, −5.3	−7.8, −5.0	−7.8, −5.3	−7.8, −5.3
Electric utilities	−3.0, −5.7	−3.0, −5.4	−3.0, −5.5	−2.9, −5.0	−3.0, −5.4	−3.0, −5.4
Average for other industries	−0.2, −0.3	−0.1, 0.1	−0.1, 0.1	−0.1, 0.1	−0.1, 0.1	−0.1, 0.1
After-tax profits (2002, 2025)						
Coal mining	−32.5, −25.8	−32.3, −25.5	−32.0, −25.1	−19.9, −12.0	−32.3, −25.5	−16.6, −10.4
Oil and gas	−2.3, −3.5	−2.3, −3.9	−0.3, −0.4	−6.6, −9.1	−2.3, −3.8	1.3, −1.8
Petroleum refining	−9.2, −3.9	−9.1, −3.6	−8.1, −2.7	−5.5, −0.9	−9.1, −3.6	−9.1, −3.6
Electric utilities	−7.7, −5.2	−7.4, −4.8	−7.1, −4.2	−5.2, −2.7	−7.4, −4.8	−7.4, −4.8
Average for other industries	−0.9, −1.1	−0.7, −0.7	0.2, 0.2	−0.7, −0.7	−0.7, −0.7	−0.7, −0.8

	Inframarginal Exemptions			Tax Payment Neutrality Imposed via	
	Exempt 100% of Actual Emissions (100% Grandfathering of Emissions Permits) (7)	Exempt 50% of BAU Emissions (8)	Exempt 90% of BAU Emissions (9)	Industry-Specific CIT Rate Cut (10)	Industry-Specific Lump-Sum Payment (11)
Gross of tax output price (2002, 2025)					
Coal mining	54.5, 57.0	54.5, 57.0	54.5, 57.0	54.3, 56.0	54.5, 57.0
Oil and gas	13.2, 8.3	13.2, 8.3	13.2, 8.3	13.2, 3.0	13.2, 8.3
Petroleum refining	6.4, 5.1	6.4, 5.1	6.4, 5.1	6.4, 2.0	6.4, 5.1
Electric utilities	2.4, 5.5	2.5, 5.5	2.5, 5.6	2.5, 5.6	2.5, 5.6
Average for other industries	-0.6, -0.6	-0.6, -0.7	-0.6, -0.6	-0.6, -0.4	-0.6, -0.6
Output (2002, 2025)					
Coal mining	-19.0, -23.3	-19.1, -23.4	-19.1, -23.4	-19.4, -23.3	-19.1, -23.5
Oil and gas	-2.0, -4.2	-2.1, -4.3	-2.1, -3.5	7.5, 23.9	-2.0, -4.1
Petroleum refining	-7.8, -5.4	-7.8, -5.4	-7.8, -5.4	-8.0, -2.8	-7.9, -5.5
Electric utilities	-3.0, -5.5	-3.0, -5.4	-3.0, -5.5	-3.0, -5.0	-3.0, -5.6
Average for other industries	-0.1, -0.1	-0.1, 0.0	-0.1, -0.1	-0.2, -0.2	-0.2, -0.2
After-tax profits (2002, 2025)					
Coal mining	542.7, 526.9	555.9, 351.5	957.2, 653.0	-19.9, -12.8	-32.4, -25.7
Oil and gas	21.4, 9.4	18.0, 5.5	34.3, 13.7	25.7, 45.7	-2.3, -3.6
Petroleum refining	-9.1, -3.8	-9.1, -3.7	-9.2, -3.8	-10.5, -2.6	-9.2, -3.8
Electric utilities	-7.5, -5.0	-7.4, -4.9	-7.5, -5.0	-8.6, -4.7	-7.6, -5.1
Average for other industries	-0.7, -0.9	-0.7, -0.8	-0.8, -0.9	-1.0, -1.0	-0.9, -1.0

Table 2.3   Equity-Value and Efficiency Impacts

	No Distributional Adjustments Revenue-Recycling via			Equity-Value Neutrality Imposed via		
	Lump-Sum Transfer (1)	PIT Rate Reduction (2)	CIT Rate Reduction (3)	Industry-Specific CIT Rate Cut (4)	Industry-Specific Lump-Sum Payment (5)	Partial Grandfathering of Emissions Permits[a] (6)
Equity values of firms (year 2000) (percentage changes from reference case)						
Agriculture and noncoal mining	−0.7	0.2	1.4	0.1	0.1	0.1
Coal mining	−28.4	−27.8	−27.2	0.0	0.0	0.0 (0.043)
Oil and gas	−4.8	−5.0	−0.7	0.0	0.0	0.0 (0.150)
Petroleum refining	−5.2	−4.5	−3.3	0.0	0.0	−4.5
Electric utilities	−6.3	−5.4	−4.4	0.0	0.0	−5.4
Natural gas utilities	−1.4	−0.4	0.6	−0.3	−0.4	−0.4
Construction	−0.3	1.5	2.1	1.8	1.5	1.5
Metals and machinery	−1.2	−0.4	1.0	−0.6	−0.4	−0.4
Motor vehicles	−0.6	0.2	1.6	0.2	0.2	0.2
Miscellaneous manufacturing	−0.8	−0.1	1.4	−0.2	−0.1	−0.1
Services (except housing)	−0.8	0.2	1.2	0.2	0.1	0.1
Housing services	0.0	0.4	0.1	0.4	0.4	0.4
Total	−0.8	−0.1	0.9	0.1	0.1	0.0
Efficiency cost						
Absolute (billions of year 2000 dollars)	−817	−471	−374	−345	−482	−506
Per ton of CO_2 reduction	102.6	60.0	47.7	46.9	61.4	64.4
Per dollar of carbon tax revenue	.73	.42	.34	.30	.43	.50

compared with coal. There are significant increases in prices and reductions in output in the petroleum refining and electric utilities industries as well, in keeping with the significant use of fossil fuels in these industries. The reductions in output are accompanied by reductions in annual after-tax profits. Associated with these output reductions is a reduction in CO_2 emissions of approximately 18 percent.[34]

34. This is the reduction in emissions associated with domestic consumption of fossil fuels. It accounts for the carbon content of imported fossil fuels, and excludes the carbon content of exported fossil fuels. These figures do not adjust for changes in the carbon content of imported or exported refined products. The percentage change in emissions is the percentage change between the reference case and policy-change case in the present value of emissions,

Table 2.3   (continued)

| | Inframarginal Exemptions | | | Tax-Payment Neutrality Imposed via | |
	Exempt 100% of Actual Emissions (100% Grandfathering of Emissions Permits) (7)	Exempt 50% of BAU Emissions (8)	Exempt 90% of BAU Emissions (9)	Industry-Specific CIT Rate Cut (10)	Industry-Specific Lump-Sum Payment (11)
Equity values of firms (year 2000) (percentage changes from reference case)					
Agriculture and noncoal mining	0.0	0.0	−0.1	−0.2	−0.4
Coal mining	1,005.4	709.1	1,284.0	1,283.3	4,283.7
Oil and gas	29.2	22.3	43.1	43.2	117.2
Petroleum refining	−4.7	−4.7	−4.8	−4.8	−12.1
Electric utilities	−5.7	−5.6	−5.7	−5.8	−12.9
Natural gas utilities	−0.8	−0.6	−0.8	−0.9	−1.1
Construction	1.0	1.1	0.8	0.7	0.2
Metals and machinery	−0.5	−0.6	−0.7	−0.7	−0.9
Motor vehicles	0.1	0.0	−0.1	−0.2	−0.4
Miscellaneous manufacturing	−0.2	−0.3	−0.4	−0.4	−0.6
Services (except housing)	−0.1	−0.1	−0.2	−0.3	−0.5
Housing services	0.1	0.3	0.3	0.2	0.1
Total	1.1	0.7	1.3	1.2	4.5
Efficiency cost					
Absolute (billions of year 2000 dollars)	−751	−549	−611	−355	−713
Per ton of CO$_2$ reduction	95.1	69.7	77.4	61.1	89.9
Per dollar of carbon tax revenue	—	.79	1.64	.32	.64

[a]Numbers in parentheses are proportion of emssions permits required to achieve equity-value neutrality.

The reductions in after-tax profits are associated with reductions in equity values. As shown in table 2.3, the largest equity-value impacts are in the coal industry, where such values fall by approximately 28 percent. The reductions in equity values in the oil and gas, petroleum refining, and electric utilities industries are also substantial, in the range of 4.8 to 6.3 percent. As indicated in the table, the impacts on equity values of other industries are relatively small.

---

where the emissions stream is discounted using the after-tax interest rate. If marginal environmental damages from emissions are constant, the percentage changes in discounted emissions will be equivalent to percentage changes in damages.

Table 2.3 also indicates efficiency impacts. We employ the equivalent variation to measure these impacts. This is a gross measure because our model does not account for the benefits associated with the environmental improvement from reduced emissions. As indicated in the table, the policy implies a gross efficiency loss of approximately $103 per ton of emissions reduced, or $0.73 per dollar of discounted gross revenue from the carbon tax.

Policy 2: Personal Income Tax Recycling

Policy 2 recycles the revenues through personal income tax (PIT) cuts rather than via lump-sum payments. As shown in column (2) of table 2.2, such recycling does not alter the impacts of the carbon tax on prices and output very much. Furthermore, such recycling only slightly attenuates the impacts on profits and equity values in the most affected industries. However, as indicated in table 2.3, this form of recycling reduces the economy-wide efficiency costs by over 40 percent. The equivalent variation is approximately $60 per ton of emissions reduced, and $0.42 per dollar of discounted carbon tax revenues. The reason for the smaller efficiency losses with PIT recycling is that, by lowering the marginal rates of the PIT, this recycling helps lower the distortionary costs of the PIT. This efficiency consequence has been termed the *revenue-recycling effect.* Despite the lower distortionary taxes, the carbon tax package still imposes gross efficiency costs because it tends to raise output prices and thereby reduce real returns to labor and capital. This *tax-interaction effect* tends to dominate the revenue-recycling effect. Hence the carbon tax still involves an overall economic cost (abstracting from the environmental benefits), even when the revenues are devoted to cuts in the PIT.[35]

Policy 3: Corporate Income Tax Recycling

The carbon tax revenue can be used instead to reduce the corporate income tax (CIT). This type of recycling further reduces the gross efficiency costs of emission reductions to 34 percent of discounted carbon tax revenues. Thus, CIT recycling appears to be more efficient than PIT recycling. This indicates that in the model the CIT is more distortionary than the PIT in the initial equilibrium. Although there is considerable disagreement as to the distortionary impacts of the CIT, these results are consistent with the prevailing results from other applied general equilibrium analyses.[36] In

---

35. This exemplifies the now-familiar result that, abstracting from the value of the environmental improvement they generate, green taxes tend to be more costly than the ordinary taxes they replace. Although this is the central result, the opposite outcome can arise when the preexisting tax system is suboptimal along nonenvironmental dimensions (e.g., involves overtaxation of capital relative to labor) and the introduction of the environmental reform helps alleviate the nonenvironmental inefficiency. For analysis and discussion of this issue see Bovenberg and de Mooij (1994), Parry (1995), Goulder (1995b), Bovenberg and Goulder (1997, 2000), and Parry and Bento (2000).

36. See, for example, Jorgenson and Yun (1991), and Goulder and Thalmann (1993).

the U.S. economy, taxes on capital investment such as the CIT appear to be more distortionary than labor income taxes or the PIT (a tax on both capital and labor income).

A cut in the CIT rate benefits the owners of capital because it reduces the tax burden on earnings from the installed capital stock. As a result, compared to the cases with lump-sum recycling or PIT recycling, in the most affected industries (coal, oil and gas, petroleum refining, and electric utilities), firms' equity values fall less. Indeed, the value of equity in the oil and gas sector is almost unaffected by this policy package. While CIT recycling significantly changes the impact on equity values, it makes relatively little difference to output patterns. This indicates that changes in industry output are an unreliable measure of the impact of environmental policy on the real earnings of industry-specific production factors.

2.6.2   Policies Achieving Equity-Value Neutrality

We now consider policies that introduce an additional instrument to alter the distributional impacts. We start with results from policies that impose the requirement of equity-value neutrality.

Policy 4: Industry-Specific Cuts in the Corporate Income Tax

Policy 4 achieves equity-value neutrality through industry-specific adjustments to CIT rates (with the remainder of the revenues recycled via cuts in the PIT). This appears to be an efficient way to attain such neutrality. In fact, as indicated in table 2.3, the gross efficiency losses are smaller than in the case where this constraint is not imposed (policy 2), suggesting that there is no trade-off between efficiency and distributional neutrality in this case. Two factors help explain this result. First, in our model the CIT is more distortionary than the PIT, as indicated by the difference in efficiency costs of policies 2 and 3. Under policy 4, the efficiency benefit from cutting the CIT rate is offset by the need to finance these cuts through the PIT—that is, the PIT cannot be lowered as much under this policy as under policy 2, which involves no CIT cuts. Since the CIT is more distortionary than the PIT, there is an overall efficiency benefit from this tax swap.

A second reason for the relatively low efficiency cost relates to tax distortions in the fossil fuel industries. The carbon tax significantly raises the overall taxation of energy industries relative to other industries. On nonenvironmental grounds, these industries are overtaxed relative to other industries. Targeted CIT cuts undo some of the nonenvironmental distortions attributable to the carbon tax. We have verified this effect by performing an additional simulation experiment that (like policy 3) recycles some carbon tax revenues through reductions in the overall CIT rate, but also (like policy 4) recycles sufficient revenues in the form of additional CIT cuts for the energy industries to preserve equity values in those industries. This policy involves an efficiency cost of $40.9 per ton, significantly lower than the efficiency cost per ton under policy 3. The difference between this policy and policy 3 is that this policy includes the additional targeted CIT cuts. Thus there is a (gross) efficiency benefit from reducing the "excess" taxation of energy industries under the carbon tax.[37] This is a second reason why policy 4's efficiency cost is lower than that of policy 2.

Policy 5: Industry-Specific Lump-Sum Transfers

Policy 5 produces equity-value neutrality through industry-specific lump-sum transfers. This appears to be an inexpensive way to ensure distributional neutrality: Relatively small lump-sum transfers ensure that the real value of equity is not affected by the carbon tax. These small transfers absorb relatively little revenue, allowing PIT rates to be cut almost as much as under policy 2. The cost of emissions abatement (relative to policy 2) is raised by only a small amount, from $60.0 to $61.4 per ton.

Policy 6: Grandfathered Emissions Permits

As discussed in section 2.2, emissions quotas or permits, by forcing firms to restrict output, can create rents for the regulated industries. At the same time, there are costs to these industries connected with the reduction in output and the associated need to remove capital or retire capital prematurely.[38] The question arises whether the policy-induced rents might be sufficient to compensate firms for the other costs associated with abatement.[39]

We consider this issue with policy 6. Here we introduce a carbon tax, but grant firms exemptions to this tax for a certain percentage of their actual emissions. It deserves emphasis that the value of the exemption, although tied to actual emissions in the industry (in the aggregate), is exogenous from the firm's point of view. As discussed in section 2.5, this experiment can also be interpreted as a policy in which the government introduces a tradable permits program, but grandfathers a percentage of the permits. The special case (considered later) in which the firm enjoys a 100 percent (inframarginal) exemption from the carbon tax corresponds to the case of fully grandfathered emissions permits.[40]

---

37. We are grateful to Ruud de Mooij for suggesting this diagnostic experiment.

38. There are other transition costs, such as the unemployment costs that may result from reduced output. These are not captured in our analysis.

39. This policy imposes equity-value neutrality only for the coal and oil and gas industries, since the carbon tax (and its exemptions) or emissions permits apply only to these industries. The government would need to invoke additional instruments to achieve equity-value neutrality in other industries.

40. Three modeling assumptions underlying this correspondence should be noted. First, the equivalence between a carbon tax policy and a carbon emissions-permits policy would not hold in a more general model in which regulators faced uncertainty. In the presence of uncertainty, taxes and permits policies intended to lead to a given level of emissions will generally yield different aggregate emissions ex post. Second, we assume that a cost-effective allocation of emissions responsibilities is achieved under the permits policy. This implicitly assumes that any differences in abatement costs (associated with heterogeneity in firms' pro-

Perhaps surprisingly, only a very small percentage of emissions permits need to be grandfathered in order to achieve equity-value neutrality. Only 15 percent need to be grandfathered in the oil and gas industry, and an even smaller percentage—4.3 percent—must be grandfathered in the coal industry! As a result, the goal of distributional neutrality can be achieved at a small cost in terms of efficiency. Earlier research has made clear that there is a trade-off between efficiency and political feasibility associated with the implementation of an emissions permit program: Efficiency requires the auctioning of permits (no grandfathering), whereas political feasibility calls for grandfathering. These results indicate that the trade-off may be relatively benign; it only takes a small amount of grandfathering, and a small sacrifice in efficiency, to preserve equity values in the industries that otherwise would suffer most from CO_2 abatement regulations.[41]

Why does a small amount of grandfathering go a long way? As suggested by the discussion in section 2.3, the gain offered to regulated firms by exemptions to a carbon tax or by the free allocation of emissions permits is enhanced to the extent that elasticities of supply are high and elasticities of demand are low. In this model, the elasticity of supply is determined by the share of cash flow (payments to owners of the quasi-fixed factor, capital) in overall production cost, along with the specification of adjustment costs. We find that for the coal and oil and gas industries, cash flow in the unregulated situation is quite small relative to production cost, which contributes to a higher supply elasticity. In addition, although adjustment costs restrict the supply elasticity in the short run, under our central values for parameters the average elasticity (taking into account the medium and long runs) is fairly high. Indeed, the long-run elasticity in the coal industry is infinite because of the assumption of constant returns to scale.[42] These conditions imply that most of the cost from abatement policies is shifted onto demand.

---

duction methods) are ironed out through trades of permits. Third, our model does not distinguish between new and old firms (although it does distinguish between installed and newly acquired capital). The model's treatment of grandfathering is most consistent with a situation in which only established firms enjoy the freely offered emissions permits, where these permits are linked to the (exogenous) initial (or old) capital stock.

41. Table 2.3 reveals that it is more costly to achieve equity-value neutrality through partial grandfathering of emissions permits (policy 6) than through lump-sum payments to firms (policy 5). The difference can be attributed to differences in the treatment of importers of fossil fuels under the two policies. Both policies can be interpreted as a carbon tax plus an inframarginal lump-sum rebate. Under policy 5, the rebate or lump-sum payment is offered only to domestic fossil fuel producers. In contrast, under policy 6, the rebate (via grandfathering) is offered both to domestic fossil fuel producers and to importers of fossil fuels. Under policy 6, the government is somewhat more generous and forgoes more tax revenue; hence the added efficiency cost. Another difference between the policies is that the petroleum refining and electric utilities industries receive direct compensation under policy 5, but enjoy no protection under policy 6. This would tend to raise the costs of policy 5 relative to policy 6. However, this cost impact is more than offset by the differences just described in the treatment of importers.

42. In the oil and gas industry, the presence of a fixed factor implies decreasing returns even in the long run.

Table 2.4   **Price Responses under Carbon Tax**

	2000	2001	2002	2004	2010	2025	2050
Coal industry							
Gross of carbon tax output price	1.1769	1.360	1.546	1.551	1.560	1.570	1.570
Net of carbon tax output price	0.986	0.978	0.973	0.978	0.995	0.995	0.998
Crude petroleum and natural gas industry							
Gross of carbon tax output price	1.046	1.090	1.132	1.125	1.109	1.083	1.059
Net of carbon tax output price	1.000	1.000	1.000	1.000	1.000	1.000	1.000

Note: Table shows ratio of price under policy change to reference-case price. Results are for policy 2, a $25 per ton carbon tax with revenues recycled through reductions in personal income tax rates. Coal and oil and gas price responses are very similar under the other carbon tax policies.

Table 2.4 bears this out. It displays the impact of a revenue-neutral carbon tax policy (policy 2) on gross and net output prices in different years. In the coal industry, the net-of-tax coal price falls a bit (relative to the reference-case price) in the short run, but the carbon tax is fully shifted in the long run. Even in the short run, over 90 percent of the tax is shifted onto consumers of coal. In the oil and gas industry, the tax is entirely forward-shifted in all years, reflecting the fact that the United States is regarded as a price-taker with respect to oil and gas.[43]

In terms of the analysis of section 2.3, the ability to shift forward the costs of regulation means that most of the R rectangle lies above the initial price. When the initial producer surplus or cash flow is small in relation to production cost, owners of the quasi-fixed factor (capital) can be fully compensated for the costs of regulation if they are given just a small piece of the R rectangle through grandfathering.

This is confirmed in table 2.5. The table shows the dynamic equivalent of the R rectangle under policy 2 (carbon tax with recycling through personal income tax cuts) and policy 6. To enhance comparisons between policies 2 and 6, we will interpret each policy as involving emissions permits; policy 2 is the case of emissions permits with no grandfathering, while policy 6 involves partial grandfathering to yield equity-value neutrality. (As mentioned, one could also interpret these as carbon tax policies, where policy 2 includes no inframarginal exemption and policy 6 involves sufficient exemptions to preserve equity-value neutrality.) Column (2) of table 2.5 shows that policy 2 causes equity values to fall by $4.9 billion in the coal industry. This policy requires the firm to purchase $119 billion of

43. The real net-of-tax price of oil and gas is the only exogenous price in the model. This price is assumed to increase at a rate of 2.7 percent per year (in keeping with baseline assumptions employed by the Energy Modeling Forum at Stanford University). Hence, the ratio of the (constant) real carbon tax rate to the (rising) net-of-tax price declines through time. This explains why, in table 2.4, the percentage increase in the gross-of-tax price of oil and gas declines after 2004.

Table 2.5   **Carbon Payments and Equity Values**

| | Industry Equity Value | | Present Value of Potential Carbon Payments | α − (Fraction of Tax Payments Exempted or Permits Grandfathered) | Present Value of Actual Carbon Payments $[(1-\alpha)\,(3)]$ | Present Value of Inframarginal Exemption or Grandfathered Permits $[\alpha\,(3)]$ |
	Level (1)	Difference from Reference Case (2)	(3)	(4)	(5)	(6)
Coal industry						
Reference case	17.6	—	—	—	—	—
Standard carbon tax (policy 2)	12.7	−4.9	119.5	—	119.5	—
Carbon tax with inframarginal exemption, carbon permits with partial grandfathering (policy 6)	17.6	—	119.6	0.043	114.5	5.1
Oil and gas industry						
Reference case	187.5	—	—	—	—	—
Standard carbon tax (policy 2)	178.0	−9.5	65.8	—	65.8	—
Carbon tax with inframarginal exemption, carbon permits with partial grandfathering (policy 6)	187.5	—	66.0	0.150	56.1	9.9

Note: Values in billions of year 2000 dollars. Values columns (4)–(6) are net of deductions to corporate and personal income taxes, as indicated in equation (12).

carbon emissions permits. "Potential carbon payments" (col. [3]) is the analog to the R rectangle: It is the permit price times the number of permits employed in the coal industry. For the coal industry, this value is $119 billion. Under policy 2, all of the permits are in fact auctioned, and thus the actual carbon payments are the same as the potential payments. Note that this number is very large in relation to the reduction in equity values— $4.9 billion—suffered by the industry. If only a small fraction of the $119 billion could be retained by firms, they would be compensated for the $4.9 billion loss. Under policy 6, firms can in fact retain a fraction of R. Enabling firms to retain just 4.3 percent of the potential carbon payments is worth an amount comparable to the $4.9 billion and prevents any reduction in equity values.

Thus, in the coal industry, a very small amount of grandfathering preserves equity values. This reflects the fact that the potential tax payment (the R rectangle) is large relative to the loss in equity value in the absence of grandfathering. This, in turn, reflects the small share of cash flow in production cost and the large elasticity of supply, as discussed previously. The result is similar in the oil and gas industry. However, in this industry the potential carbon payments are not as large in relation to the loss of equity value. Hence, the firm must be relieved of a somewhat larger fraction (15 percent) of these payments to suffer no loss of equity value.

While policy 6 preserves profits in the fossil fuel industries, it does not insulate all industries from negative impacts on profits. The petroleum refining and electric utilities industries—which use fossil fuels (carbon) most intensively—also endure noticeable losses of profit and equity values, as indicated by table 2.3. Protecting these industries would require expanded policies involving additional instruments.[44]

Policies 7–9: Other Policies Involving Inframarginal Exemptions

The emissions permits policy just analyzed is one of many potential policies that offer inframarginal exemptions to the regulated firms. Before investigating policies imposing tax-payment neutrality, it seems worthwhile to examine some related policies that grant inframarginal exemptions from carbon taxation. Under policy 7 we consider the limiting case where 100 percent of actual emissions are (inframarginally) exempt from the carbon tax or, equivalently, where all emissions permits are grandfathered. The results are shown in tables 2.2 and 2.3. Full grandfathering leads to very large increases in equity values in the regulated industries—especially the coal industry. These large increases are consistent with the predicted magnitudes of the rents these policies generate and the associated increases in the discounted value of after-tax cash flow.

We also consider some policies that offer exemptions based on business-as-usual (BAU) emissions. In particular, we examine the case where firms receive exemptions from emissions corresponding to 50 or 90 percent of their first-period emissions in the unregulated situation. These simulations differ in two ways from the earlier experiments involving exemptions. First, the exemptions are tied to BAU emissions rather than to the actual emissions occurring under the new policy. Second, the exemptions are constant through time. (In the earlier experiments, actual emissions tend to grow with time, which means that the number of permits in circulation grows as well.) Since we model a growing economy in which outputs of all industries tend to increase through time (even under a carbon tax), the exemption represents a diminishing percentage of actual output.

These policies (8 and 9) are less generous to firms than the policy involving 100 percent grandfathering, but more generous than policy 6, which grandfathered just enough permits to assure equity-value neutrality. As shown in table 2.3, offering a permanent exemption equal to 50 percent of

44. One possible extension is to employ input subsidies for selected downstream industries. Such a subsidy could insulate downstream users from higher fuel prices. We are grateful to Ruud de Mooij for suggesting this option to us. Another possibility would be to give downstream users some of the carbon (fossil fuel) permits. This effectively is a lump-sum transfer to downstream users; such users could sell the permits to fossil fuel suppliers and earn revenues that compensate them for the higher costs of fossil fuels.

first-year BAU emissions is enough to increase equity values of firms relative to the unregulated situation: Equity values rise by a factor of 7 in the coal industry and by approximately 22 percent in the oil and gas industry. A permanent exemption equal to 90 percent of first-year BAU emissions raises equity values by more than a factor of 12 in the coal industry and by approximately 43 percent in the oil and gas industry. The government forgoes more revenues under these policies than under policy 6; accordingly, the efficiency costs of these policies are somewhat higher than under policy 6.

2.6.3   Policies Achieving Tax-Payment Neutrality

Policy 10: Industry-Specific Cuts in the Corporate Income Tax

Policy experiments 10 and 11 invoke additional instruments (relative to policy 2) to yield tax-payment neutrality for the coal, oil and gas, petroleum refining, and electric utilities industries. In experiment 10, we introduce industry-specific CIT cuts to achieve such neutrality, with a constraint that the CIT cuts cannot bring the industry's tax rate below 0. It turns out that this constraint is binding for the coal industry; under this policy, the CIT rate for this industry reaches 0 before tax-payment neutrality is achieved. This reflects the fact that this industry's carbon tax payments are very large relative to CIT payments under the status quo. In the oil and gas industry, tax-payment neutrality is achieved when the CIT rate is lowered to 0.17 from its initial value of 0.42.

The CIT reductions that move toward tax-payment neutrality (in the coal industry) or achieve such neutrality (in the oil and gas industry) are much larger than the reduction necessary to achieve equity-value neutrality (policy 4), and they imply extremely large increases in equity values relative to the unregulated situation. The reason is as follows. As mentioned earlier, the "average" of short-, medium-, and long-run elasticities of supply in the model is fairly high. Thus, overall, firms are able to shift onto demanders a large fraction of the carbon tax. Because producers bear only a small share of the tax burden, only a small CIT cut is needed to undo the potential impact of a carbon tax on profits and equity values. In contrast, a CIT cut that achieves tax-payment neutrality vastly overcompensates firms in terms of the real burden of the carbon tax to producers— most of the tax is shifted onto consumers.

Policy 11: Industry-Specific Lump-Sum Transfers

When we maintain tax-payment neutrality through lump-sum recycling, the efficiency costs are fairly high. As in the case with industry-specific CIT cuts, lump-sum recycling substantially overcompensates firms in terms of profits and equity values. Again the reason is that most of the carbon tax's burden falls on demanders rather than suppliers. But the case

Table 2.6   Sensitivity Analysis

	Policy 1: Carbon Tax with Recycling via Lump-Sum Transfer			Policy 2: Carbon Tax with Recycling via PIT Rate Reduction			Policy 6: Carbon Tax with Partial Grandfathering of Emissions Permits for Equity-Value Neutrality[a]			Policy 7: Carbon Tax with 100% Inframarginal Exemption (Emissions Permits with 100% Grandfathering)		
A. Carbon Tax Rate ($)	12.50	25.00	50.00	12.50	25.00	50.00	12.50	25.00	50.00	12.50	25.00	50.00
Change in equity value of firms, year 2000 (%)												
Coal mining	-17.0	-28.4	-42.8	-16.6	-27.8	-41.9	0.0 (0.037)	0.0 (0.043)	0.0 (0.053)	554.9	1,005.4	1,749.6
Oil and gas	-2.6	-4.8	-9.3	-2.8	-5.0	-9.4	0.0 (0.151)	0.0 (0.150)	0.0 (0.141)	14.7	29.2	56.7
Petroleum refining	-2.7	-5.2	-9.6	-2.3	-4.5	-8.4	-2.4	-4.5	-8.4	-2.5	-4.7	-8.9
Electric utilities	-3.4	-6.3	-11.2	-2.8	-5.4	-9.5	-2.9	-5.4	-9.6	-3.1	-5.7	-10.3
Efficiency cost per ton of CO_2 reduction	93.1	102.6	126.0	51.7	60.0	77.9	60.2	64.4	83.0	82.5	95.1	109.4
B. Elasticity of Demand	Low	Medium	High	Low	Medium	High	Low	Medium	High	Low	Medium	High
Change in equity value of firms, year 2000 (%)												
Coal mining	-19.0	-28.4	-43.2	-18.2	-27.8	-42.9	0.0 (0.032)	0.0 (0.043)	0.0 (0.058)	1,104.9	1,005.4	840.5
Oil and gas	-5.1	-4.8	-4.5	-5.3	-5.0	-4.5	0.0 (0.151)	0.0 (0.150)	0.0 (0.154)	28.7	29.2	28.9
Petroleum refining	-4.6	-5.2	-5.8	-3.8	-4.5	-5.3	-3.8	-4.5	-5.3	-4.2	-4.7	-5.5
Electric utilities	-4.4	-6.3	-8.3	-3.3	-5.4	-7.6	-3.3	-5.4	-7.6	-3.7	-5.7	-7.9
Efficiency cost per ton of CO_2 reduction	151.9	102.6	60.7	86.6	60.0	38.3	94.5	64.4	38.8	140.0	95.1	48.8
C. Adjustment Costs	Low	Medium	High	Low	Medium	High	Low	Medium	High	Low	Medium	High
Change in equity value of firms, year 2000 (%)												
Coal mining	-26.2	-28.4	-31.1	-25.5	-27.8	-30.5	0.0 (0.040)	0.0 (0.043)	0.0 (0.067)	1,065.4	1,005.4	901.3
Oil and gas	-4.8	-4.8	-5.7	-5.0	-5.0	-5.9	0.0 (0.136)	0.0 (0.150)	0.0 (0.158)	30.3	29.2	26.2
Petroleum refining	-5.0	-5.2	-5.4	-4.3	-4.5	-4.7	-4.3	-4.5	-4.7	-4.6	-4.7	-5.0
Electric utilities	-6.1	-6.3	-6.5	-5.2	-5.4	-5.4	-5.2	-5.4	-5.5	-5.6	-5.7	-5.9
Efficiency cost per ton of CO_2 reduction	101.6	102.6	102.1	57.9	60.0	59.8	62.0	64.4	65.5	90.2	95.1	89.5

[a] Numbers in parentheses are proportion of emissions permits required to achieve equity-value neutrality.

with lump-sum recycling is considerably more costly than the previous case because it lacks the beneficial influence on efficiency associated with the cut in marginal CIT rates. The efficiency costs, however, do not become as large as with full lump-sum recycling to households (policy 1). The reason is that part of the lump-sum transfers to firms are taxed away by the PIT on dividend income when firms pay out these lump-sum transfers as dividends. Since the PIT rate endogenously balances the government budget, the additional tax revenue associated with the higher dividend income is returned to households in the form of a lower PIT rate. In contrast, under full lump-sum recycling to households (policy 1), the PIT rate stays constant.

It may be noted that ensuring tax neutrality through lump-sum transfers does not have much consequence for industry-specific output, employment, and investment. Lump-sum transfers decouple interindustry allocation from interindustry distribution.

These experiments indicate that imposing tax-payment neutrality substantially overcompensates firms. While equity concerns might justify compensating firms for lost profits or equity values, they seem to offer little justification for tax-payment neutrality. The real-world backing for tax-payment neutrality may stem from the misperception that leaving firms' tax revenues unchanged is necessary to neutralize the real burden from regulation.

2.6.4   Sensitivity Analysis

Table 2.6 indicates how key parameters affect results under policies 1, 2, 6, and 7. Panel A of the table illustrates the implications of the size of the carbon tax (or extent of abatement). Here we consider carbon tax rates of $12.50 and $50.00, as well as the previously considered central-case value of $25.00. Impacts on equity values increase with the size of the carbon tax, but somewhat less than linearly. The results for the 100 percent-exemption case indicate that considerable rents are generated even when the carbon tax is $12.50—in this situation, equity values in the coal mining industry rise by almost a factor of six. Under all three policies, efficiency costs per ton (or average abatement costs) increase as the carbon tax rises from $12.50 to $50.00, attesting to rising marginal costs of abatement. The efficiency rankings of the three policies do not change as the carbon tax rate changes.

Panel B reveals the significance of alternative values for elasticity of demand for energy. The high-elasticity case involves a doubling in each industry of the elasticities of substitution between the energy composite E and the materials composite M, as well as the elasticity of substitution among the specific forms of energy. The low-elasticity case halves each of these elasticities in each industry. For the coal industry, the high-, central-, and low-elasticity of substitution cases yield general equilibrium elasticities of demand of 0.26, 0.41, and 0.64, respectively, under policy 2. Under every policy, the efficiency costs per ton of abatement are lower, the larger the elasticity of demand. The efficiency rankings of policies remain invariant across the various demand-elasticity scenarios. The numbers in parentheses in the columns associated with policy 6 are the proportion of emissions permits required to achieve equity-value neutrality. In keeping with the analysis of section 2.2, this proportion tends to rise with the elasticity of demand.

Panel C examines the implications of alternative assumptions for the elasticity of supply. We regulate this elasticity by altering the parameter ξ in the adjustment cost function for each industry; see equation (A2) in appendix A. This parameter determines the marginal adjustment cost, which is inversely related to the elasticity of supply. The low-adjustment-cost (high-elasticity-of-supply) case reduces this parameter by 25 percent in all industries; the high-adjustment-cost (low-elasticity-of-supply) case doubles it everywhere.[45] Efficiency costs per ton do not vary substantially with this parameter. As predicted by the analysis in section 2.2, the proportion of emissions permits required to achieve equity-value neutrality rises with adjustment costs (or falls with the elasticity of supply).

2.7   Conclusion

This paper indicates that the government can implement carbon abatement policies and at the same time protect profits in key industries without substantial costs to the budget and thus to economy-wide efficiency. In particular, the government has to grandfather only a small fraction of tradable pollution permits or exempt a small fraction of inframarginal emissions from a carbon tax to protect the value of capital in industries that are especially vulnerable to impacts from carbon taxes.

In the coal and oil and gas industries, initial cash flow is fairly small in relation to total cost. This small cost share contributes to a high supply elasticity and the ability to shift a large share of the tax burden onto demand. The ability to shift a significant portion of the tax burden implies that only a very small fraction of the potential rents associated with CO_2 policies needs to be earmarked for the fossil fuel industries in order to preserve profits and equity values. These equity values can be safeguarded through policies that depart only slightly from the most efficient carbon tax (or auctioned carbon permits) policies. The efficiency sacrifice is small.

The simplest programs involving freely allocated emissions permits or inframarginal exemptions to the carbon tax protect only the fossil-fuel-supplying industries; downstream industries are not protected. In our model, the downstream industries that suffer the largest proportionate re-

---

45. We were unable to obtain a solution for the model when adjustment cost parameters were reduced by more than 25 percent.

ductions in equity values are the electric utilities and petroleum refining industries. To protect these industries, other policy instruments would need to be invoked. We find that the potential carbon revenues are very large in relation to the revenue that would be required to protect profits in the fossil fuel industries and the electric utilities and petroleum refining industries. As a result, in our simulations the profits of this broader group of energy industries can be protected at relatively small efficiency cost.

Some caveats are in order. First, our model's aggregation may mask significant losses in some industries (such as aluminum manufacturing) that are not explicitly identified. To protect these industries, additional compensation methods would be required.

Second, our model assumes pure competition. If, in contrast, industries producing fossil fuels already exercise considerable market power before the carbon tax is introduced, they may have already enjoyed much or all of the potential rents. In this case, CO_2 abatement policies may be unable to generate significant additional rents, and thus the opportunities for achieving equity-value neutrality at low cost may be considerably more limited.

Third, in our model capital is the only factor that is not perfectly mobile. To the extent that labor also is imperfectly mobile, there can be serious transition losses from policy changes, and such losses may have significant political consequences. Overcoming barriers to political feasibility requires attention to these losses.

Finally, it is worth emphasizing that the forces underlying the political feasibility of CO_2 abatement policies are complex. Protecting the profits of key energy industries may not be sufficient to bring about political feasibility.

Notwithstanding these qualifications, the present analysis offers significant hope that some major distributional concerns related to CO_2 abatement policies can be eliminated at low cost. Hence the price tag on removing key political obstacles to domestic CO_2 abatement policies may be lower than previously thought.

Appendix A
Structure of the Numerical Model

Production

Technology: General Features

Equations (1)–(3) of the text indicate the nested production structure. The second term in equation (1) represents the loss of output associated with installing new capital (or dismantling existing capital). Per-unit adjustment costs, ϕ, are given by

(A1)
$$\phi(I/K) = \frac{(\xi/2)(I/K - \delta)^2}{I/K},$$

where I represents gross investment (purchases of new capital goods) and ξ and δ are parameters. The parameter δ denotes the rate of economic depreciation of the capital stock. The production function for the oil and gas industry, equation (4), contains the additional element γ_f, which is a decreasing function of cumulative oil and gas extraction:

(A2)
$$\gamma_{g,t} = \varepsilon_1[1 - (Z_t/\overline{Z})^{\varepsilon_2}],$$

where ε_1 and ε_2 are parameters, Z_t represents cumulative extraction as of the beginning of period t, and $\overline{Z}$ is the original estimated total stock of recoverable reserves of oil and gas (as estimated from the benchmark year).

The following equation of motion specifies the evolution of Z_t:

(A3)
$$Z_{t+1} = Z_t + X_t.$$

Equation (A2) implies that the production function for oil and gas shifts downward as cumulative oil and gas extraction increases. This addresses the fact that as reserves are depleted, remaining reserves become more difficult to extract and require more inputs per unit of extraction.

Behavior of Firms

In each industry, managers of firms are assumed to serve stockholders in aiming to maximize the value of the firm. The objective of firm-value maximization determines firms' choices of input quantities and investment levels in each period of time.

The equation of motion for the firm's capital stock is

(A4)
$$K_{s+1} = (1 - \delta)K_s + I_s.$$

Household Behavior

Consumption, labor supply, and saving result from the decisions of an infinitely lived representative household maximizing its intertemporal utility with perfect foresight. The model employs a nested utility function. In year t the household chooses a path of full consumption C to maximize

(A5)
$$U_t = \sum_{s=t}^{\infty}(1 + \omega)^{t-s}\frac{\sigma_U}{\sigma_U - 1}C_s^{(\sigma_U - 1)/\sigma_U},$$

where ω is the subjective rate of time preference and σ_U is the intertemporal elasticity of substitution in full consumption. C is a CES composite of consumption of goods and services $\tilde{C}$ and leisure ℓ:

(A6) $$C_s = [\tilde{C}_s^{(v-1)/v} + \alpha_C^{1/v}\ell_s^{(v-1)/v}]^{v/(v-1)},$$

where v is the elasticity of substitution between goods and leisure; and α_C is an intensity parameter for leisure.

The variable $\tilde{C}$ in equation (A6) is a Cobb-Douglas aggregate of 17 composite consumer goods:

(A7) $$\tilde{C}_s = \prod_{i=1}^{17} \overline{C}_{i,s}^{\alpha_{\overline{C}i}},$$

where the $\alpha_{\overline{C},i}(i = 1, \ldots, 17)$ are parameters. The 17 types of consumer goods are shown in table 2.1.

Consumer goods are produced domestically and abroad. Each composite consumer good $\overline{C}_i$, $i = 1, \ldots, 17$, is a CES aggregate of a domestic and foreign consumer good of a given type:

(A8) $$\overline{C} = \gamma_{\overline{C}}[\alpha_{\overline{C}}CD^{\rho_{\overline{C}}} + (1 - \alpha_{\overline{C}})CF^{\rho_{\overline{C}}}]^{1/\rho_{\overline{C}}}.$$

In this equation, CD and CF denote the household's consumption of domestically produced and foreign-made consumer good of a given type at a given point in time. The parameter $\rho_{\overline{C}}$ is related to $\sigma_{\overline{C}}$, the elasticity of substitution between CD and CF, $\rho_{\overline{C}} = (\sigma_{\overline{C}} - 1)/\sigma_{\overline{C}}$. For simplicity, we have omitted subscripts designating the type of consumer good and the time period.

The household maximizes utility subject to the intertemporal budget constraint given by the following condition governing the change in financial wealth, WK:

(A9) $$WK_{t+1} - WK_t = \bar{r}_t WK_t + YL_t + GT_t - \tilde{p}_t \tilde{C}_t,$$

where $\bar{r}$ is the average after-tax return on the household's portfolio of financial capital, YL is after-tax labor income, GT is transfer income, and $\tilde{p}$ is the price index representing the cost to the household of a unit of the consumption composite, $\tilde{C}$.

Government Behavior

A single government sector approximates government activities at all levels—federal, state, and local. The main activities of the government sector are purchasing goods and services (both nondurable and durable), transferring incomes, and raising revenue through taxes or bond issues.

Components of Government Expenditure

Government expenditure, G, divides into nominal purchases of nondurable goods and services (GP), nominal government investment (GI), and nominal transfers (GT):

(A10) $$G_t = GP_t + GI_t + GT_t.$$

In the reference case, the paths of real GP, GI, and GT all are specified as growing at the steady-state real growth rate, g. In simulating policy changes, we fix the paths of GP, GI, and GT so that the paths of real government purchases, investment, and transfers are the same as in corresponding years of the reference case. Thus, the expenditure side of the government ledger is largely kept unchanged across simulations. This procedure is expressed by

(A11a) $$GP_t^P / p_{GP,t}^P = GP_t^R / p_{GP,t}^R,$$

(A11b) $$GI_t^P / p_{GI,t}^P = GI_t^R / p_{GI,t}^R,$$

(A11c) $$GT_t^P / p_{GT,t}^P = GT_t^R / p_{GT,t}^R,$$

where the superscripts P and R denote policy change and reference case magnitudes, while p_{GP}, p_{GI}, and p_{GT} are price indexes for GP, GI, and GT. The price index for government investment, p_{GI}, is the purchase price of the representative capital good. The price index for transfers, p_{GT}, is the consumer price index. The index for government purchases, p_{GP}, is defined in the next section.

Allocation of Government Purchases

GP divides into purchases of particular outputs of the 13 domestic industries according to fixed expenditure shares:

(A12) $$\alpha_{G,i} GP = GPX_i p_i \quad i = 1,\ldots,13,$$

where GPX_i and p_i are the quantity demanded and price of output from industry i, and $\alpha_{G,i}$ is the corresponding expenditure share. The ideal price index for government purchases, p_{GP}, is given by

(A13) $$p_{GP} = \prod_{i=1}^{13} p_i^{\alpha_{G,i}}.$$

Appendix B

Parameter Values of the Numerical Model

Table 2B.1   **Elasticities of Substitution in Production**

	Parameter					
	σ_f	σ_{g1}	σ_{g2}	σ_E	σ_M	σ_x
Substitution margin	$g_1 - g_2$	$L-K$	$E-M$	E components	M components	Domestic-foreign inputs
Producing industry						
1. Agricultural and noncoal mining	0.7	0.68	0.7	1.45	0.6	2.31
2. Coal mining	0.7	0.80	0.7	1.08	0.6	1.14
3. Oil and gas extraction	0.7	0.82	0.7	1.04	0.6	(Infinite)
4. Synthetic fuels	0.7	0.82	0.7	1.04	0.6	(Not traded)
5. Petroleum refining	0.7	0.74	0.7	1.04	0.6	2.21
6. Electric utilities	0.7	0.81	0.7	0.97	0.6	1.0
7. Gas utilities	0.7	0.96	0.7	1.04	0.6	1.0
8. Construction	0.7	0.95	0.7	1.04	0.6	1.0
9. Metals and machinery	0.7	0.91	0.7	1.21	0.6	2.74
10. Motor vehicles	0.7	0.80	0.7	1.04	0.6	1.14
11. Miscellaneous manufacturing	0.7	0.94	0.7	1.08	0.6	2.74
12. Services (except housing)	0.7	0.98	0.7	1.07	0.6	1.0
13. Housing services	0.7	0.80	0.7	1.81	0.6	(Not traded)

Table 2B.2   **Parameters of Stock Effect Function in Oil and Gas Industry**

	Parameter			
	Z_0	$\overline{Z}$	ε_1	ε_2
Value	0	450	1.27	2.0

Note: This function is parameterized so that γ_f approaches 0 as Z approaches $\overline{Z}$; see equation 8. The value of $\overline{Z}$ is 450 billion barrels (about 100 times the 1990 production of oil and gas, where gas is measured in barrel-equivalents). $\overline{Z}$ is based on estimates from Masters et al. (1988). Investment in new oil and gas capital ceases to be profitable before reserves are depleted. The values of ε_1 and ε_2 imply that, in the baseline scenario, oil and gas investment becomes 0 in the year 2050.

Table 2B.3   **Utility Function Parameters**

	Parameter			
	ω	σ_u	ν	η
Value	0.007	0.5	0.69	0.84

References

Armington, P. S. 1969. A theory of demand for products distinguished by place of production. *IMF Staff Papers,* 159–76.

Bovenberg, A. Lans, and Ruud A. de Mooij. 1994. Environmental levies and distortionary taxation. *American Economic Review* 84 (4): 1085–89.

Bovenberg, A. Lans, and Lawrence H. Goulder. 1996. Optimal environmental taxation in the presence of other taxes: General equilibrium analyses. *American Economic Review* 86 (4): 985–1000.

———. 1997. Costs of environmentally motivated taxes in the presence of other taxes: General equilibrium analyses. *National Tax Journal* 50 (1): 59–87.

———. 2000. Environmental taxation and regulation in a second-best setting. In *Handbook of public economics,* ed. A. Auerbach and M. Feldstein, 2nd ed. Amsterdam: North-Holland, forthcoming.

Buchanan, James M., and Gordon Tullock. 1975. Polluters, profits and political response: Direct controls versus taxes. *American Economic Review* 65: 139–47.

Bull, N., K. Hassett, and G. Metcalf. 1994. Who pays broad-based energy taxes? Computing lifetime and regional incidence. *Energy Journal* 15:145–64.

Bureau of Economic Analysis. U.S. Department of Commerce. 1994. *Survey of Current Business,* August.

Cruz, Miguel, and Lawrence H. Goulder. 1992. An intertemporal general equilibrium model for analyzing U.S. energy and environmental policies: Data documentation. Stanford University. Unpublished manuscript.

Farrow, Scott. 1999. The duality of taxes and tradeable permits: A survey with applications in central and eastern Europe. *Environmental and Development Economics* 4:519–35.

Felder, Stefan, and Reto Schleiniger. 1999. Environmental tax reform: Efficiency and political feasibility. Faculty of Economics, Institute for Social Medicine, Otto-Von-Guericke-Universität Magdeburg. Working paper. January.

Fullerton, Don, and Gilbert Metcalf. 1998. Environmental controls, scarcity rents, and pre-existing distortions. University of Texas at Austin. Working paper.

Gaskins, Darius, and John Weyant, eds. 1996. *Reducing global carbon dioxide emissions: Costs and policy options.* Stanford, Calif.: Energy Modeling Forum, Stanford University.

Goulder, Lawrence H. 1994. Energy taxes: Traditional efficiency effects and environmental implications. In *Tax policy and the economy,* vol. 8, ed. James M. Poterba. Cambridge, Mass.: MIT Press.

———. 1995a. Effects of carbon taxes in an economy with prior tax distortions: An intertemporal general equilibrium analysis. *Journal of Environmental Economics and Management* 29:271–97.

———. 1995b. Environmental taxation and the "Double Dividend": A reader's guide. *International Tax and Public Finance* 2 (2): 157–83.

Goulder, Lawrence H., Ian W. H. Parry, and Dallas Burtraw. 1997. Revenue-raising vs. other approaches to environmental protection: The critical significance of pre-existing tax distortions. *RAND Journal of Economics* 28:708–31.

Goulder, Lawrence H., and Phillipe Thalmann. 1993. Approaches to efficient capital taxation: Leveling the playing field vs. living by the golden rule. *Journal of Public Economics* 50:169–86.

Hall, Robert. 1988. Intertemporal substitution in consumption. *Journal of Political Economy* 96 (2): 339–57.

Jorgenson, Dale W., Daniel T. Slesnick, and Peter J. Wilcoxen. 1992. Carbon taxes and economic welfare. *Brookings Papers on Economic Activity, Microeconomics,* 393–431.

Jorgenson, Dale W., and K. Y. Yun. 1991. *Tax reform and the cost of capital.* Oxford: Oxford University Press.

Keohane, Nathaniel, Richard Revesz, and Robert Stavins. 1998. The choice of regulatory instruments in environmental policy. *Harvard Environmental Law Review* 22 (2): 313–67.

Lawrance, Emily. 1991. Poverty and the rate of time preference: Evidence from panel data. *Journal of Political Economy* 99 (1): 54–77.

Masters, Charles, Emil Attanasi, William Dietzman, Richard Meyer, Robert Mitchell, and David Root. 1988. World resources of crude oil, natural gas, natural bitumen, and shale oil. In *Proceedings of the twelfth World Petroleum Congress,* vol. 5, 3–27. New York: Wiley.

Metcalf, Gilbert E. 1998. A distributional analysis of an environmental tax shift. NBER Working Paper no. 6546. Cambridge, Mass.: National Bureau of Economic Research.

Olson, Mancur. 1965. *The logic of collective action.* Cambridge, Mass.: Harvard University Press.

Parry, Ian W. H. 1995. Pollution taxes and revenue recycling. *Journal of Environmental Economics and Management* 29:S64–S77.

Parry, Ian W. H., and A. Bento. 2000. Tax deductions, environmental policy, and the "double dividend" hypothesis. *Journal of Environmental Economics and Management* 39 (1): 67–96.

Parry, Ian W. H., Roberton C. Williams III, and Lawrence H. Goulder. 1999. When can CO_2 abatement policies increase welfare? The fundamental role of pre-existing factor market distortions. *Journal of Environmental Economics and Management* 37:52–84.

Pezzey, John. 1988. Market mechanisms of pollution control: "Polluter pays," economic and practical aspects. In *Sustainable environmental management: Principles and practice,* ed. R. K. Turner. London: Belhaven Press.

———. 1992. The symmetry between controlling pollution by price and controlling it by quantity. *Canadian Journal of Economics* 25 (4): 983–91.

Pezzey, John C. V., and Andrew Park. 1998. Reflections on the double dividend debate: The importance of interest groups and information costs. *Environmental and Resource Economics* 11 (3/4): 539–55.

Poterba, James M. 1991. Tax policy toward global warming: On designing a carbon tax. In *Economic policy responses to global warming,* ed. R. Dornbusch and J. Poterba. Cambridge, Mass.: MIT Press.

Poterba, James M., and Lawrence H. Summers. 1985. The economic effects of dividend taxation. In *Recent advances in corporate finance,* ed. E. Altman and M. Subramanyam. Homewood, Ill.: Irwin.

Russek, Frank S. 1996. Labor supply and taxes. Macroeconomic Analysis Division, Washington, D.C. Congressional Budget Office. Working paper.

Schillo, Bruce, Linda Giannarelli, David Kelly, Steve Swanson, and Peter Wilcoxen. 1996. The distributional impacts of a carbon tax. In *Reducing global carbon dioxide emissions: Costs and policy options,* ed. D. Gaskins and J. Weyant. Stanford, Calif.: Energy Modeling Forum, Stanford University.

Summers, Lawrence H. 1981. Taxation and corporate investment: A *q*-theory approach. *Brookings Papers on Economic Activity,* no. 1, 67–127.

Vollebergh, Herman R. J., Jan L. de Vries, and Paul R. Koutstaal. 1997. Hybrid carbon incentive mechanisms and political acceptability. *Environmental and Resource Economics* 9:43–63.

Williamson, Oliver E. 1996. The politics and economics of redistribution and efficiency. In *The mechanisms of governance.* Oxford: Oxford University Press.

Comment   Ruud A. de Mooij

This paper reveals that the efficiency costs of neutralizing the adverse industry impact of CO_2 abatement policies are rather low. The analysis should be the starting point of a new research agenda to explore the sensitivity of this result with respect to some important assumptions.

The Efficiency-Equity Trade-Off

The literature on environmental policy reveals that instruments that charge a price for pollution—such as taxes or tradable pollution permits—induce smaller efficiency costs than non-revenue-raising instruments. In addition to the fact that economic instruments equalize marginal abatement costs, there is a second-best argument that underpins this result. In particular, environmental policy that raises revenue induces a revenue-recycling effect: The revenues from the environmental policy can be used to reduce other distortionary taxes. Accordingly, revenue-raising instruments are attractive from an efficiency point of view (Goulder, Parry, and Burtraw 1997).

In contrast, revenue-raising instruments are typically unattractive from a political perspective. Indeed, by assigning the property rights of the environment to the government, these instruments redistribute income away from polluters and toward the government. This reduces the political viability of environmental policy. To avoid adverse distributional effects, the government may compensate the regulated agents through its revenue-recycling strategy, provided that it has the appropriate instruments available. Recent literature has indeed emphasized how the revenues from environmental taxes can be used to neutralize the distributional consequences of environmental taxes on households (Metcalf 1998). The study by Bovenberg and Goulder (henceforce BG) focuses on the distributional implications of CO_2 abatement policies applied to firms. In particular, BG explore the implications for industry profits and equity values that seem to be the most natural indicators for the distributional impact of CO_2 abatement policies on firms.

Availability of Instruments

BG consider different instruments to neutralize distributional effects. First, they analyze industry-specific corporate income taxes and industry-specific lump-sum transfers. It is straightforward to see how these instruments can neutralize the disproportional equity effects on specific industries. However, one may doubt the relevance of this approach since such instruments are typically not available in practice. For instance, industry-

Ruud A. de Mooij is a senior research fellow of the Research Center for Economic Policy (OCFEB) at the Erasmus University Rotterdam and heads the European Comparative Analysis unit at CPB Netherlands Bureau for Economic Policy Analysis.

specific lump-sum transfers can be viewed as state aids, which are forbidden in many countries, while corporate income taxes are generally not differentiated across industries. A more promising strategy, therefore, is the third instrument explored by BG, namely, the grandfathering of pollution permits (or the introduction of tax exemptions). A highly relevant and surprising result from the BG analysis is that the government has to grandfather only a very small fraction of tradable pollution permits to protect the value of capital in the most vulnerable industries. Furthermore, this neutralization involves only small efficiency costs. This is an important conclusion for policymakers. Many countries consider the introduction of environmental taxes or tradable pollution permits. However, governments are generally reluctant to impose these policies due to the disproportional impact on particular industries.

Grandfathering and Entry Barriers

The intuition behind the BG result can be summarized as follows. Restricting output of a particular industry creates scarcity rents. Indeed, whereas a single firm in the industry may not have market power, a government that restricts total sector output exploits the monopoly power of the industry as a whole. Accordingly, all firms in the sector benefit. The efficiency costs associated with this output restriction are borne by the consumers of the commodity in terms of a higher consumer price.

Scarcity rents can be created by grandfathered emissions permits or pollution-control instruments. Notice that scarcity rents can be maintained only if the policy applies to existing firms alone. Without this restriction, new firms would enter the market to share the rents from freely issued permits, thereby causing economy-wide pollution to rise to the level before the regulation was imposed. Hence, the scarcity rents originate in a barrier for new firms to enter the market.[1]

As explained by Fullerton and Metcalf (1999), revenue-raising environmental-policy instruments tax away the scarcity rents from the regulated sector. BG propose a hybrid system in which part of the scarcity rents are held privately while the residual part flows to the government (to be used for revenue recycling).

Input versus Output Taxes

BG consider a tax that is imposed on the suppliers of fossil fuels. This choice is attractive if one seeks a solution for the adverse effects on equity values in the coal and the oil and gas industries. Indeed, a tax exemption (or grandfathered emissions permits) compensates the industry on which

---

1. BG (n. 33) mention that environmental tax revenues from firms in Denmark and the Netherlands are earmarked for technology subsidies to the same industries. These targeted subsidies do not apply to existing firms alone. However, by phasing out these subsidies in subsequent years, the risk of entry is limited.

the tax is imposed. However, other industries that suffer a decline in equity value, such as energy-intensive firms that face higher factor prices, cannot be compensated through such exemptions. In the BG simulations, these industries are therefore compensated through industry-specific corporate income tax reductions or lump-sum transfers. If such instruments are not available, the adverse impact on the equity value of the energy-intensive industries may impose an important political problem.

BG could alternatively consider a tax imposed on energy inputs, rather than a tax on energy supply. In that case, exemptions or grandfathered emissions permits would create rents for the industries that demand fossil fuels since, in aggregate, these industries would exploit their monopsony power and reduce the before-tax price of energy. Accordingly, the government would have access to an instrument that directly compensates industries that demand energy, although it would lose its instrument to compensate the energy-supplying industries. This may be an interesting case. Indeed, political debates in Europe—an energy importer—have been dominated by concerns about the energy-intensive industries rather than energy suppliers. Exemptions or grandfathered emissions permits on energy inputs may result in a politically acceptable CO_2 policy for the bulk of heterogeneous industries that demand energy. For energy-importing countries, this compensation may even be sufficient since energy-supplying firms—who bear the burden of this policy—are located abroad.

Border-Tax Adjustments

BG assume that exports of energy are exempt from the CO_2 tax, while imports are taxed. In the presence of these border-tax adjustments, tax exemptions create rents because the energy supplier can differentiate its output price between domestic and foreign consumers. Indeed, the scarcity rents originate in a higher before-tax output price in the domestic market. If there were no border-tax adjustments, energy-producing firms could no longer differentiate between domestic and foreign consumers; both would be charged the world-market price. In that case, non-revenue-raising instruments will hardly raise the output price and thus do not create scarcity rents. Accordingly, the exemptions necessary to compensate the energy-supplying firms would be much higher than in the BG simulations.

For energy-intensive industries, it would be more difficult to impose border-tax adjustments since the government would require information about the carbon content of products with foreign origin. In case of an input tax on energy without border-tax adjustments, however, it is difficult to compensate the energy-intensive industries through tax exemptions. Indeed, domestic industries will generally exert only little monopsony power on the international market for energy inputs. My guess is that exemptions (or the fraction of grandfathered permits) would thus have to be very large to compensate these industries for the loss in equity value.

A Future Research Agenda

The paper by BG should be the starting point of a new research agenda on how the disproportional industry impact of CO_2 abatement policies can be neutralized at the least cost to efficiency. In my view, this is a promising research area that is highly relevant for the policy debate. To better understand the impact of different neutralization options, it may be helpful to develop small stylized models, in addition to the applied model used by BG. In addition, researchers may perform similar analyses using different models. Accordingly, they may test the robustness of the results and understand the crucial parameters that determine the magnitude of the effects. In this connection, I would also like to encourage BG to explore alternative policy proposals, in addition to the experiments analyzed in this paper. Furthermore, an extensive sensitivity analysis may reveal the most important elasticities that require further empirical investigation.

References

Fullerton, D., and G. E. Metcalf. 1999. Environmental controls, scarcity rents, and pre-existing distortions. University of Texas at Austin. Working paper.

Goulder, L. H., I. W. H. Parry, and D. Burtraw. 1997. Revenue-raising vs. other approaches to environmental protection: The critical significance of pre-existing tax distortions. *RAND Journal of Economics* 28:708–31.

Metcalf, G. E. 1998. A distributional analysis of an environmental tax shift. NBER Working Paper no. 6546, Cambridge, Mass.: National Bureau of Economic Research.

Green Taxes and
Administrative Costs
The Case of Carbon Taxation

Sjak Smulders and Herman R. J. Vollebergh

3.1 Introduction

Implementing environmental policies—through standards, tradable permits, or environmental taxes alike—is far from costless. For instance, when implementing an environmental tax, the tax department has to run a special unit to enforce and collect taxes and to monitor compliance. In practice, the costs of implementing environmental policies play a significant role in the choice between policy options. The proposals of the European Commission for a European-wide energy/CO_2 tax provide clear examples (Vollebergh 1995). Instead of proposing a totally new tax on CO_2 emissions, the European Commission employed the close linkage between CO_2 emissions and the implicit taxation of carbon by the existing taxes on energy products (which are usually intermediate inputs). Indeed, using existing instruments rather than introducing new ones to address new policy areas may save considerably on administrative costs.

However, just the minimization of transaction costs might come at a cost for society. A strategy based on input taxes, for example, forgoes the gains that are potentially reaped by a more direct way of taxing the externality through emissions taxes. Any deviation from the principle of taxing externalities at the point where they arise introduces an incentive to misallocate resources. Thus a trade-off arises between minimizing transaction costs and directly inducing incentive effects. The optimal tax structure has

Sjak Smulders is a fellow of the Royal Netherlands Academy of Arts and Sciences and is affiliated with CentER, Tilburg University, The Netherlands. Herman R. J. Vollebergh is assistant professor in public economics and a research fellow of the Research Centre for Economic Policy (OCFEB) at Erasmus University Rotterdam.

The authors thank conference participants as well as Dallas Burtraw, Jan Pieters, and especially Don Fullerton for their detailed comments on an earlier draft.

to balance the burden of complex and expensive-to-run tax systems against the incentives it induces to internalize the externality that it aims to address.

This paper investigates the potential trade-off between administrative costs and incentives of environmental regulation, in particular if the government aims to reduce CO_2 emissions. We analyze how the optimal choice of carbon taxes is affected by the administrative costs incurred by the regulator (government). Using a simple model, we determine the optimal rates for emissions and input taxes in the presence of administrative costs and which of these taxes should optimally be introduced. Moreover, we explore and interpret the scarce empirical evidence on administrative costs of taxation in the light of optimal carbon taxation. Because empirical information on the role of implementation costs in the design of environmental policy is almost entirely lacking, we concentrate on what factors might be expected to determine those costs, based on studies of administrative costs outside the environmental policy area.

Although most formal analysis of environmental regulation ignores administrative costs, compliance costs, or transaction costs in general, a growing literature takes these issues seriously (see the overview in Krutilla 1999).[1] Several papers recognize that administrative costs may be important and rule out the use of emissions taxes on these grounds. Typically investigated is which taxes could best replace or "approximate" emissions taxes (Smith 1992). Moreover, under some circumstances other taxes or tax combinations are equivalent to perfect emissions taxes (i.e., emissions taxes in a world without transaction costs). For instance, Xepapadeas (1999) reviews the conditions under which input taxes and emissions taxes are equivalent. Eskeland and Devarajan (1996) show how the combination of mandated technology and output taxes approaches the ideal emissions tax. Fullerton and Wolverton (1997) propose to combine output taxes and subsidies on clean goods, or more general two-part instrument systems of a deposit-refund nature, to replace the emissions tax that involves costly monitoring.

The implicit assumption in these papers, however, is that emissions taxes are prohibitively costly to administer and that other taxes have negligibly low administrative costs. We extend this approach by more explicitly taking into account the administrative costs of all types of taxes, without assuming beforehand that emissions taxes are always the most costly type of tax from the administrative point of view. In particular we allow different tax instruments to feature differences in administrative costs, which, in addition, are endogenously dependent on the tax rates (cf. Yitzhaki 1979). Once other taxes, as well as emissions taxes, are subject to signifi-

---

1. The relation between taxation in general and transaction costs is more widely analyzed; see Slemrod and Yitzhaki (1998) for an overview.

cant administrative costs, it becomes unlikely that the first-best optimum can be reached. Hence, alternative tax systems should be considered that are no longer equivalent to perfect emissions taxes.

Shortle, Horran, and Abler (1998) research to what extent input taxes can approach perfect emissions taxes if not all inputs that directly affect emissions can be taxed. We extend their analysis by explicitly taking into account administrative costs and allowing for the simultaneous use of emissions and input taxes. We find that a mixed tax system might be (second-best) optimal. Schmutzler and Goulder (1997) arrive at a similar result using a model of mixed output and emissions taxation that incorporates monitoring. We complement their analysis by investigating input taxation and by exploring in more detail how optimal tax rates in the presence of administrative costs differ from Pigouvian taxes. Administrative costs in our model mainly represent costs stemming from monitoring, and thus our paper is related to the literature on monitoring and enforcement of environmental policy (see Cohen 1998 for a survey). Because we are primarily interested in optimal taxation rather than optimal monitoring, we do not model monitoring in an explicit way.

The theoretical part of this chapter is also closely related to Fullerton, Hong, and Metcalf (chap. 1 in this volume). The two chapters complement one another in various respects. Both chapters compare ideal emissions taxes with alternative taxation, but they differ with respect to the production structure and the government budget constraint. First, Fullerton, Hong, and Metcalf analyze a model in which there is a one-to-one correspondence between input use and emissions. Input taxes and emissions taxes are therefore equivalent, but output taxes provide an (imperfect) substitute form of taxation. In contrast, our model separates input use from emissions, and considers abatement explicitly. Accordingly, we allow for three ways to reduce pollution: output reduction, input reduction, and abatement. We study input taxation as an (imperfect) substitute for emissions taxes. Second, whereas Fullerton, Hong, and Metcalf consider a second-best world with a distortionary labor tax for revenue-raising purposes, the second-best nature of the policies considered here arises because of administrative costs. Thus, the present chapter abstracts from tax-interaction effects due to recycling effects.

The structure of our paper is as follows. First, we explain the nature of the trade-off involved if the implementation costs of corrective taxes, in particular administrative costs, are considered explicitly. Second, we analyze a stylized model that incorporates both emissions and input taxes to sort out critical determinants that shape this trade-off. Finally, we evaluate both explicit and implicit carbon taxation in Organization for Economic Cooperation and Development (OECD) countries in terms of the trade-off and suggest some opportunities for welfare-improving carbon tax policies. Note in advance that taxing carbon inputs is not equivalent to taxing CO_2

emissions, as is sometimes suggested. Although a close linkage exists between the carbon content of energy products and CO_2 emissions, this is not a fixed chemicotechnological relationship because several opportunities for carbon abatement or removal exist (Okken et al. 1992).

3.2   The Trade-off between Incentives and Administrative Costs

In this section we argue that the administrative costs argument per se is not sufficient to rule out the implementation of emissions taxes. In the presence of administrative costs, the costs and benefits associated with each specific type of tax should be compared. First, we hypothesize which factors influence the shape of the administrative-costs curve. Next, we show why administrative costs introduce such a general trade-off between the costs and benefits of various implementation strategies. We also develop some useful terminology.

3.2.1   Administrative Costs

We define *transaction costs* as the costs associated with tax assessment, collection, and enforcement; and all other costs incurred by any party to enable, facilitate, and ensure transactions from taxpayers to tax authorities (Vollebergh 1995). An alternative term that we use is *implementation costs.* The terms include ex ante costs (e.g., costs of exclusion) and ex post costs (e.g., monitoring costs). It is common to categorize these costs further into costs for the government (tax receiver), or *administrative costs,* to handle forms and enforce compliance, and the costs for the tax-liable agent (taxpayer), or *compliance costs,* to carry out the obligations of calculating and paying the tax (see Sandford, Godwin, and Hardwick 1989). In our analysis we concentrate on administrative costs.[2]

Administrative costs of a particular tax are closely related to the base to which the tax is applied. The tax base usually varies with the type of tax. For example, an emissions tax taxes the physical volumes of hazardous substances, while an input tax taxes such substances indirectly, for instance through their use as (intermediate) inputs. In turn, these differences induce both tax authorities and taxpayers to set up and maintain various systems for collecting and processing information about the tax, that is, to record how much is emitted or how much input is used, in order to be able to calculate the total tax payments due.

One important characteristic of the tax base that determines (differences in) administrative costs is the number of agents liable for the tax. A large number of taxable legal units implies a large implementation cost for the

---

2. Section 3.4.2, however, shows that administrative and compliance costs turn out to move together in practice; that is, taxes for which compliance costs are relatively important are also associated with relatively high administrative costs.

tax agency, since each unit requires separate treatment. Taxing a particular pollutant that is emitted by many producers may be associated with large administrative costs. Taxing the inputs from which the pollutant arises as a by-product may be associated with significantly lower administrative costs. For instance, inputs need no longer be taxed at the points of consumption, but can also be taxed at the point of delivery, such as gas stations or distributors of electricity. Hence, switching from emissions to inputs as the tax base could change administrative costs.

Note that the difference in administrative costs is independent of the induced regulatory effect. It is a difference in the fixed-cost component of administrative costs, that is, the setup cost and part of the cost to run the information system. Each liable unit submits its own tax form. The cost of processing forms depends on the number of forms rather than the tax amount due. Nevertheless, this still leaves the possibility of economies of scale for a given type of tax. If the tax base can be broadened across a larger number of taxpayers, the overall administrative costs per taxpayer can be reduced.

A second important determinant of administrative costs is measurability of the base. In most cases emissions levels are likely to be more difficult to measure, report, and record than input or output levels. Heterogeneity across industries and their technologies compounds the complexity of a tax system. For instance, a tax base in terms of weighted units of measurement, rather than in terms of a single unit, may be expected to create higher administrative costs if firms use highly firm-specific technologies. One well-known example is NO_x emissions from road transportation, which depend on vehicle type, equipment, fuel type, driving patterns, and so forth (see also Hoel 1998, 89).

Administrative costs are also likely to vary with the tax rates and the revenue raised. The possibility of evasion by taxpayers requires monitoring expenditures by regulators. The remark by Fullerton (1996, 7) that many of the administrative costs "are 'fixed' costs of calculating the tax base, not marginal costs of collecting more revenue by raising the rate of tax on a given tax base" seems to call for a qualification in this respect. The larger the tax bill, the larger are the incentives to evade tax payment and the more attractive it is for the regulator to spend resources to reduce tax evasion.

Regulators usually have various strategies for monitoring and need to sort out the efficient choice of monitoring levels and techniques. A large literature on monitoring and enforcement studies this policy in detail (Cohen 1998). Here, we do not need this level of detail. With respect to environmental monitoring, we can safely assume that when the optimal mix of monitoring instruments is chosen, total cost of monitoring increases with the number of polluters, the variety of production and abatement techniques used, the importance of stochastic influences on actual pollution, and the difficulty of measuring emissions.

To sum up, no general shape can be assumed ex ante for different types of taxes. However, it seems fruitful to assume that both fixed and variable costs (varying with the tax rate) play a significant role. Both in theory and practice, we need a case-by-case approach to study the nature and implications of administrative costs.

3.2.2   The Role of Linkage

The efficiency of instruments to reach a certain policy goal is usually defined in terms of the extent to which the instrument increases social welfare. The most efficient instrument to hit a given target has the smallest gross welfare cost, where *gross welfare cost*[3] refers to the change in welfare apart from that arising from the reduction in the externality.[4]

In a first-best world without transaction costs, different instruments can be ranked in terms of efficiency by investigating their effect on private welfare. Things become more complicated in a world with transaction costs because both administrative costs and the linkage between regulatory aim, emissions reduction, and the type of regulatory tax used play a role (Smith 1992).[5] First of all, different types of taxes usually differ with respect to the directness of the incentive they provide to reduce emissions (assuming emissions reduction reflects the goal of the government). Less-direct taxation of the marginal damages caused by an individual polluter causes an efficiency loss, but may lower administrative costs. Furthermore, different instruments distort private welfare not only directly, but also indirectly through their implications for transaction costs. The usual gross welfare cost of taxation has to be supplemented by the transaction costs of the tax.

Before turning to how welfare analysis of environmental taxation is influenced by transaction costs, it is useful to clarify our terminology and make it precise. We explicitly separate the transaction costs from the total change in welfare associated with the use of a certain (tax) instrument. Hence, in our case of environmental taxation, we distinguish (1) administrative (transaction) costs, (2) the welfare gain from an improvement in the environment, (3) the residual welfare change, that is the gross welfare cost ignoring transaction costs. The third component is called here *private gross welfare cost*. An instrument that has relatively low private gross welfare

---

3. The term "gross welfare cost" is from Goulder (1995).

4. This definition applies to corrective taxes. The gross welfare cost in the case of revenue raising can be similarly defined as the change in welfare apart from that arising from relaxing the government budget constraint.

5. There is an interesting analogue between the current paper and the long-standing issue in environmental economics of selecting instruments to improve ambient quality directly or indirectly through the reduction of emissions. It is well known that the linkage between emissions and ambient quality is often indirect, but the cost of ambient-quality regulation can be prohibitive. Thus an interesting trade-off exists between the utility loss in terms of the directness of linkage, on the one hand, and the cost of regulation, on the other hand. We owe this point to Dallas Burtraw.

costs is called relatively privately efficient. Of course, in a world without transaction costs, efficiency just coincides with this notion of private efficiency, since gross welfare costs do not contain transaction costs.[6] Thus, the relative efficiency of different types of taxes can be measured with the following formula:

$$(1) \qquad\qquad U = Y - T - D(E),$$

where U is social welfare of the representative agent, Y is gross private welfare, T is the welfare loss due to transaction (administrative) costs, and D is the damage from pollution. Let t_1 and t_2 be two distinct tax regimes that yield the same aggregate emissions: $E(t_1) = E(t_2)$. The private costs of t_1 are lower than those of t_2 if $Y(t_1) > Y(t_2)$.

We do not need to discuss extensively the determinants of private efficiency here, since they are well known from analyses without transaction costs. For example, the efficiency of a tax to internalize pollution externalities is larger if the individual's tax bill is more directly linked to the externality. Hence, emissions taxes are more (privately) efficient than input taxes. Also, efficiency requires that the effective tax rate on marginal contributions to damage (D) is equal across polluters. Hence, an emissions tax that applies to all polluters is (privately) more efficient than an emissions tax with exemptions or a nonuniform emissions tax.

As noted in section 3.1, it is often argued that emissions taxes are too costly to implement and that administrative costs provide a basic motivation for other (tax) solutions.[7] However, instead of simply assuming that such a shift away from emissions taxes is optimal, we aim at explicitly deriving such a conclusion within a comprehensive welfare framework. A first step in this direction has been taken by McKay, Pearson, and Smith (1990), who hypothesize that a clear trade-off exists between shifts in the tax system to save on transaction costs, on the one hand, and tax reforms that harness incentives and promote (private) efficiency, on the other hand. They assume that regulation that is linked less directly to the externality does indeed save on administrative costs, but that it comes at a cost to society by distorting private decisions more.

Figure 3.1 illustrates this. The horizontal axis measures various tax systems with respect to the directness with which they address incentives to reduce damage; for example, an emissions tax ranks high and an input tax

6. We realize that this term might be misleading, since transaction costs also affect (ultimately) private welfare. However, the term captures the fact that we focus on administrative costs that first affect the tax authority (and not private agents directly). Indeed, of the three terms in equation (1), only the first (Y) captures direct changes in private welfare. The third term, the environmental gain, is a "public" component of the welfare change if the environment is assumed to be a public good. Alternatively, we could have used the terms "frictionless gross welfare cost" and "frictionless efficiency."

7. In fact, Smith (1992) shows that the basic idea can be traced back to the seminal paper of Diamond (1973).

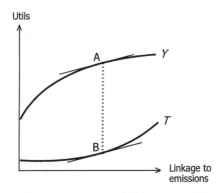

Fig. 3.1   The basic trade-off between transaction costs and the efficiency of internalization of tax systems

ranks low. Taxes on different inputs rank differently, depending on the closeness of the linkage between input use and emissions subject to regulation. The vertical axis measures two components of utility. The figure compares a continuum of tax systems. It is assumed that all of them yield the same level of damage D by appropriate choice of tax rates. The two curves represent the other two components of overall utility, transaction costs T and private utility Y, for each of the tax systems. Administrative costs T increase when taxation is better linked to emissions. The idea behind this is that more-direct taxation implies fewer links to already existing procedures of the existing tax system. Gross private welfare Y also increases with the linkage of taxation to emissions. The more direct the taxation, the larger the (private) utility is for a given level of emissions. The optimal tax system balances transaction costs and efficiency. In the figure, welfare is maximized by an indirect tax that corresponds to points A and B. The complete switch to emissions taxes is too costly: The associated increase in administrative costs would outweigh the gains from having a more-direct tax with better incentives.

Figure 3.1 is hypothetical and suggestive. As noted before, we have to assess tax proposals case by case. For example, if marginal administrative costs increase only slowly, emissions taxes may be optimal despite the presence of administrative costs. Moreover, it is not at all guaranteed that the curves T and Y have nice convex and concave shapes, respectively. Smulders and Vollebergh (1998), for instance, represent the linkage to pollution by the fraction of (symmetrical) sectors that is liable to an emissions tax and find that in a very simple setting the Y curve first declines and then increases. In general, administrative costs introduce nonconvexities because of their fixed-cost nature, and the conventional marginal approach to optimal taxation has to be extended.

Administrative costs have many dimensions. The government may affect administrative costs by varying the number of firms or sectors subject to the tax, the tax rates chosen for input and emissions taxes, the accuracy of measurability aimed for, the enforcement spending to reduce the (probability of) tax evasion, and so on. Each of these dimensions can be measured along the horizontal axis in a figure similar to figure 3.1. Needless to say, each of these factors directly influences the overall welfare effect of implementing environmental taxes.

It is not only the multidimensionality of administrative costs that makes the simple diagram in figure 3.1 problematic. As Feldstein (1976) pointed out long ago, a distinction should be made between the design of a tax system de novo and the reform of an existing tax system. This is true for its associated administrative system as well. Indeed, in practice every tax reform starts from a given tax and administrative system inherited from the past. This system determines the (short-run) scope of welfare-improving tax reform at low administrative cost (Smith 1992; Vollebergh 1995).

For instance, increasing existing taxes rather than introducing new taxes might save on the fixed costs of administration and therefore on total administrative costs. It is also attractive to exploit such economies of scale and scope when designing environmental taxes. Levying environmentally motivated taxes on a base that is already taxed for other purposes, rather than introducing an entirely new emissions tax, would certainly save on administrative costs. Furthermore, economies of scope with the administrative system used for other regulatory instruments may also arise. When implementing environmental taxes, the regulator could benefit from experience in related administrative procedures for operations already undertaken. As does Smith (1992), we label this use of existing administrative procedures and experience for new purposes "piggybacking."

3.3   Critical Determinants Shaping the Trade-Off

This section develops a simple model along the lines of Kaplow (1990) and Shortle, Horan, and Abler (1998) to compare emissions taxes and input taxes in the presence of administrative costs. The aim of the regulator is to correct externalities from pollution. The presence of administrative costs implies that the regulator should deviate from the first-best Pigouvian tax. Hence, administrative costs in themselves cause policies to be second best. We abstract from other second-best issues. In particular, we assume that lump-sum taxes and transfers are available to the government, so that there is no revenue requirement that affects tax rates and we can ignore labor taxes.[8]

---

8. We also abstract from output taxes and abatement subsidies. See Smulders and Vollebergh (1999) for the interaction between these instruments and administrative costs.

3.3.1 The Model

We assume a given number of heterogenous sectors, indexed i. The production of one unit of final output q_i requires labor l_i and a single homogenous intermediate input x (in amount x_i). Moreover, firms can spend labor services on abatement a_i, which reduces emissions per unit of output e_i. The minimum labor requirement per unit of output equals $l_i(x_i)$. Labor and inputs are substitutes: $l_i' \equiv \partial l_i / \partial x_i < 0$. Emissions per unit of output depend negatively on abatement effort and positively on inputs: $e_i(a_i, x_i)$ with $e_{ai}' = \partial e_i / \partial a_i < 0$ and $e_{xi}' = \partial e_i / \partial x_i < 0$.[9]

Final-good producers face a (sector-specific) emissions tax τ_i and a (per-unit) input tax (t_{xi}). Perfect competition prevails, and firms take the output price p_i as given. They maximize profits by choosing output, abatement, and input levels. We normalize the wage to unity. The first-order conditions can be written as

$$(2) \qquad p_i = l_i + a_i + e_i\tau_i + (p_x + t_{xi})x_i,$$

$$(3) \qquad 1 \geq (-e_{ai}')\tau_i \quad \text{with equality if } a_i > 0,$$

$$(4) \qquad p_x + t_{xi} + \tau_i e_{xi}' \geq -l_i' \quad \text{with equality if } x_i > 0.$$

Equation (2) says that price equals cost, which in turn equals labor cost for production, labor cost for abatement, and taxes due per unit of output. Condition (3) states that with positive abatement levels, the marginal cost of abatement (on the left-hand side) equals the marginal benefits in the form of a reduction of emissions tax payments (on the right-hand side). Condition (4) equates the marginal cost to the marginal benefits of input use. Marginal input costs consist of the price of the input p_x, the sector-specific input tax t_{xi}, and the induced additional emissions tax payments. Marginal benefits consist of the labor saving in production.

The intermediate good is produced with labor only and subject to constant returns to scale. We choose units such that one unit of labor produces one unit of the intermediate good. For simplicity we assume that the production of the intermediate input is nonpolluting (but this can be easily modified in a way that is completely analogous to pollution in the final-goods sector). Intermediate-good producers face a price p_x, which they take as given. Hence, their first-order condition for profit maximization simply states that the price equals the wage, which is normalized to 1:

$$(5) \qquad p_x = 1.$$

---

9. Furthermore, $e_{xx}'' > 0$, $e_{aa}'' > 0$, and $l'' > 0$. We ensure concavity by assuming $[l'' + (-e_a')] e_{xx}'' e_{aa}'' - (-e_a')(e_{ax}'')^2 > 0$.

Equilibrium in the market for the input requires[10]

$$X = \sum x_i q_i,$$

where X is total supply of the intermediate good.

We impose a very simple demand structure by choosing a quasi-linear utility function with no cross-demand effects, and where the opportunity cost of labor is constant (and normalized to 1). The utility function is

(6) $$u = \sum_i u_i(q_i) + l_0 - D(E),$$

where l_0 is leisure, D is damage from emissions, and E is aggregate emissions defined as

(7) $$E = \sum e_i q_i.$$

Consumers take prices and emissions as given and maximize utility, subject to their budget constraint $\sum p_i q_i = L - l_0 + Z$, where Z are transfers from the government. The first-order conditions read

(8) $$u_i' = p_i.$$

The government collects tax revenue, pays civil servants for the tax administration (T), and rebates the remainder of tax revenue to households in a lump-sum fashion (Z). The tax administration employs T units of labor at wage $w = 1$. The required administrative costs are sector specific and depend on sectoral taxes and output levels.[11]

(9) $$T = \sum F_i(I_{\tau i}, I_{txi}) + \sum V_i(\tau_i, t_{xi}, q_i),$$

where F represents the fixed costs of the tax system, and V represents the administrative costs varying with the size of the rates and bases of the tax system. Fixed costs are determined only by certain taxes being implemented or not. This is modeled by the dependence of F on indicator functions $I_{\hat{t}}$, each of which takes the value 1 if tax $\hat{t}$ (e.g., τ_i) is positive and the value 0 if the tax is 0. The natural restrictions we impose are *sign* $V_t' = $ *sign* $\hat{t}$ for any tax $\hat{t}$, that is, both taxes and subsidies are costly to implement; and $V_i(0, 0, q_i) = 0$, that is, all fixed costs are excluded from $V(\cdot)$.

The labor market clears. Labor endowment is fixed and given by L. Hence, we write

(10) $$L = l_0 + \sum(l_i + a_i + x_i)q_i + T.$$

---

10. To simplify notation, all summation signs refer to summation over all final goods sectors, unless stated otherwise.
11. Note that by assuming linear sectoral separability we ignore economies of scope as discussed in section 3.2.

Substituting equations (7) and (10) into (6), we may write utility as

(11) $\quad U = \sum u_i(q_i) + L - \sum (l_i + a_i + x_i)q_i - T - D\left(\sum e_i q_i\right).$

Totally differentiating utility, and substituting the first-order conditions for firms' and households' maximization problems (2), (3), (4), (5), and (8), we obtain

(12) $\qquad dU = \sum t_{xi} dX_i - dT - \sum (D' - \tau_i) dE_i,$

where $E_i \equiv e_i q_i$ is total emissions in sector i and $X_i \equiv x_i q_i$ is total input use in sector i. Equation (12) shows the welfare effects associated with changes in input demands, transaction costs, and environmental quality. The first term on the right-hand side of (12) stands for the distortionary effect of excises on the goods market associated with input taxes. The last term reveals that a reduction in emissions ceteris paribus improves utility as long as the marginal damage is larger than the emissions tax.

3.3.2   Optimal Taxation

We can rewrite equation (12) to separate the three components of welfare, as in equation (1):

(13) $\qquad dU = \sum [\tau_i dE_i + t_{xi} dX_i] - dT - D' dE.$

Equation (13) categorizes the welfare effect of any policy in the three components mentioned in section 3.2.2. The bracketed term on the right-hand side is the private gross welfare effect of the policy, denoted by dY, in line with equation (1); dT is the transaction costs of the policy; and $-D' dE = -dD$ is the environmental welfare gain. Note that the private gross welfare cost is a tax-base effect; the change in each tax base times the tax rate corresponding to that tax base together determine this effect.[12]

In the presence of administrative costs, a necessary condition for optimality of the tax system is that the expression in (13) be 0. The government maximizes welfare, taking as given the reactions of households and firms to changes in taxes. It faces a two-stage decision problem: (1) deciding which taxes to use (tax-base decision), and (2) setting the appropriate tax level (tax-rate decision).

Concerning the tax-rate decision, we find conditions for optimal taxation by rewriting equation (12) in terms of the total derivatives with respect to each of the taxes and setting these expressions equal to 0.[13] For any tax $\hat{\imath}$ this condition reads

---

12. See the analysis in Bovenberg and Goulder (1998, sec. 3.1).

13. Note that equations (2), (3), (4), (5), and (8) allow us to determine how a_i, x_i, q_i, p_i, and p_x—and hence also $l_i(x_i)$, $e_i(a_i, x_i)$, E_i, X_i, T, and U—depend on the tax rates.

(14) $\dfrac{dU}{d\hat{\imath}} = \sum\left[t_{xi}\dfrac{dX_i}{d\hat{\imath}} - (D' - \tau_i)\dfrac{dE_i}{d\hat{\imath}}\right] - \dfrac{dT}{d\hat{\imath}} = 0.$

This equation binds only if the tax is implemented; that is, equation (14) guides the tax decision, conditional on the tax being implemented.

Concerning the tax-base decision, the regulator should compare utility levels associated with any combination of taxes implemented at the rate implied by equation (14). The optimal tax system may include nonzero taxes, set at the level implied by equation (14), as well as zero taxes, that is, taxes that are not implemented. For the latter taxes, equation (14) may be violated, that is, utility may marginally increase in this tax. Yet it is optimal not to implement these taxes. The reason is that, by construction, in an optimally designed tax system setting any zero tax at the level implied by (14)—and adjusting all nonzero tax rates such that they satisfy (14)—decreases welfare (nonmarginally) because of fixed administrative costs. Similarly, in an optimally designed tax system, switching the rate of any nonzero tax from the rate implied by (14) to a zero rate—and adjusting all other nonzero taxes such that they satisfy (14)—decreases welfare (nonmarginally). Since fixed administrative costs play a role, the tax-base decision is subject to nonconvexities and no simple smooth optimality condition can be written.

Instead of optimizing the overall tax system, a more practical issue is to find a welfare improving tax reform. Such an approach takes into account the fact that actual changes to the tax system are usually slow and piecemeal due to the role of the existing tax system (Feldstein 1976) and, as we like to add, its associated administrative costs. A change in an existing tax system is worth pursuing if this change entails an increase in welfare even if the maximum level of welfare is not reached. In particular, we are interested in the welfare effects of the introduction of a new tax, if some taxes already exist (as well as their associated tax administration). The obvious rule for a welfare-improving introduction of a new tax is that the net welfare gain from exploiting the newly introduced tax should exceed the fixed costs of introducing the tax. For any tax $\hat{\imath}$, this condition can be written as[14]

(15) $\hat{\imath}^* = \hat{\imath}^o \quad \text{if} \quad \left[\displaystyle\int_0^{\hat{\imath}^o} \dfrac{dU}{d\hat{\imath}}\,d\hat{\imath}\right] \geq F_t,$

where $\hat{\imath}^o$ is the level of the tax that corresponds to equation (14) (i.e., the solution to $dU/d\hat{\imath} = 0$, or the corner solution 0), $F_t \equiv dT/dI(\hat{\imath})$ is the fixed cost (administrative setup cost) associated with introducing tax $\hat{\imath}$, $\hat{\imath}^*$ is the

14. This condition can be called the "entry condition," analogous to industrial organization models where firms enter if the operating profits (cf. welfare), measured at the optimal price (cf. tax), exceed the entry cost (cf. tax introduction/setup cost).

(second-best) optimal tax rate,[15] and we evaluate all total derivatives, taking into account changes in other taxes so as to satisfy (14) for all other taxes.

As a benchmark, consider the (first-best) case without transaction costs, that is, $T = dT = 0$. As is well known, the optimal emissions tax then equals the marginal damage D' in each sector and all other taxes should be 0.[16] This can be immediately seen from equation (12). Indeed, equation (14) is satisfied for these tax rates. Under the usual conditions on utility and production functions, the tax-base optimality condition is automatically met since fixed costs do not play a role and the maximization problem is convex. Starting from a situation without any taxes, introducing the emissions tax improves welfare.[17]

The first-best outcome may be realized in some special cases even if transaction costs play a role. Obviously, if transaction costs are associated with other taxes, but not with emissions taxes, the Pigouvian tax should still be implemented. The other way around, if transaction costs apply to emissions taxes only and other taxes can be implemented without such costs, a first-best outcome may arise provided that other taxes (or tax combinations) are equivalent to emissions taxes with respect to their incentive effects (private efficiency). For example, if the emissions-input ratio is fixed, an input tax can bring about the first-best outcome.[18]

A second-best situation arises when other taxes also involve transaction costs or when other instruments are privately less efficient than emissions taxes. Once transaction costs play a role, it is no longer guaranteed that emissions taxes should be uniform, nor that output or input taxes should be excluded. Most of the literature on second-best optimal environmental taxation concentrates on cases in which other taxes (taxes on outputs or inputs) can replace emissions taxes without loss of incentives and without administrative costs (e.g., see the double dividend literature; de Mooij 2000).

If administrative costs are mentioned as a reason not to use emissions

---

15. To be precise, $\hat{\imath}^*$ is the tax that maximizes welfare given the set of taxes employed; $\hat{\imath}^* = 0$ if equation (15) is violated.

16. Solving the social-planner problem for the case without transaction costs, we find the following optimality conditions: (1) $u'_i = l_i + a_i + x_i + e_i D'$, (2) $1 \geq -e'_{ai} D'$ with equality if a_i is positive, (3) $1 \geq -l'_i - e'_{xi} D'$ with equality if x_i is positive. Comparing these conditions to equations (2), (3), (4), (5), and (8), we find that $\tau_i = D'$, $t_{xi} = 0$ implements the first-best outcome. As a special case, if $e'_{ai} = 0$ and $e'_{xi} = e_i / x_i \; \forall \; i$, any combination of taxes that satisfies $\tau_i + (x_i / e_i) t_{xi} = D' \; \forall \; i$ also implements the first-best social optimum (input taxes and emissions taxes are equivalent, cf. sec. 3.3.4).

17. For this case $dU/d\tau$ reduces to $(\tau - D') dE_i / d\tau$, which is positive for $\tau < D'$. Hence, the left-hand side of the inequality in (15) is positive while the right-hand side is 0, and (15) is satisfied.

18. Similarly, two-part instruments may do the job. If only one pollutant causes an externality and if all other outputs and inputs can be taxed at zero transaction costs, the first-best outcome can be reached (see Fullerton and Wolverton 1997). In the present model, this would require (sector-specific) taxes on output and input use and a (sector-specific) subsidy on abatement. Note, however, that optimality breaks down when more pollutants play a role.

taxes, the most common case in the literature is the one in which emissions taxes are too costly to implement because of the transaction costs associated with emissions taxes but not with other taxes (the most discussed case is nonpoint pollution; see Xepapadeas 1999). Our model allows for more subtle impacts of administrative costs by considering administrative costs throughout the entire tax system and taking into account that administrative costs may endogenously vary with tax rates. To investigate these in more detail, we consider some special cases.

3.3.3   Pure Emissions Taxes

Let us first focus on emissions taxes by considering the case in which all other taxes are ruled out. Note that we cannot simply suppose that only emissions taxes are used; we have to explain within the model why this is so. We give this explanation in subsection 3.3.4 and concentrate here on the optimality conditions for emissions taxes only.

Evaluating equation (14) for an emissions tax in sector i, we find that the following optimality conditions should hold:

$$(16) \qquad \frac{dU}{d\tau_i} = -\frac{dT}{d\tau_i} - (D' - \tau_i)\frac{dE_i}{d\tau_i} \leq 0 \quad \text{with equality if } \tau_i > 0.$$

Hence, if implemented, the optimal emissions tax reads

$$(17) \qquad \tau_i^o = \max\left(0, D' - \frac{dT/d\tau_i}{-dE_i/d\tau_i}\right).$$

This tax should be implemented if the total welfare gain exceeds the fixed administrative costs; see equation (15). We approximate the welfare gain by a second-order Taylor expansion, evaluated at $\tau_i = \tau_i^o$. The optimal tax τ_i^* is given by

$$(18) \qquad \tau_i^* = \tau_i^o \quad \text{if} \quad \frac{1}{2}\left[-\frac{dE_i}{d\tau_i}(\tau_i^o)^2 + (\eta_{Ei} + \eta_{T\tau i})\frac{dT_i}{d\tau_i}\tau_i^o\right] \geq F_{\tau i},$$

$$\tau_i^* = 0 \quad \text{otherwise},$$

where η_E and $\eta_{T\tau}$ are the positively defined elasticities of $dE/d\tau$ and $dT/d\tau$ with respect to τ.

Conditions (17) and (18) reveal two cases in which it is optimal not to use emissions taxes in a particular sector because of administrative costs. The first case is the case in which the fixed costs of administering the tax are large relative to the total potential gains; see equation (18). The gains are small indeed if emissions are insensitive to the emissions tax, that is, if abatement *and* changes in the input mix are expensive ($dE_i/d\tau_i$ small), if the marginal damage (D') is small, and if marginal administrative costs ($dT/$

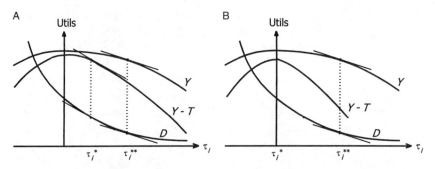

Fig. 3.2   The optimal second-best tax rate for an emissions tax with (*A*) low and (*B*) high transaction costs

$d\tau_i$) are large.[19] A second case in which a sector should be optimally exempted from an emissions tax is the case in which marginal administrative costs for the sector are relatively large, such that, for any small increase in the sector-specific emissions tax, higher administrative costs more than offset gains from the induced emissions reduction ($dU/d\tau_i < 0$ for any τ_i so that $\tau_i^o = 0$).

Figure 3.2 illustrates the case of emissions taxes in terms of the trade-off between efficiency and administrative costs (see section 3.2). Private gross welfare, Y, is maximized for zero emissions taxes, since, loosely speaking, emissions taxes impede free-market forces. However, they reduce damage D and hence improve social welfare. In a first-best world without administrative costs, the Pigouvian tax τ_i^{**} maximizes welfare $Y - D$. In the presence of administrative costs, T, the gross welfare cost of emissions taxation (i.e., the effect on $U - D$) is higher and rises more steeply with tax rates. The (second-best) optimal tax maximizes $Y - T - D$, and it can be easily seen that this tax is below the first-best tax. In panel B of figure 3.2, transaction costs rise steeply with the tax rate, and the fixed-cost component is large. As a result, the second-best optimal emission tax is 0.

How emission taxes should be optimally differentiated across sectors is also revealed by condition (17), conditional on being implemented. Note that the optimal tax equals marginal damage minus a correction term that is proportional to marginal administrative costs. The optimal tax equals the Pigouvian tax if marginal administrative costs are 0 ($dT/d\tau_i = 0$). The gap between optimal taxes and the Pigouvian tax widens if administrative costs rise steeply with tax levels and if emissions are not very sensitive to emissions taxation. The second may arise because of a low elasticity of demand (it is hard to accomplish emissions reductions by cutting demand) or because the emissions intensity is not very sensitive to emissions-tax changes (steeply rising abatement and input substitution costs). To clarify

19. To see this, substitute equation (17) into (18).

this, we decompose the emissions-reduction effect of the tax, which appears as the denominator in (17), into these three effects:

$$(19) \qquad -\frac{dE_i}{d\tau_i} = \varepsilon_i + \alpha_i + \lambda_i \xi_i,$$

where

$$(20) \qquad \varepsilon_i \equiv -\frac{dq_i}{d\tau_i} e_i = \left(-\frac{dq_i}{dp_i}\right) e_i e_i,$$

$$(21) \qquad \alpha_i = -q_i e'_{ai} \frac{da_i}{d\tau_i},$$

$$(22) \qquad \xi_i = -q_i \frac{e_i}{x_i} \frac{dx_i}{d\tau_i},$$

$$(23) \qquad \lambda_i = \frac{e'_{xi} x_i}{e_i}.$$

That is, ε represents the effect of emissions taxes on emissions through changes in demand, α measures the direct effect of emissions taxation on emissions through abatement, and $\xi\lambda$ measures the analogous effect through input reduction (the reason to separate ξ and λ becomes clear in subsection 3.3.4).

So far, we have assumed that administrative costs rise with tax rates because incentives to evade increase with the tax rate, thus raising the cost for the tax authority to administer the tax. The opposite, however, is possible as well. Using a partial equilibrium model, Polinsky and Shavell (1982) find that the optimal emissions tax in the presence of administrative costs may be larger than the Pigouvian tax. The argument is that a higher emissions tax saves on transaction costs if administrative costs depend on the number of taxpaying firms and if an increase in the emissions tax reduces market demand and the number of firms. In our setup, the number of firms is indeterminate because of the constant-returns-to-scale production functions, but the equation immediately shows that Polinsky and Shavell's result also applies here if administrative costs decrease with the tax rate, that is, if $dT/d\tau_i < 0$.

3.3.4   Input Taxes: The Role of Linkage

To investigate the trade-off between emissions taxes and input taxes, we first consider sector-specific taxes on emissions (τ_i) and on the use of input x (t_{xi}). Evaluating equation (14) for these taxes, we find[20]

---

20. Note that $(x_i/e_i)^2(\varepsilon_i + \xi_i\beta_i) = -dX_i/dt_{xi}$ and $(x_i/e_i)(\varepsilon_i + \xi_i) = -dE_i/dt_{xi} = -dX_i/d\tau_i$.

$$(24) \quad \tau_i^o = \max\left\{0, D' - \frac{dT/d\tau_i}{\varepsilon_i + \xi_i\lambda_i + \alpha_i} - \left(\frac{\varepsilon_i + \xi_i}{\varepsilon_i + \xi_i\lambda_i + \alpha_i}\right)\frac{x_i}{e_i}t_{xi}\right\},$$

$$(25) \quad t_{xi}^o = \frac{e_i}{x_i}\left[(D' - \tau_i)\left(\frac{\varepsilon_i + \xi_i}{\varepsilon_i + \xi_i\beta_i}\right) - \frac{dT/dt_{xi}}{(x_i/e_i)(\varepsilon_i + \xi_i\beta_i)}\right],$$

where

$$(26) \quad \beta_i = \frac{e_i}{x_i}\frac{dx_i/dt_{xi}}{dx_i/d\tau_i} = \left[\lambda_i + \left(\frac{-e'_{a_i}x_i}{e'_{aa_i}e_i}\right)e''_{axi}\right]^{-1}.$$

Note that ξ measures the direct effect of input taxation on emissions,[21] λ measures the elasticity of the emissions function with respect to input use, and β measures how much input use is more sensitive to input taxation than to emissions taxes.

According to equation (24), input taxes can serve as environmental taxes and reduce the need for explicit emissions taxes. Note that the first two terms are the same as in equation (17) after the substitution of equation (19). The smaller the direct emissions-tax effect $\varepsilon + \lambda\xi + \alpha = -dE/d\tau$, the larger is not only the effect of marginal administrative costs on optimal emissions taxes, but also the scope for input taxes to replace emissions taxes, as is clear from the third term in equation (24). Indeed, with high marginal administrative costs of emissions taxes, input taxes only should be used as environmental taxes and should be set according to equation (25), with $\tau_i = 0$, which can be written as

$$(27) \quad t_{xi} = \left(\frac{dE_i/dt_{xi}}{dX_i/dt_{xi}}\right)D' - \left(\frac{dT/dt_{xi}}{-dX_i/dt_{xi}}\right).$$

Note that inputs should then be taxed according to their marginal emissions content dE_i/dX_i times marginal damage D' corrected for administrative costs as a result of changes in input use. (Of course, we must make the provision that in the presence of large fixed administrative costs, such that equation [15] is violated for t_{xi}, the input tax should not be implemented.)

Replacing emissions taxes by input taxes reduces efficiency. Input taxes distort the input mix and fail to provide direct incentives for abatement. Only if the input-to-emissions ratio is constant and there are no abatement possibilities are the input taxation and emissions taxation equivalent in the absence of transaction costs. This corresponds to e_i/x_i = constant, $\lambda = \beta = 1$, and $\alpha = 0$. With an interior solution, conditions (24) and (25) can then be rewritten as

---

21. It can be derived, from equations (3) and (4), that $dx/d\tau_i = e'_{ai}(da_i/dt_{xi}) + e'_{xi}(dx_i/dt_{xi})$.

$$(28) \quad D' - \tau_i - t_{xi}x_i/e_i = \left(\frac{1}{\varepsilon_i + \xi_i} \right) \frac{dT}{d\tau_i} = \left(\frac{1}{\varepsilon_i + \xi_i} \right) \frac{dT}{d(t_{xi}x_i/e_i)}.$$

With a fixed emissions-input ratio, input and emissions taxes would be equivalent in the absence of administrative costs (as is well known; see, e.g., Xepapadeas 1999). Indeed, according to condition (25), without marginal administrative costs, any combination of taxes such that $\tau_i + t_{xi}x_i/ e_i = D'$ would achieve the first-best optimum. This implies that the two taxes are equally efficient in terms of the sum of gross private welfare and the environmental benefit (see section 3.2). Hence, transaction costs considerations entirely determine the choice between the two taxes.

Differences in (fixed and/or variable) administrative costs across tax instruments remove the indeterminacy in the optimal tax choice. First, if fixed administrative costs differ across the two taxes, but administrative costs are not affected by tax-rate levels, to satisfy the entry condition only the tax with lowest fixed administrative costs should be introduced, either $\tau_i = D'$ or $t_{xi}x_i/e_i = D'$. Note that the effective tax on pollution equals marginal damage (the Pigouvian tax). Second, when both tax rates increase administrative costs, the effective tax on pollution $(\tau_i + t_{xi}x_i/e_i)$ should be smaller than marginal damage D'. When, in addition, the sum of fixed administrative costs for the two taxes are sufficiently small to justify the introduction of both taxes, the taxes should be set so as to minimize variable administrative costs, as appears in the second equality in (28).

In the general case of variable and sector-specific emissions per unit of input, input taxes are less efficient than emissions taxes. Hence, if at the same time administrative costs for emissions taxes are higher, efficiency and administrative costs may be optimally traded off by choosing a mixed system of input and emissions taxes. Solving equations (24) and (25) for an interior solution, and for simplicity assuming that abatement and input use separately affect emissions $(e''_{ax} = 0$ so that $\beta = 1/\lambda)$, we obtain

$$(29) \quad \tau_i^o = D' - \frac{1}{\Delta_i} \left(\frac{\lambda_i}{\xi_i} + \frac{1}{\varepsilon_i} \right) \frac{dT}{d\tau_i} + \frac{\lambda_i}{\Delta_i} \left(\frac{1}{\xi_i} + \frac{1}{\varepsilon_i} \right) \frac{dT}{dt_{xi}},$$

$$(30) \quad t_{xi}^o = \frac{e_i}{x_i} \left[\frac{\lambda_i}{\Delta_i} \left(\frac{1}{\xi_i} + \frac{1}{\varepsilon_i} \right) \frac{dT}{d\tau_i} - \frac{\lambda_i}{\Delta_i} \left(\frac{1}{\xi_i} + \frac{\lambda_i}{\varepsilon_i} + \frac{\alpha_i}{\xi_i \varepsilon_i} \right) \frac{dT}{dt_{xi}} \right],$$

where

$$(31) \quad \Delta_i = (\lambda_i - 1)^2 + \left(\frac{1}{\varepsilon_i} + \frac{\lambda_i}{\xi_i} \right) \alpha_i > 0.$$

In the above expressions, Δ_i measures the "efficiency edge" of emissions taxes over input taxes. Indeed with a constant emissions input ratio ($\alpha = 0$ and $\lambda = 1$), we have $\Delta = 0$, and equations (29)–(30) collapse to equation (28). The efficiency edge of emissions taxes increases in abatement possibilities α and in $|\lambda - 1|$. We call this latter expression the extent of linkage between emissions and inputs. The closer to unity the elasticity of emissions is with respect to inputs (λ), the closer is the correspondence between inputs and emissions, and the more efficiently input taxes mimic emissions taxes. Equations (29) and (30) reveal that marginal administrative costs are less important in determining the optimal tax rates if the efficiency of emissions taxes relative to input taxes (Δ) is larger, that is, if more abatement possibilities abound (α is larger) and emissions are more closely linked to inputs (λ is closer to 1).

3.3.5   Lessons from the Model

To internalize environmental externalities in the presence of administrative costs, pure emissions taxes are optimal only under specific conditions. These conditions include (1) low fixed administrative costs, (2) not-too-steeply-rising administrative costs (as a result of increases in emissions taxes) relative to marginal damage and direct emissions-reduction effect of emissions taxes, and (3) relatively low incentive effects from alternative environmental taxes (taxes on polluting inputs) to reduce emissions. The optimal second-best rate of emissions taxes falls short of marginal damage.

Input taxes may indeed serve as (optimal) environmental taxes. With the close linkage between input use and emissions, and if abatement of emissions (as an alternative means of reducing the pollution intensity of production, rather than changing the input mix) is relatively costly, taxes on polluting inputs may supplement emissions taxes that fall short of marginal damage to internalize pollution externalities more fully. In this case a mixed system of emissions taxes and input taxes is optimal, essentially because it saves on administrative costs while only moderately affecting incentives to reduce emissions. If linkage is close and abatement expensive, and if administrative costs associated with input taxation are sufficiently low relative to administrative costs associated with emissions taxation, input taxes should fully replace emissions taxes.

3.4   Carbon Taxation and Administrative Costs

In this section we assess existing and potential environmental taxes relevant for climate change policy, in particular through carbon taxation. We argue that current policy (proposals) can be substantially improved if the trade-off between incentive regulation and administrative costs is explicitly taken into account. We concentrate on the explicit carbon taxes introduced in a number of European countries since the beginning of the 1990s.

Table 3.1   **Characteristics of Main Fossil Fuels**

Fuel	Energy Content (GJ)	Carbon Content (ton)	Tons of Oil Equivalent (TOE)	Normalized Carbon Content (ton/TOE)
Coal (metric ton)	25–30	0.61	0.6	0.96–1.00
Crude oil (barrel)	6.1	0.12	1	0.76–0.84
Natural gas (1,000 m³)	9.6–10.7	0.17	8.0[a]	0.56–0.64

Source: OECD/IEA (1993).
[a]Average for Gronings gas, based on the upper bound of 8.37381 and lower bound of 7.535714.

We first review relevant facts on existing carbon taxes, then present evidence on administrative costs, next assess current carbon taxes, and finally discuss the scope for improvement.

3.4.1   Carbon Taxes in Practice

Since the early 1990s, taxes have been seriously considered to combat climate change, in particular carbon taxes to curb CO_2 emissions (e.g., Pearce 1991; Cnossen and Vollebergh 1992; Poterba 1992). The debate in Europe was strongly influenced by a proposal of the European Commission, COM(92) 226 (European Commission 1992), for a hybrid European Union (EU) tax on energy/CO_2 to be implemented at the European level. The basic idea behind this proposal is to bring the (minimum) rate structure more in accordance with the carbon content across currently taxed energy products, mainly hydrocarbon fuels, as well as to extend the carbon tax base to energy products that are not yet subject to an excise. The same idea is also behind the carbon taxes actually implemented in several European countries.

Thus the aim is to raise the implicit taxation of carbon at the margin. As is well known, the amount of CO_2 emitted per kind of fuel differs considerably (see table 3.1). Clearly, oil emits less carbon than coal does. Natural gas, in turn, is cleaner than oil. The obvious implication is that emissions intensities also can be reduced by internalizing the respective carbon contents in the price of each kind of fossil fuel. By differentiating the fossil fuel excise by the carbon emissions coefficient instead of energy content coefficient, or even a hybrid coefficient, the consumption of carbon is put at a disadvantage at the margin. Thus, users would be induced to substitute oil for coal and natural gas for coal and oil, and, further, nonfossil fuels for fossil fuels.

However, the EU proposal was never implemented due to considerable resistance from industry and specific countries such as the United Kingdom. Despite this failure to implement an EU-wide carbon tax, several individual European countries have introduced explicit carbon taxes (see table 3.2). Finland, at that time not a member state, was the first country to impose a CO_2 tax in 1990. This environmental tax is additional to an excise tax (basic duty) and is calculated according to the carbon and energy content of the energy products. Furthermore, it is imposed on primary energy inputs, including heavy fuel oil, liquified petroleum gas (LPG), coal, and natural gas.

Other Nordic countries soon followed: Norway and Sweden in 1991, and Denmark in 1992. The CO_2 tax in Norway affects the use of mineral oils, coal, natural gas, and petroleum on the continental shelf. Interestingly, CO_2 tax rates differ among these products, with petroleum and natural gas [*sic!*] taxed most heavily (per unit CO_2) and heavy fuel oil and coal at a much lower level. Also, electricity production and consumption are taxed. Sweden's CO_2 tax applies to primary energy inputs, such as natural gas and coal, but also includes heavy fuel oil and gas oil. The Danish tax is levied on all energy products with the exception of petrol and amounts to a tax-rate reform from dollars per liter to dollars per unit carbon. A tax reform in 1996 explicitly distinguishes energy consumption in industry according to the categories room heating, light processes, and heavy processes, with tax rates varying accordingly.

The Netherlands has had an environmental tax on fuels (hydrocarbon oils) since 1988, with the CO_2 component added in 1990. However, only the regulatory tax on energy from 1996 was specifically aimed at achieving carbon emissions reduction by households and small firms. The tax base included primary energy products, while the tax rates correspond to the proposed EU CO_2 energy tax. Austria also imposed an energy tax on electricity and natural gas in 1996.

In a recent analysis of these carbon taxes, Ekins and Speck (1999) show how exemptions for industry are used to provide considerable tax relief for certain sectors facing considerable competitive pressure. Tax relief is usually established by applying lower or zero carbon tax rates or systems of rebate for specific industries that use these products as inputs (often in addition to exemptions already provided for already existing energy excises). Sometimes a maximum is set to the tax liability for specific energy-intensive industries, such as the steel industry, usually in terms of a percentage of sales value (this provision was also envisaged in the hybrid EU tax). Finally, improvements in energy efficiency are promoted by explicitly targeted tax reliefs. As a result, nominal and effective tax rates for specific industries tend to differ considerably.

Table 3.3 shows that for several energy products Sweden, Denmark, and Norway apply much lower effective rates for specific industries. Only Finland does not apply lower rates, although this is now heavily debated in Finland. Furthermore, it is remarkable that considerable differences exist in tax rates per ton of CO_2 across energy products, especially in Norway.

Table 3.2   Excise Taxation of Energy Products in Countries Applying Carbon Taxes, 1997

Country	Petrol (ECU/1000 liters)	Diesel (ECU/1000 liters)	Gas Oil (ECU/1000 liters)	Heavy Fuel Oil (ECU/1000 liters)	Coal (ECU/ton)	Natural Gas (ECU/m³)	Electricity (ECU/kWh)
Denmark	533	321	236	266	160	0.03091	0.06719
Finland	616	307	50	38	29	0.02443	0.00533
Netherlands	579	302	47	16	0	0.00962	0
Norway	658	485	56	79	56	0.10897	0.00397
Sweden	597	337	210	217	144	0.12031	0.01316
EU, minimum	337	245	18	13	0	0	0
EU, proposed 2000	450	343	37	23	13	0.01400	0.00200

Source: Ekins and Speck (1999, 371, table 1).

Note: CO_2 taxes and existing energy excises per unit of fuel are included.

Table 3.3   Effective Tax Rates of Explicit CO$_2$ Taxes for Some Industries in the Nordic Countries (% of nominal tax rates)

Energy Products	Sweden, Manufacturing Industry	Denmark, Heavy Processes	Norway, Pulp/Paper Industry	Finland, All Industry
Gas oil (heating)	0.50	0.24	0.50	1.0
Heavy fuel oil	0.50	0.23	0.50	1.0
LPG	0.50	0.25	0	0
Coal	0.50	0.25	1.0	1.0
Natural gas	0.50	0.24	1.0	1.0

Source: Calculations based on Ekins and Speck (1999, 380).

Norway, like Finland, exempts LPG, while coal and natural gas are taxed (much) more heavily than is oil.

The carbon taxes in the Nordic countries are quite similar to the original proposal for a common carbon tax within the EU jurisdiction (see European Commission 1992 and its evaluation by Smith and Vollebergh 1993). This tax is aimed at lowering the use of fossil fuels in proportion to their carbon content. The European carbon/energy tax, the first explicit uniform EU-wide tax, was proposed as an additional tax on top of the (non) existing taxes. Since the tax base would include several energy products that were not subject to tax before, the proposal also broadens the tax base of current energy taxes. Thus, an incentive would be provided for industry and consumers to reduce their use of carbon-based energy, and hence for CO$_2$ emissions to be reduced.

Because this EU proposal was never implemented, a later proposal was more closely linked to the existing drafts on Mineral Oil Excise Harmonization, COM(95) 172 (European Commission forthcoming) and therefore concentrated its effort on a much smaller carbon tax base (see table 3.2 and its evaluation in Vollebergh 1995). In 1997, the European Commission came up with a new proposal to use the directive on excise harmonization across EU countries more specifically for the purpose of a carbon tax policy (see Ekins and Speck 1999 for further details). According to this proposal the minimum target levels for the existing excise taxes on mineral oils should be raised in three steps; small minimum rates on primary energy products, such as coal and natural gas, are also proposed, as well as a tax on electricity (see table 3.2 for the proposed rates for 2000).

All EU proposals allow for exemptions. In the 1992 draft directive, an exemption would depend on a case-by-case assessment of the degree of competitive pressure faced from countries not taking equivalent measures. Member states could grant firms a reduction in the carbon tax payable (through an exemption or an equivalent refund) if energy costs (minus the value-added tax) amount to at least 8 percent of value-added. In addition, the proposed directive in 1992 also allows for reductions or refunds if firms invest in energy-efficiency improvements or carbon abatement.

To summarize, the recently introduced (unilateral) carbon taxes in several European countries indeed broaden the existing (implicit) carbon tax base by including specific primary energy products, such as coal and natural gas. These products were usually not taxed before. The agents who pay the tax are mainly (downstream) distributors of final fuel products or electricity at the point of delivery to households and to small and large businesses. Furthermore, with the exception of Norway, the tax rate is equal per unit carbon across energy products and is interwoven with (existing) energy excise rates, if available. Finally, with the exception of Finland, all Nordic countries choose to exempt specific agents, mainly energy-intensive industries, by applying (much) lower or even zero carbon tax rates.

3.4.2   Evidence on Administrative Costs

An empirical estimation of the administrative costs of different environmental tax policies does not, to our knowledge, exist. The same holds for compliance costs with only a few exceptions, such as Fullerton's (1996) analysis of the Superfund's corporate environmental tax. Direct estimates of the administrative costs of carbon taxes are also lacking. This section reviews the existing evidence on the administrative costs of taxation in general, and the factors that appear from this literature as relevant for the level of these costs.

The lack of evidence on administrative costs is not surprising because only a few explicit environmental taxes exist in practice (see, e.g., Fullerton 1996). Explicit environmental taxes are those for which the legislator has explicitly expressed the aim that this tax should serve some environmental purpose. However, the analysis of environmental taxation and administrative costs would be severely restricted if one were to limit the analysis to explicit environmental taxes only. As shown in section 3.3, input taxes are also important for environmental purposes. Indeed, taxes such as excise taxes and value-added taxes (VAT) affect the environment (e.g., taxes on gasoline and driving), as well as provisions in income taxes (tax allowances for commuting expenses, mine exploration, pollution control equipment, etc.).[22] For carbon taxation, current energy taxes, such as excises on hydrocarbon oils, are the most important because they are likely to have an impact on emissions through changes in input mix and changes in demand for energy.

Unfortunately, empirical information on the administrative costs of

---

22. Barthold (1994) mentions 51 federal tax-code provisions for the United States, and the OECD in more recent inventories mentions a much larger number of relevant taxes.

Table 3.4   **Relative Administrative and Compliance Costs of Different Types of Taxes (% of total revenue)**

Tax or Group	Administrative Costs	Compliance Costs	Total Operating Costs
Income tax	1.53	3.40	4.93
VAT	1.03	3.69	4.72
Corporation tax	0.52	2.22	2.74
Petroleum revenue tax	0.12	0.44	0.56
Excise duties (hydrocarbon oils; tobacco, alcoholic drinks)	0.25	0.20	0.45
Minor taxes (stamp duty, cars, betting and gambling)	0.85	1.48	2.33

Source: Sandford, Godwin, and Hardwick (1989, 192).

other taxes is scarce as well. Only a few studies exist.[23] Many problems exist regarding how to measure these costs, especially their absolute levels. One issue is the significant element of transferability between compliance costs and administrative costs (Sandford, Godwin, and Hardwick 1989, 203). Also, difficulties arise in categorizing operating costs. For instance, the (marginal) cost of transferring forms is highly influenced by the level of integration with existing administration.

Table 3.4 summarizes the results of Sandford, Godwin, and Hardwick (1989). Both the administrative and compliance costs of each tax are expressed as a percentage of the revenue raised by the tax. Administrative costs vary from 0.12 percent for the petroleum revenue tax to 1.53 percent for the income tax. The overall picture is clear: income tax and VAT are relatively expensive to administer, while excise duties are especially inexpensive in terms of administrative costs. This finding is also in accordance with findings in other studies; although OECD (1988) provides lower estimates on the total cost of VAT (between 0.40 and 1.09 percent), this study also ranks income taxes as being relatively most expensive and excises (interpreted as a single-stage general consumption tax) as being least expensive to implement (total cost around 0.5 percent).[24]

Because Sandford, Godwin, and Hardwick (1989) also include compliance costs, we can test whether we bias our analysis by focusing on administrative costs only. On average, compliance costs are three times higher than administrative costs. Compliance costs are relatively higher only for VAT. It is more important for our purposes, however, that the ranking of different types of taxes according to implementation costs is the same, whether we use administrative costs or total operating costs. Hence, the

23. Sandford, Godwin, and Hardwick (1989) analyze administrative and compliance costs of different taxes in the United Kingdom in 1986–87. OECD (1988) discusses operating costs for consumption taxes relative to other taxes.
24. See the discussion in Cnossen (1994).

basic picture is not influenced by adding compliance costs. The similar relative importance of compliance and administrative costs across different taxes suggests that administrative costs can be taken as being representative of both.

We now turn to the factors that determine the level of the administrative costs (see also section 3.2.1). Administrative costs as a percentage of the total revenue raised by a tax are not very relevant for the choice between different types of taxes. It is more important to know their fixed- and variable-costs characteristics, and how they are affected by the choice of tax base and rate. Unfortunately, such information is available only in a very limited way. As far as the role of the number of taxpayers is concerned, empirical information on the administrative costs of VAT indeed suggests the existence of economies of scale. In that case, costs per registered business should be relatively lower in countries with a low small-business exemption than in countries with a high exemption; broadening the tax base across a larger number of taxpayers reduces overall administrative costs per taxpayer. Cnossen (1994, 1652) notes that the data observed by OECD (1988), with the exception of Denmark, indeed fit this observation.

Another important determinant of administrative costs is the measurability of the tax base. One factor here is the differences among taxpayers. Some taxpayers will be more expensive to tax due to specific characteristics that have to be checked. Again, an interesting example is the small-business exemption from VAT. The larger the exemption, the smaller the number of registered businesses and the lower the absolute levels of administrative costs (see Cnossen 1994, 1652). Usually exemptions will be responsible for higher administrative costs. For instance, to give a tax rebate to a particular industry requires extra excise officers to handle and check such claims. Of course, exemptions for specific agents can also lower administrative costs if the agent is liable neither for the tax payment nor for a rebate. We do not find evidence for the assumption that more complex forms for calculating the tax base would raise administrative costs. Also, empirical studies have not tried to quantify the precise shape of the fixed- and variable-costs components of administrative costs of different tax types in relation to the use of differences in tax rates.

No decisive empirical information exists on the (general) shape of the transaction-costs curve for different types of taxes, especially environmental taxes. Moreover, as observed by Cnossen (1994, 1663), the findings of Sandford, Godwin, and Hardwick (1989) on the comparatively high VAT compliance costs are in clear contrast with evidence on VAT compliance costs in Germany. Here the estimated costs are only a fraction of the costs observed for the United Kingdom, which is mainly explained by the much longer tradition and experience in Germany and the integration of VAT with the administration of the business income tax. Thus, even if some information exists, the evidence seems to be dependent on local circumstances and institutional settings.

The implementation and enforcement of environmental taxes, however, has much in common with the operation of the age-old excises on alcohol, tobacco, and petroleum products (Cnossen 1977). Generally, these excises rely on quantitative measurement for assessment purposes, with compliance ensured through physical controls. Similar close controls should be exercised at points of import.[25] Thus, it seems safe beforehand not to always expect prohibitively high administrative costs for environmental taxes. This might be different only if the regulatory tax base asks for the monitoring of emissions that are difficult to measure and that therefore require costly metering technology.

Furthermore, the change in administrative costs depends heavily on the sectors already subject to other existing taxes or environmental regulation. For instance, according to Hoornaert (1992, 87), the physical control necessary for energy excises is very closely related to carbon taxes, while administrative controls for VAT are quite different and more time-consuming. The same might hold for other regulatory procedures that are already in force. Usually direct controls for environmental purposes also reflect tight supervision of technological processes and quantitative measurement. Thus, if closely linked production processes are already subject to monitoring, administrative costs need not be very high.

Summarizing, the level of administrative costs depends much on how emissions specifically relate to the production processes, their heterogeneity, and the number of these processes included in the tax base. Using existing excises for environmental regulation might be a relatively cheap way of taxing emissions since tax officers already have a lot of information required to operate the tax system.

3.4.3   Assessment in Terms of the Trade-Off

As noted before, the overall effect on welfare of introducing new environmental taxes, such as carbon taxes, should be compared to the incentives provided by the tax. An important result of our theoretical model is that input taxes offer an interesting alternative for emissions taxes if three conditions are met (see the end of section 3.3.4 in particular). First, there should be a clear linkage between inputs and emissions. Second, only few possibilities must exist for abating carbon emissions separately. Third, the administrative costs of emissions taxes should be high. In this section, we argue that these conditions are indeed met in the case of carbon taxation that supports the strategy chosen by the various countries applying these

---

25. Note that the tax base of specific excises requires physical control due to the physical dimensions in which they are usually expressed (dollars per unit, liter, etc.). This is fundamentally different from taxes expressed on an ad valorem basis (percentage of price or [added] value).

taxes. At the same time, however, the current design of the carbon taxes in practice leaves considerable room for improvement.

The first condition is related to the linkage issue (measured through $| \lambda - 1 |$ as part of the efficiency edge, Δ_i, in section 3.3.4). In the carbon case, CO_2 emissions are indeed in a one-to-one correspondence with the carbon content in energy products used as inputs (e.g., crude or refined oil products, natural gas, and various types of coal). Moreover, (potential) harmful CO_2 emissions are mainly related to the consumption of fossil fuels in modern societies. Thus, rather than taxing each unit of carbon emitted separately, it is rational to use taxes on energy products that contain carbon to pursue climate-change objectives. Such taxes on energy products provide indirect incentives, using the relationship between the burning of these products and transactions that can more easily be taxed. Thus, instead of taxing the emissions from car exhaust, an additional tax may be levied on gasoline purchases, on the assumption that the environmental damage caused is proportional to the amount of gasoline used.

This approach is indeed largely reflected in the carbon taxes applied in practice. They all take advantage of this fact by using carbon content of fuels as its tax base (although often a hybrid tax base is applied with a combination of both carbon and energy content). Thus coal-based energy-production processes are put at a disadvantage compared to other fossil and nonfossil fuel energy products. The same holds for oil relative to natural gas and nonfossil fuels. This is entirely in alliance with the purpose of the tax: providing much better targeted incentives compared to an indirect excise tax on energy alone. However, applying differences in tax rates per unit of carbon, as in the Norwegian case, cannot be justified and the rate structure as applied considerably weakens its incentive effect (e.g., coal is taxed at a much lower rate compared to natural gas, which contains fewer units of carbon per unit of energy).

The second important condition is that only few possibilities should exist to abate carbon emissions separately (measured through α as part of the efficiency edge, Δ_i). If emissions are very sensitive to emissions taxation, that is, if agents can abate CO_2 emissions easily, input taxes might become inefficient because they do not provide appropriate incentives for reducing carbon emissions directly. In other words, a loss in the efficiency of input taxes can be expected only if direct carbon abatement is possible, although not stimulated by a tax levied on the agents who are responsible for these CO_2 emissions. With respect to the abatement of carbon emissions (carbon disposal), indeed relatively few possibilities are available and almost none is actually employed.[26] Furthermore, these possibilities can

26. Of course, many opportunities exist for savings on energy use (improvements of energy efficiency) that also implicitly reduces carbon emissions (Eskeland and Devarajan 1996). However, the condition applied here is the improvement of carbon efficiency at the margin (as measured through the efficiency edge of emissions taxes over input taxes; see section

usually be applied only on a rather large scale. Therefore they are outside the reach of small individual firms or households. Thus, the use of input taxes in the case of carbon is indeed justified in this respect, in particular because the explicit carbon tax rates are very low.[27]

The third condition is that administrative costs related to emissions taxes should be high, or, in other words, the cost of input taxation should be relatively small; see equations (29) and (30), in particular. Usually administrative cost of newly designed taxes are relatively expensive due to a fixed setup cost of monitoring activities. This also applies to excise taxes, whether they are emissions or input taxes, even though they are cheap to administer compared with other types of taxes (see section 3.4.2). For that reason, tax reform of existing taxes is very attractive for policymakers because the effect on (marginal) administrative costs can be expected to be small. A rise of the marginal carbon tax rate is simply reached by using the existing implicit energy taxes on carbon (i.e., the existing energy excises). Thus, tax-rate reform is sufficient, that is, a reform of currently existing energy-input taxes toward taxes based on emissions coefficients (see section 3.4.1).

Indeed, the strategy chosen by the Nordic countries when implementing carbon taxes basically follows this logic. We checked which products and agents were already subject to energy excises in these countries in the pre-carbon-tax period, say 1990.[28] Table 3.5 presents our results. We distinguish between three potential groups of taxpayers: households (Hh), industrial consumers (I), and electricity generators (E). It is immediately clear from this table that the most important carbon-containing energy products consumed or produced in the Nordic countries were already subject to energy excises before the introduction of the carbon tax. The basic picture is that existing excises were levied on fuels consumed by households, with the exception of natural gas in Sweden and Norway (which is a small category anyway). The inputs of electricity were usually not subject to tax, in contrast to its delivery to consumers (both households and industries). Although many energy products are subject to tax, including even the products used as inputs in industry, it turns out that the industrial sector is often exempted or pays lower tax rates, especially energy-intensive industries (e.g., refineries, and steel and aluminum producers).

Thus, the effects on administrative costs of introducing carbon taxes on

---

3.3.4). We also exclude compensation techniques, such as carbon sequestration (by planting trees) because they are not directly related to the production techniques employed for producing output.

27. This might change if carbon tax policy became stricter because technological improvements might considerably reduce the cost of existing carbon abatement potentials.

28. We only checked excises because the introduction of a carbon tax is closely related to existing energy excises. Furthermore, in terms of the fixed-cost element, it is not important whether these products are VAT exempt or not. As discussed in section 3.4.2, the administrative procedures for VAT differ considerably with the excise administration.

Table 3.5 **Energy Excises Applying to Households, Industry, and the Electricity Sector in the Nordic Countries, 1990**

Energy Product	Sweden			Denmark			Norway			Finland		
	Hh	I	E	Hh	I	E	Hh	I	E	Hh	I	E
Diesel	+	+	−	+	+	−	+	+	−	+	+	−
Heavy fuel	−	+	0	−	+	0	−	+	0	−	+	+
Coal	+	+	0	+	+	0	0	0	na	na	+	+
Natural gas	0	0	0	+	na	na	0	0	0	+	+	na
Electricity	+	+	−	+	0	−	+	+	−	+	0	−

Source: OECD (1993a, 1993b).
Notes: Abbreviations: Hh, households; I, industry; E, electricity generation; +, tax; 0, no tax; −, not used; na, not available.

fuel content in these countries is dominated by the use of the existing energy excise administration. As long as this administration is also used for the carbon tax, one can safely assume a small rise in administrative costs. The only factor that might give an upward effect is the more complicated tax-base calculations due to the integration of two instead of one indicator (both energy content and carbon content). The same holds for carbon tax exemptions, especially in the case of rebates. As noted before, rebates often complicate the tax and cause higher administrative costs. If, however, exemptions in the carbon tax also take advantage of these institutional setups, additional administrative costs still need not be high (sunk cost element).

In all Nordic countries, however, the carbon excise is also imposed on new products, especially the production of electricity (use of inputs) and natural gas. Also, coal seems to be taxed now on a more comprehensive basis. But the effects of these changes on administrative costs also seem to be limited. Like the existing excise systems for other energy products, tax administration can take advantage of the way in which final fuel products, such as diesel or electricity, is usually delivered to consumers (both industry and households). The administration of energy excise taxes saves on the number of taxpayers by using points of delivery (e.g., fuel stations and energy distributors) instead of taxing all consumers separately. This is applicable in the case of natural gas (delivery through pipelines), as well as in the case of coal (points of distribution). Thus, the broadening of the tax base implies only a small increase in the number of taxpayers.

3.4.4 Scope for Improvement

Although the current carbon tax strategy in the Nordic countries satisfies the conditions for using input taxes instead of emissions taxes, considerable scope for improvement seems to exist. The coverage of the carbon excises in the Nordic countries (as well as the Netherlands) is far from exhaustive, especially in terms of the agents subject to an effective tax. Exemptions are widely used, mainly motivated by concerns about international competitiveness. Often the energy inputs of domestic industries are taxed at lower rates or not taxed at all. Furthermore, the existing energy excises related to oil products are of the final-fuel type, which implicitly exempts the production of the fuels themselves. Also the extraction of any fossil fuel is not subject to this tax (although other type of taxes and subsidies apply).

Our theoretical results suggest that sectoral differentiations in the tax rate are justified by administrative costs, if linkage and marginal abatement cost (MAC) differ among sectors. Exemptions can also be justified by differences in fixed administrative costs. A difference between linkage and MAC, as well as fixed administrative costs among sectors, seems to apply in the carbon case. However, current differentiation is exactly the opposite of what our model suggests is optimal. Dijkgraaf and Vollebergh (1997) show that this observation generalizes across OECD countries. In general, households face much higher taxes on average compared to industry. Furthermore, most OECD countries tax final oil products (e.g., diesel and gasoline) much more heavily than primary energy products on average (e.g., heavy fuel oil, natural gas, and coal). In this respect, the countries that introduced a carbon tax, already applied a much broader (implicit) carbon tax base compared to the other countries.

Thus the industries that are usually exempted now, mainly the energy-intensive industries (both producing energy products and energy-intensive products), are also the taxpayers who can be taxed with lowest transaction costs per unit of emissions. In other words, the most important polluters (the small number of taxpayers consuming the larger part of fossil fuels) still pay only a very small amount of tax, if any. The same holds for the choice to exempt certain energy products consumed by specific sectors, such as coal by electricity generation. Finally, not taxing particular energy products that cause considerable carbon emissions, such as coal, seems to be particularly unattractive.[29] Of course, issues of carbon leakage are of considerable importance here. If a country follows a unilateral strategy without any compensation for its carbon-exposed industries, import substitution could easily reduce the effectiveness of its carbon abatement policy. However, several mechanisms are available to compensate for these effects, with small or even no negative effect on administrative costs, such as tax credits (Vollebergh, Koutstaal, and de Vries 1997).

Another issue closely linked to the selectivity of coverage is that all explicit carbon taxes are based simply on the amount of carbon contained in the actual products. This implies that carbon emitted in the process of producing those fuels is not taxed at all. As Pearson and Smith (1991, 29) note, such a scheme gives an undesirable incentive to the use of highly

---

29. OECD (1998) shows that coal is still subsidized in quite a number of OECD countries.

refined fuel products, in which as much as possible of the carbon emissions have taken place before the excise is applied. Thus, this tax will be less efficient at encouraging carbon-reducing fuel substitutions. According to Vollebergh (1995) it might be an efficient strategy in this case to use a materials-balance approach to impute the amount of upstream carbon emissions related to energy products of the final fuel type.

A third possibility for improvement is to supplement current input taxes with incentives for abatement (introducing a mixed system of input and emissions taxes). Although abatement of CO_2 emissions is very limited for small energy users, large industries and energy producers may have some opportunities for abatement that are less costly than separate abatement possibilities such as carbon sequestration. Large-scale firm-specific investments are involved in these abatement projects. Emissions taxes for energy producers may provide appropriate abatement incentives. Moreover, the administration costs for emissions taxes in the energy-production sector can be expected to be considerably lower than for small industry and households. Technologies are more homogenous, and the number of agents is small. For large energy-intensive industries, however, the competitiveness argument may prevent the implementation of emissions taxes, since these taxes increase costs and require, again, compensation schemes. Alternatively, abatement subsidies decrease costs and seem more feasible.

The most important step toward more efficient carbon policies is explicit coordination of carbon policies on the EU, OECD or, better, world scale. Carbon leakage then no longer offsets unilateral carbon policies. Thus, the exemption of large energy-intensive exporting industries to restore international competitiveness would no longer be a reasonable strategy. Only then would it be possible to initiate a full-fledged tax reform imposing carbon taxes on agents that have the most options for abatement, contribute most to CO_2 emissions, and for which the administrative costs involved are relatively smallest.

References

Barthold, T. A. 1994. Issues in the design of environmental excise taxes. *Journal of Economic Perspectives* 8 (1): 133–51.

Bovenberg, A. L., and L. H. Goulder. 1998. Environmental taxation. In *Handbook of public economics,* ed. A. Auerbach and M. Feldstein, 2d ed. Amsterdam: North-Holland, forthcoming.

Cnossen, S. 1977. *Excise systems.* Baltimore: Johns Hopkins University Press.

———. 1994. Administrative costs and compliance costs of the VAT. *Tax Notes International* 8:1649–68.

Cnossen, S., and H. R. J. Vollebergh. 1992. Toward a global excise on carbon. *National Tax Journal* 45 (1): 23–36.

Cohen, M. A. 1998. Monitoring and enforcement of environmental policy. Vanderbilt University, Nashville. Mimeo.

de Mooij, R. A. 2000. *Environmental taxation and the double dividend.* Contributions to Economic Analysis, vol. 246. Amsterdam: North-Holland.

Diamond, P. A. 1973. Consumption externalities and imperfect corrective pricing. *Bell Journal of Economics and Management Science* 4:526–38.

Dijkgraaf, E., and H. R. J. Vollebergh. 1997. Energy taxation in OECD countries. Paper presented at the International Institute of Public Finance conference, Tel Aviv.

Ekins, P., and S. Speck. 1999. Competitiveness and exemptions from environmental taxes in Europe. *Environmental and Resource Economics* 13:369–96.

Eskeland, G. S., and S. Devarajan. 1996. Taxing bads by taxing goods: Towards efficient pollution control with presumptive charges. In *Environmental taxation in a second-best world,* ed. S. Cnossen and L. Bovenberg, 61–112. Boston: Kluwer Academic Publishers.

European Commission (EC). 1992. Proposal for a council directive introducing a tax on carbon emissions and energy, COM(92) 226 final, June. *Official Journal* C 196. 3 August.

———. Forthcoming. Amended proposal for a council directive introducing a tax on carbon dioxide emissions and energy, COM(95) 172. *Official Journal.*

Feldstein, M. 1976. On the theory of tax reform. *Journal of Public Economics* 6: 77–101.

Fullerton, D. 1996. Why have separate environmental taxes? *Tax policy and the economy,* vol. 10, ed. J. M. Poterba, 33–70. Cambridge, Mass.: MIT Press.

Fullerton, D., and A. Wolverton. 1997. The case for a two-part instrument: Presumptive tax and environmental subsidy. NBER Working Paper no. 5993. Cambridge, Mass.: National Bureau of Economic Research.

Goulder, L. H. 1995. Environmental taxation and the "double dividend": A reader's guide. *International Tax and Public Finance* 2:157–83.

Hoel, M. 1998. Emission taxes versus other environmental policies. *Scandinavian Journal of Economics* 100 (1): 79–104.

Hoornaert, L. 1992. The use of taxation as a policy instrument aimed at limiting the community's CO_2 emissions: Practical dimensions of implementation. *European Economy,* special edition no. 1, 63–90.

Kaplow, L. 1990. Optimal taxation with costly enforcement and evasion. *Journal of Public Economics* 43:221–36.

Krutilla, K. 1999. Environmental policy and transactions costs. In *Handbook of environmental and resource economics,* ed. J. C. J. M. van den Bergh, 249–64. Cheltenham, U.K.: Edward Elgar.

McKay, S., M. Pearson, and S. Smith. 1990. Fiscal instruments in environmental policy. *Fiscal Studies* 11 (4): 1–20.

OECD. 1988. *Taxing consumption.* Paris: Organization for Economic Cooperation and Development.

———. 1993a. Environmental taxes in OECD countries: A survey. OECD Working Papers, Environment Directorate and Directorate for Financial, Fiscal and Enterprise Affairs 1, no. 4. Paris: Organization for Economic Cooperation and Development.

———. 1993b. *Energy prices and taxes.* Paris: Organization for Economic Cooperation and Development.

———. 1998. *Subsidies and the environment.* Paris: Organization for Economic Cooperation and Development.

OECD/IEA. 1993. *Greenhouse gas emissions—The energy dimension.* Paris: Organization for Economic Cooperation and Development/International Energy Agency.

Okken, P. A., P. Lako, D. Gerbers, T. Kram, and J. R. Ybema. 1992. *CO_2 removal* *in competition with other options for reducing CO₂ emissions.* ECN, RX 92–025. Petten: The Netherlands Energy Research Foundation.

Pearce, D. 1991. The role of carbon taxes in adjusting to global warming. *Economic Journal* 101:938–48.

Pearson, M., and S. Smith. 1991. *The European carbon tax: An assessment of the European Commission's proposals.* London: Institute for Fiscal Studies.

Polinsky, A. M., and S. Shavell. 1982. Pigouvian taxation with administrative costs. *Journal of Public Economics* 19:385–94.

Poterba, J. M. 1992. Tax policy to combat global warming: On designing a carbon tax. In *Global warming,* ed. R. Dornbusch and J. M. Poterba, 71–98. Cambridge, Mass.: MIT Press.

Sandford, C., M. Godwin, and P. Hardwick. 1989. *Administrative and compliance costs of taxation.* Bath, U.K.: Fiscal Publications.

Schmutzler, A., and L. H. Goulder. 1997. The choice between emission taxes and output taxes under imperfect monitoring. *Journal of Environmental Economics and Management* 32:51–64.

Shortle, J. S., R. D. Horan, and D. G. Abler. 1998. Research issues in nonpoint pollution control. *Environmental and Resource Economics* 11:571–85.

Slemrod, J., and S. Yitzhaki. 1998. Tax avoidance, evasion, and administration. In *The handbook of public economics,* ed. A. Auerbach and M. Feldstein. Amsterdam: North-Holland, forthcoming.

Smith, S. 1992. Taxation and the environment: A survey. *Fiscal Policy* 13 (4): 21–57.

Smith, S., and H. R. J. Vollebergh. 1993. The European carbon excise proposal: A green tax takes shape. *EC Tax Review* 3 (4): 207–21.

Smulders, S., and H. R. J. Vollebergh. 1998. Environmental taxation and administrative costs: Should we optimally exempt some firms from emission taxes? Paper presented at the World Congress of Environmental and Resource Economists, Venice.

———. 1999. The choice between upstream and downstream taxation in the presence of administrative costs. Tilburg University. Mimeo.

Vollebergh, H. R. J. 1995. Transaction costs and European carbon tax design. In *Environmental standards in the European Union in an interdisciplinary framework,* ed. M. Faure, J. Vervaele, and A. Weale, 135–54. Antwerp: MAKLU.

Vollebergh, H. R. J., P. Koutstaal, and J. de Vries. 1997. Hybrid carbon incentive mechanisms and political acceptability. *Environmental and Resource Economics* 9:43–63.

Xepapadeas, A. 1999. Non-point source pollution control. In *Handbook of environmental and resource economics,* ed. J. van den Bergh, 539–50. Cheltenham, U.K.: Edward Elgar.

Yitzhaki, S. 1979. A note on optimal taxation and administrative costs. *American Economic Review* 69:475–80.

Comment   Dallas Burtraw

The insight of the thorough and well-written paper by Sjak Smulders and Herman Vollebergh is in identifying an important trade-off when choosing

Dallas Burtraw is a senior fellow in the Quality of the Environment Division at Resources for the Future in Washington, D.C.

among potential policy instruments. This is the trade-off between transaction costs and incentives, which indeed has not been developed adequately in the environmental literature. In this application, transaction costs include "administrative costs" incurred by the regulator, and "compliance costs" incurred by the agent. "Incentives" refer to the signal to change behavior received by the agent. Focusing on this trade-off informs the choice of policy instruments and also draws attention to the cost of environmental policy, which forces one away from the first-best (Pigouvian) level of regulation.

Drawing on the sketchy evidence available from previous studies, the authors make some key observations. The first is that the administrative costs incurred by the government (which are the focus of paper, rather than compliance costs incurred by agents) are often greater for emissions taxes than for other types of environmental taxes, especially input taxes. Why might this be true? One reason has to do with the measurability of the tax base. It may be more difficult to measure emissions from distributed pollution sources than it is to measure the economic activity of those sources because the purchase of inputs is already accounted for financially. The second reason may have to do with the number of agents that are covered under various tax schemes. Typically one might imagine inputs (such as fuel) to come from a small number of sources while emissions come from a large number of distributed sources, and one might expect administrative costs to vary with the number of agents that have to be monitored.

The second observation recognizes a distinction between fixed and variable components, reminiscent of Stavins's (1995) treatment of transactions costs for tradable permits. Further, it is noted that variable costs may rise with tax burden, or in this case with the quantity of emissions. One reason this may be true has to do with the incentive that agents have for tax evasion.

The authors use these observations as a point of departure to craft broad policy guidance. Their proposition is that other environmental taxes, especially input taxes, are preferable to emissions taxes when

1. Emissions taxes have high fixed administrative costs, that is, the cost associated with setting up a new tax on emissions. For many types of input taxes, it is recognized that these costs are already incurred because these taxes are already in place.

2. Emissions taxes have high elasticity of variable costs; that is, variable costs rise sharply with the level of emissions or the tax revenue collected from the emissions tax (as a result of increases in the tax rate).

3. The incentive effects from input taxes are high with respect to the production of emissions. For example, this occurs when there is little opportunity to abate emissions, so emissions have a high correlation with taxable inputs.

Table 3C.1   **Comparison of a Variety of Environmental Problems**

Environmental Problems	High Fixed Administrative Costs for Emissions Taxes?	Elastic Variable Administrative Costs for Emissions Taxes?	Little Opportunity for Abatement (Inputs and Emissions Linked)?
CO_2	Strong–weak	Strong–weak	Strong
NO_x	Strong–weak	Strong–weak	Weak
SO_2	Weak	Moderate–weak	Moderate
Mercury	Strong	Moderate–weak	Moderate
Nonpoint water pollution	Strong	Strong	Strong–moderate

This comparison, between environmental (input) taxes and emissions taxes, supplements a yet more fundamental trade-off guiding environmental policy analysis. Emissions taxes as a strategy to address environmental problems are already second-best approaches. Environmental economics ideally would recommend a first-best strategy that would target ambient concentrations of pollution and the marginal damages that result, the class of direct instruments. However, in many settings it is clear that direct instruments are not practical because of the administrative costs of implementing such a system. And so we are already in the second-best world of considering emissions taxes, from which these authors take us still further, toward consideration of input taxes.

The authors take this broad advice and apply it to one problem in particular, controlling emissions of carbon dioxide (CO_2). They find that in this case, at least, the shoe fits. Implicitly the paper provides an endorsement of the approach taken recently by several governments in Europe that have sought to implement various environmental taxes on energy inputs as an indirect means of regulating CO_2 emissions.

The finding is likely to survive in the context of CO_2 policy because, for the foreseeable future, a technology that allows for postcombustion abatement of CO_2 emissions will be prohibitively expensive. Therefore, taxing inputs is identical to taxing emissions, with respect to the incentives for emissions reduction. However, one must use care in seeking to generalize from this example, and it is interesting to attempt to do so.

Table 3C.1 provides a subjective assessment of the authors' criteria applied to a number of environmental problems in the United States. In the cases of CO_2, NO_x, and SO_2, I suggest that a range of values may apply to the first criterion, which asks whether high fixed administrative costs with emissions taxes exist. The reason this criterion may apply in a weak way is that a significant fraction of emissions, especially SO_2, comes from the electric utility sector, where continuous emissions monitors already are required on all large fossil-fired power plants under the 1990 Clean Air Act Amendments. These monitors track the emissions of these three pollutants, making administrative costs of enforcing a tax relatively low. In other sectors the administrative costs could be quite high, and in general that would apply to mercury emissions and nonpoint water pollution.

The second criterion asks whether administrative costs vary with the magnitude of the policy or quantity of emissions. Again, the electricity sector may be a special case because the monitors are thought to be accurate. Outside this sector, however, the incentive and opportunity for avoiding monitoring may be high. However, the magnitude of potential tax revenues is probably not substantial for most affected industries, with the particular exceptions of CO_2 and nonpoint water pollution.

The third criterion addresses the linkage between inputs and emissions. This provides the strongest argument for an input tax for CO_2 policies. Other air pollutants offer some opportunity for abatement, which severs the link between inputs and emissions. Nonetheless, for SO_2 and mercury, mass-balance principles apply, linking inputs and residuals (either abated or emitted). However, emissions of NO_X in large part are variable with respect to combustion temperature and other factors. Nonpoint water pollution provides perhaps the second strongest link (second to CO_2) since nonpoint pollution largely is a function of inputs, although there exist some management practices that affect nonpoint pollution.

Even after taking the contribution of this paper fully into account, environmental economists should remain sensitive and perhaps reluctant to back away from more direct instruments for environmental control. Experience teaches us that important changes in technology and in social organizations occur as we try to build institutions that put strong incentives in place regarding environmental performance. New technologies are emerging for monitoring emissions sources, including continuous emissions monitors on smokestacks, pollution markers that can allow the regulator to trace concentrations to individual pollution sources, and remote sensors that can be used to identify pollutant emissions from individual vehicles. As technologies emerge, administrative costs of more finely tuned environmental instruments fall. Hence, it is important to think about the design of instruments in the dynamic context. For example, in the United States the emissions reduction credit program of the early 1980s was a mixed success. Anecdotal evidence is that the transaction costs of trades under the program were approximately 30 to 40 percent of the value of the trades, on average. However, under the SO_2 emissions allowance trading program in the United States in the 1990s, transaction costs have fallen to less than 1 percent of the value of the trades. This comes from organizational learning, new institutions, and new technologies. In general, one should give consideration in the design of environmental programs to the enhancement of these innovations.

The authors go beyond development of the criteria discussed here to offer another proposition. They suggest that when fixed administrative costs of both environmental (input) taxes and emissions taxes are low, relative to variable costs, a hybrid policy involving both types of taxes may avoid the problem of rising variable administrative costs. This may occur because the tax rate on emissions can be kept low because part of the emissions reduction is occurring through substitution among inputs in response to an input tax, and vice versa.

I remain skeptical about the efficiency characteristics of a hybrid of multiple taxes to achieve a desired level of emissions reduction. Each tax gives rise to the opportunity for political horse-trading, and I fear a mix of taxes to achieve a single policy goal would seemingly give rise to unmodeled coordination costs. I conjecture that, in the broad scheme of things, administrative costs as they are addressed in this paper are typically less than the political costs of navigating the legislative process. The difficulty of fine-tuning the tax system would seem to be even more daunting when multiple tax instruments are considered. In fact, toward the end of the paper the authors provide a useful summary of existing input taxes and identify myriad deviations from consistent or efficient policy. In Europe, I understand, there is increasing use of hybrid tax instruments to achieve environmental tax reform. It may be that the hybrid tax instrument enhances the political acceptability of environmental action or that it is a way to slip the camel's nose (control of CO_2 emissions) under the tent of economic growth. However, it is not obvious that the outcome is likely to be efficient in the sense addressed by these authors.

These thoughts provide some cautionary notes for why we may not necessarily want to move away from a more direct approach to environmental policy making for an approach that in the short run has lower administrative costs. However many of these issues are beyond the scope of the present paper. So to close, let us return to the present paper to reconsider its major contributions and guidance for policy.

The conclusions offered for CO_2 emissions reduction by the authors align with the direction currently being taken by governments as they try to grapple with the greenhouse gas problem. This analysis tells us that this direction of policy is probably the right one. And also this analysis gives us reason to focus on a critical issue—the trade-off between first-best incentives and administrative costs—that is an important consideration in actual policy design and that is often missing from economic models.

Reference

Stavins, Robert N. 1995. Transaction costs and tradeable permits. *Journal of Environmental Economics and Management* 29 (2): 122–48.

4

An Industry-Adjusted Index of State Environmental Compliance Costs

Arik Levinson

4.1 Introduction

This paper describes a new industry-adjusted index of state environmental compliance costs that can be used to compare regulations both across states in a given year and within states over time.[1] It compares that index to others used in the environmental economics literature and uses the index to answer several often-raised questions about the pattern of environmental regulations in the United States and how that pattern has changed over time. Finally, the paper describes an application of the index, as used to assess the effect of environmental regulatory stringency on foreign direct investment to U.S. states.

There are three key motivations for creating this index. First, the Environmental Protection Agency (EPA) has worried publicly that some states are laggards in enforcing federal standards.[2] The index described here documents the variation across states in their environmental compliance costs. Second, since 1980 responsibility for monitoring and enforcement of environmental regulations has been devolving from the federal government to state and local regulators. In theory, this could cause states to become less or more similar in their standard stringency as they are freed to set their

Arik Levinson is associate professor of economics at Georgetown University and a faculty research fellow of the National Bureau of Economic Research.

This paper has benefited from funding provided by the National Science Foundation; helpful comments by Wayne Gray, Kevin Hassett, John List, and Domenico Siniscalco; and research assistance by Victor Davis and Joe Hendrickson.

1. Interested readers can find a Stata file containing the index at http://www.georgetown.edu/faculty/am16/index2.htm.

2. See, e.g., the articles by J. H. Cushman, Jr., "States Neglecting Pollution Rules, White House Says," *New York Times,* 15 December 1996, p. 1.1; and "E.P.A. and States Found to be Lax on Pollution Law," *New York Times,* 7 June 1998.

own levels of stringency and to compete with their neighbors to attract industry or clean their environments. This index provides data on the degree of convergence in state standard stringency over time. Third, analysts studying the effects of environmental regulations on local and national economies have been hampered by the difficulty of accurately measuring and comparing the stringency of those regulations (Jaffe et al. 1995). In particular, studies of the effect of regulations on local economies rely almost exclusively on cross-sectional data,[3] subjective indexes of state standards or cost-based measures that do not control for industrial composition.

Existing measures of environmental regulatory stringency take two forms. First, there are the environmental groups' rankings of states. These are subjective and typically only measure perceptions of states' efforts at one time, so intertemporal comparisons are not possible. Most analysts have therefore relied on the Census Bureau's Pollution Abatement Costs and Expenditures (PACE) survey data to construct measures of statewide compliance costs per unit of output. These measures, however, fail to control for states' industrial compositions. Consequently, states with a lot of polluting industry have relatively high environmental compliance costs, regardless of their regulations.

To address these concerns, section 4.2 describes the new industry-adjusted index and reports findings about relative stringency and how it has changed over time. Section 4.3 describes existing subjective measures of environmental standard stringency and compares them to the industry-adjusted index. Section 4.4 describes an application of the index to assess the effect of environmental stringency on foreign direct investment to U.S. states.

4.2   An Industry-Adjusted Index of State Environmental Compliance Costs

Many researchers have relied on the Census Bureau's PACE survey to construct indexes of state environmental regulatory stringency. The PACE survey collected data from manufacturing establishments about their pollution abatement operating and capital costs from 1977 to 1994, when it was discontinued.[4] Most commonly, studies use these costs divided by some measure of state economic activity, such as total employment or

3. Because most studies examine differences among jurisdictions at one time, they cannot distinguish between the simultaneous effects of regulations on economic growth and that of economic growth on regulations. Notable exceptions include Gray (1997), Greenstone (1998), and Becker and Henderson (1997).

4. Recently, it appears that the EPA and the Census Bureau have agreed on plans to again collect the PACE data. Unfortunately, there will have been a minimum of 5 years during which the data were not collected. The PACE data collected from 1973 to 1976 are incompatible with later surveys in their treatment of small plants. Also, the PACE data were not collected in 1987.

gross state product (GSP).[5] The most significant problem with such measures is that they fail to adjust for industrial composition. States that have pollution-intensive industrial compositions will incur high pollution abatement costs, whether or not they have stringent regulations. Ideally, one would use the pollution abatement costs in the relevant industry as an index of regulatory stringency. While abatement costs by state and industry are published annually by the Census Bureau, so many of the observations are censored to prevent disclosure of confidential information that the data are not comparable year-to-year or state-to-state.[6] Therefore, this paper proposes an alternative index.

The index compares the *actual* pollution abatement costs in each state, unadjusted for industrial composition, to the *predicted* abatement costs in each state, where the predictions are based solely on nationwide abatement expenditures by industry and each state's industrial composition.[7] Let the actual costs per dollar of output be denoted

(1)
$$S_{st} = \frac{P_{st}}{Y_{st}},$$

where P_{st} is pollution abatement costs in state s in year t, and Y_{st} is the manufacturing sector's contribution to the GSP of state s in year t. S_{st} is the type of unadjusted measure of compliance costs commonly used. By failing to adjust for the industrial composition of each state, it probably overstates the compliance costs of states with more pollution-intensive industries and understates the costs in states with relatively clean industries.

To adjust for industrial composition, compare equation (1) to the predicted pollution abatement costs per dollar of GSP in state s:

(2)
$$\hat{S}_{st} = \frac{1}{Y_{st}} \sum_{i=20}^{39} \frac{Y_{ist} P_{it}}{Y_{it}},$$

5. See, e.g., Crandall (1993), Friedman, Gerlowski, and Silberman (1992), or List and Co (2000). Consulting firms specializing in industrial siting decisions have also relied on such simple indexes of environmental regulatory stringency (Alexander Grant & Co. 1985).

6. Several papers have used the confidential plant-level PACE data to construct such indexes (Levinson 1996; Gray 1997). However, those data are unavailable to most researchers, and the purpose here is to construct an easily accessible resource for analysts. Later, I do compare the index created from the confidential data to that compiled from the published data.

7. For two reasons, I use pollution abatement operating expenses (as opposed to capital expenses) in the index. First, operating expenses for pollution abatement equipment are easier for PACE survey respondents to identify separately. Abatement capital expenses may be difficult to disentangle from investments in production process changes that have little to do with pollution abatement. Second, abatement capital expenditures are highest when new investment takes place. So states that have thriving economies and are generating manufacturing investment tend to have high levels of abatement capital expenses, regardless of the stringency of those states' environmental laws. Operating costs are more consistent from year to year.

where industries are indexed from 20 through 39 following the two-digit manufacturing Standard Industrial Classification (SIC) codes,[8] Y_{ist} is the contribution of industry i to the GSP of state s at time t, Y_{it} is the nationwide contribution of industry i to the national GDP, and P_{it} is the nationwide pollution abatement operating costs of industry i. In other words, S_{st} is the weighted average pollution abatement costs (per dollar of GSP), where the weights are the relative shares of each industry in state s at time t.

To construct the industry-adjusted index of relative state stringency, S^*_{st}, I compute the ratio of actual expenditures in equation (1) to the predicted expenditures in equation (2),[9]

$$(3) \qquad\qquad S^*_{st} = \frac{S_{st}}{\hat{S}_{st}}.$$

When S^*_{st} is greater than 1, industries in state s at time t spent more on pollution abatement than those same industries in other states. When S^*_{st} is less than 1, industries in state s at time t spent less on pollution abatement. By implication, states with large values of S^*_{st} have relatively more stringent regulations than states with small values of S^*_{st}.[10]

Table 4.1 presents the average values of various environmental indexes. The first column contains the average unadjusted index, S_{st}, from 1977 to 1994 (omitting 1987, when the PACE data were not collected). The second column contains the industry-adjusted index, S^*_{st}.[11] Table 4.2 contains the rankings of these indexes. Several striking facts can be seen from comparing the indexes. First, the ranking of state regulatory stringency according to the industry-adjusted index (S^*) is often quite different from the ranking according to the unadjusted index (S). For example, New Jersey manufacturers spent a relatively large amount on pollution abatement, causing the state to be ranked 20th in terms of the average unadjusted index in column (1) of table 4.2. However, when New Jersey's relatively pollution-intensive industrial composition is accounted for, the state's ranking falls to 34th. In contrast, when Oregon's relatively clean industrial composition is

---

8. SIC code 23 (apparel) is omitted because it is relatively pollution-free, and as a result no data for that industry are collected by the PACE survey.

9. Note that the state's GSP is in both the numerator and the denominator of equation (3). Equation (3) can thus be expressed as $S^*_{st} = P_{st}/P_{st}$, where P_{st} is the summation term in equation (2).

10. I have also calculated the index described by equations (1), (2), and (3) using the number of production workers in each two-digit SIC code to control for industrial composition, instead of using each industry's share of GSP. The broad conclusions are similar, although the rankings of some states do change. Also, annual employment totals by state and industry are more often censored to prevent disclosure of confidential information.

11. Appendix table 4A.1 presents annual values of the industry-adjusted index (S^*) and its ranking of states. Appendix table 4A.2 presents annual values of the unadjusted index (S) and its rankings.

Table 4.1 Indexes of State Environmental Effort

	Unadjusted Cost (S_s) Average[a] (1)	Adjusted Cost (S_s^*) Average[a] (2)	Conservation Foundation (3)	FREE Index (4)	Green Index (5)	Southern Studies (6)	LCV Average[a] (7)	Levinson (1996) (8)
AL	0.0219	1.19	10	16	8,658	681	24.4	−0.035
AZ	0.0148	1.39	24	27	7,342	567	29.2	−0.232
AR	0.0168	1.17	27	18	8,353	579	43.7	−0.072
CA	0.0121	0.90	46	48	4,931	423	57.4	−0.149
CO	0.0113	1.01	26	24	6,110	377	48.0	−0.384
CT	0.0079	0.67	32	44	5,483	442	74.0	−0.001
DE	0.0344	1.30	29	24	6,821	518	67.8	−0.273
FL	0.0138	1.21	31	41	6,320	461	50.3	0.022
GA	0.0127	0.91	25	26	7,488	544	43.3	−0.194
ID	0.0181	1.66	16	18	6,513	425	17.2	−0.004
IL	0.0132	0.91	28	41	7,052	563	60.3	0.055
IN	0.0196	1.14	36	36	7,939	687	45.9	0.013
IA	0.0106	0.96	29	39	6,541	491	54.5	−0.034
KS	0.0115	0.76	23	29	7,732	625	35.5	−0.330
KY	0.0146	0.99	34	28	7,694	594	32.3	0.065
LA	0.0538	1.51	21	21	8,383	708	25.7	−0.102
ME	0.0237	1.55	32	36	4,892	331	77.1	−0.041
MD	0.0185	1.17	37	34	5,585	413	70.6	0.148
MA	0.0067	0.67	44	41	5,076	389	86.6	−0.109
MI	0.0121	1.01	30	43	6,297	541	67.9	0.084
MN	0.0092	0.66	47	38	5,000	381	64.9	−0.209
MS	0.0213	1.47	15	14	8,299	612	20.1	−0.255
MO	0.0104	0.79	14	31	7,006	530	42.6	−0.195
MT	0.0341	1.49	37	23	6,546	559	49.9	0.110

(*continued*)

Table 4.1 (continued)

	Unadjusted Cost (S_S) Average[a] (1)	Adjusted Cost (S_S^*) Average[a] (2)	Conservation Foundation (3)	FREE Index (4)	Green Index (5)	Southern Studies (6)	LCV Average[a] (7)	Levinson (1996) (8)
NE	0.0088	0.83	22	31	7,001	520	40.9	-0.196
NV	0.0072	0.63	22	23	6,670	434	38.1	-0.239
NH	0.0072	0.75	21	32	5,803	310	58.6	-0.276
NJ	0.0158	0.82	45	47	5,790	464	78.6	0.117
NM	0.0306	1.64	18	23	6,998	533	30.0	-0.500
NY	0.0087	0.77	37	43	5,419	424	64.4	0.000
NC	0.0088	0.82	25	42	6,772	578	34.2	-0.144
ND	0.0105	0.77	22	16	6,833	458	43.8	-0.566
OH	0.0139	0.82	30	36	7,411	586	61.5	0.056
OK	0.0103	0.58	19	29	7,644	588	27.5	-0.396
OR	0.0139	1.22	42	35	4,583	395	53.9	0.122
PA	0.0169	0.91	28	32	6,905	511	55.1	0.022
RI	0.0075	0.72	26	30	5,105	397	79.8	-0.245
SC	0.0160	0.99	25	31	7,407	611	41.4	-0.184
SD	0.0056	0.68	30	23	6,965	396	51.1	-0.020
TN	0.0165	1.10	23	29	8,151	698	43.2	-0.078
TX	0.0311	1.39	22	26	8,197	703	28.3	-0.151
UT	0.0164	0.93	23	16	7,122	556	17.5	-0.494
VT	0.0065	0.66	32	28	4,921	282	83.8	-0.111
VA	0.0118	0.96	28	33	7,055	521	33.8	-0.097
WA	0.0196	1.37	39	29	5,473	430	56.3	-0.182
WV	0.0433	1.58	23	15	8,117	652	54.4	-0.115
WI	0.0110	0.89	37	49	5,478	379	69.8	-0.186
WY	0.0259	0.72	23	16	7,445	601	16.8	-0.412

[a] For 1977–94; averages omit 1987, when the PACE survey was not collected.

Table 4.2 Rankings of State Environmental Effort

	Unadjusted Cost (S_S) Average[a] (1)	Adjusted Cost (S_S^*) Average[a] (2)	Conservation Foundation (3)	FREE Index (4)	Green Index (5)	Southern Studies (6)	LCV Average[a] (7)	Levinson (1996) (8)
AL	9	14	48	44.5	48	44	44	19
AZ	21	9	31	31	33	33	40	38
AR	16	15	25	41.5	46	35	28	21
CA	29	29	2	2	4	12	16	29
CO	32	20	26.5	34.5	15	4	25	44
CT	42	44	14	4	11	17	6	15
DE	3	11	20.5	34.5	23	23	10	2
FL	25	13	16	9	17	19	23	11.5
GA	27	28	29	32.5	37	29	29	34
ID	14	1	45	41.5	18	14	47	16
IL	26	26	23	9	30	32	14	10
IN	12	17	11	14	41	45	26	13
IA	34	24	20.5	11	19	21	19	18
KS	31	38	34	26.5	40	42	35	43
KY	22	22	12	29.5	39	38	38	8
LA	1	5	41.5	40	47	48	43	24
ME	8	4	14	14	2	3	5	20
MD	13	16	8.5	17	12	11	7	3
MA	46	43	4	9	6	7	1	25
MI	28	19	18	5.5	16	28	9	7
MN	38	46	1	12	5	6	11	37
MS	10	7	46	48	45	41	45	41
MO	36	35	47	22	29	26	31	35
MT	4	6	8.5	37.5	20	31	24	6

(*continued*)

Table 4.2   (continued)

	Unadjusted Cost (S_S) Average[a] (1)	Adjusted Cost (S_S^*) Average[a] (2)	Conservation Foundation (3)	FREE Index (4)	Green Index (5)	Southern Studies (6)	LCV Average[a] (7)	Levinson (1996) (8)
NE	40	31	38.5	22	28	24	33	36
NV	44	47	38.5	37.5	21	16	34	39
NH	45	39	41.5	19.5	14	2	15	42
NJ	20	34	3	3	13	20	4	5
NM	6	2	44	37.5	27	27	39	47
NY	41	36	8.5	5.5	8	13	12	14
NC	39	33	29	7	22	34	36	28
ND	35	37	38.5	44.5	24	18	27	48
OH	23	32	18	14	35	36	13	9
OK	37	48	43	26.5	38	37	42	45
OR	24	12	5	16	1	8	21	4
PA	15	27	23	19.5	25	22	18	11.5
RI	43	41	26.5	24	7	10	3	40
SC	19	21	29	22	34	40	32	32
SD	48	42	18	37.5	26	9	22	17
TN	17	18	34	26.5	43	46	30	22
TX	5	8	38.5	32.5	44	47	41	30
UT	18	25	34	44.5	32	30	46	46
VT	47	45	14	29.5	3	1	2	26
VA	30	23	23	18	31	25	37	23
WA	11	10	6	26	9	15	17	31
WV	2	3	34	47	42	43	20	27
WI	33	30	8.5	1	10	5	8	33
WY	7	40	34	44.5	36	39	48	1

Note: Equal observations receive the average rank.

[a] For 1977–94; averages omit 1987, when the PACE survey was not collected.

accounted for, that state's ranking improves from 24th to 12th. Similar reordering takes place for other states, supporting the conclusion that using abatement costs without adjusting for industrial composition yields a misleading picture of states' relative regulatory compliance costs.

A second fact that emerges from the industry-adjusted index is that while most state rankings are relatively stable, a few change significantly over time. Appendix tables 4A.1 and 4A.2 present the annual figures. From 1977 to 1991, Florida dropped from the 4th most costly state to the 25th most costly. By contrast, during the same time period Illinois rose from the 32nd most costly state to the 23rd.

Third, each of these statements should be tempered by the observation that there is considerable noise in the data, both in the adjusted and unadjusted indexes especially for the smaller states. For example, it is hard to imagine that Rhode Island leapt from the 42nd most costly state in 1986 to 4th in 1988. Most likely, some of the year-to-year variation in the indexes results from sampling error in the PACE survey or from the small size of some states.

Despite the noisiness of the data for small states, some consistent patterns emerge. To study trends in the data, I regressed S^* on year dummies and a time trend and plotted the residuals. As an example, figure 4.1 plots

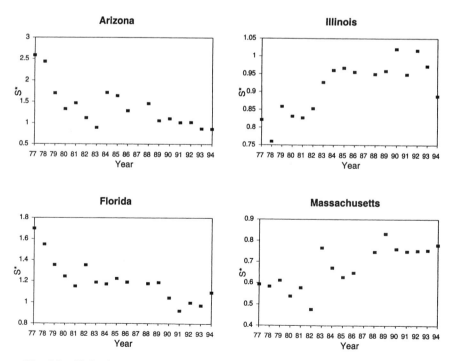

Fig. 4.1 S^* for four states

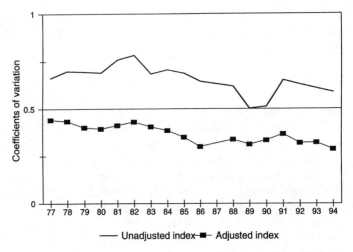

Fig. 4.2   Coefficients of variation: unadjusted and industry-adjusted indexes

S^* for four large states. Compliance costs in Arizona and Florida declined between 1977 and 1994, relative to the changing compliance costs in other states. Note that because S^* is already normalized on an annual basis, the downward trend in S^* does not mean that Arizona and Florida have become less expensive in absolute terms, only less expensive relative to other states. By contrast, over the same period relative compliance costs rose in Illinois and Massachusetts.

Another important fact discernible from this index is that the variation among states in their regulatory stringency is decreasing. It has often been speculated that pressure from federal regulators and national attention is forcing a convergence in state regulatory stringency. This index provides the first simple evidence of that convergence. Figure 4.2 depicts the coefficient of variation in both time series from 1977 to 1994. The coefficient of variation of the industry-adjusted index drops from 0.44 in 1977 to 0.29 in 1994. Meanwhile variation in the unadjusted index falls much less, from 0.66 to 0.59. Taken together, these two time series suggest that while states' industrial compositions have become more dissimilar over time, their regulatory stringency has become more similar.

Appendix table 4A.2 makes clear that despite the convergence of states' stringency, there remains substantial variation among states, even as late as 1994. Expenditures on pollution abatement in 1994 ranged from 0.5 percent of GSP in Nevada, to 6 percent in Louisiana. While much of this difference is accounted for by differences in the two states' industrial compositions, the industry-adjusted index for the most expensive state (Maine) remains 1.7 times the national average, and 4.13 times as large as for the least expensive state (Nevada).

Before comparing this index to others, it is important to note a few caveats. First, this index is not necessarily a measure of regulatory stringency alone. Other state characteristics may well drive up the cost of pollution abatement. For example, if the wages of environmental engineers vary state-to-state, they will affect the relative pollution abatement costs. Furthermore, this index is not intended to be a measure of environmental quality. Many of the nation's most polluted regions also have the strictest regulations.

Second, this industry-adjusted index makes no attempt to control for the relative age of different states' manufacturers. This is important because many state environmental standards are more strict for new sources of pollution than for existing sources. Consequently, states that have relatively new manufacturing bases also have relatively high compliance costs, even after controlling for their industrial compositions. Conversely, states that have relatively old manufacturers will experience lower compliance costs. There is, therefore, potentially a positive correlation between the amount of new investment and this industry-adjusted index of regulatory compliance costs. However, there is also reason to believe that this bias is small. In another paper (Levinson 1996), I regressed pollution abatement expenditures at the plant level on plant characteristics, including an indicator for plants built in the last 5 years. The new plant indicator, although positive, was small and statistically insignificant.

Third, this industry-adjusted index of environmental stringency, S^*, controls for states' industrial compositions at the level of two-digit SIC codes. While this surely accounts for a lot of the differences among states, there is equally certain to be heterogeneity among states within two-digit classifications. For example, SIC 26, pulp and paper, includes both pulp mills, which are among the most pollution-intensive manufacturers, and envelope assemblers, which emit very little pollution. To the extent that some states contain relatively more pulp mills and others merely assemble envelopes, if the former experience high abatement costs, that will probably be due to differing industrial compositions rather than more stringent environmental regulations. In other words, the index S^* retains some of the bias due to industrial compositions—in particular, heterogeneity of industrial compositions within any given two-digit SIC code.

4.3 Existing Indexes of State Environmental Regulatory Effort

Attempts to quantify state environmental regulations have taken numerous forms over the years. Many environmental organizations have compiled indexes for this purpose, and these indexes form a standard against which the industry-adjusted index can be compared.

Conservation Foundation Index. In 1983 the Conservation Foundation attempted to measure each state's "effort to provide a quality environment for its citizens" (Duerkson 1983, 218). They compiled an index from 23 components, including environmental and land-use characteristics such as the League of Conservation Voters' assessment of each state's congressional delegation's voting record, the existence of state environmental-impact-statement processes, and the existence of language specifically protecting the environment in state land-use statutes. Conservation Foundation staff assigned weights to each component based on subjective assessments of their importance, and the weighted sum is an index ranging from 0 to 63. Minnesota and California received the best scores, while Missouri and Alabama received the worst.

FREE Index. The Fund for Renewable Energy and the Environment (FREE 1987) published an index of the strength of state environmental programs. The components of the index include state laws regarding air quality, hazardous waste, and groundwater pollution. Wisconsin and California scored the highest, while West Virginia and Mississippi received the lowest marks.

Green Index. Hall and Kerr (1991) compiled the widely cited Green Index of state and environmental health from 256 measures of public policy and environmental quality. Oregon and Maine lie at the top of the ranking, while Louisiana and Alabama are last.

Southern Studies Index. The Institute for Southern Studies (1994) ranked the states based on 20 environmental measures such as air quality, state spending on the environment, pollution and waste generation, and energy efficiency, and then added up the 20 rankings of each state to get a composite index. Vermont and New Hampshire had the best scores, while Texas and Louisiana had the worst.

League of Conservation Voters (LCV). Each year, the LCV assigns each U.S. senator and representative a score from 1 to 100, based on his or her voting record on environmental bills chosen by the LCV. Some researchers have used these scores as a measure of the environmental sentiment in each state (Gray 1997). To compare these scores to the compliance cost index, I averaged each state's House and Senate delegation's environmental voting records. Each record is the average voting record for each member of the state's delegation. Thus, for states with more House members than Senate members, the Senate votes are weighted more heavily (and vice versa).[12]

Table 4.1 reports the values of each environmental index for each state. Table 4.2 presents each index's ranking of states. The rankings of the sub-

---

12. For further details, see the appendix.

jective indexes conform loosely to anecdotal evidence and to reports in the popular press. Alabama, Mississippi, and Louisiana consistently receive the lowest grades from environmental organizations, while Massachusetts, Wisconsin, Minnesota, and California receive the highest grades. I suspect that few policy analysts, environmental regulators, or industry representatives would be surprised by these rankings, and I therefore refer to these indexes as the "conventional wisdom" regarding states' relative environmental efforts.

In column (8) of table 4.1, I present an index calculated from the confidential plant-level Census of Manufactures, as described in Levinson (1996). Using the raw, establishment-level 1988 PACE data, I regressed the log of gross pollution abatement operating costs on the log of the book value of capital, the log of the number of production workers, the log of value-added, a dummy for new plants, dummies for four-digit SIC codes, and individual state dummies, all from the 1987 Census of Manufactures.[13] The state dummy coefficients are reported in column (8) of table 4.1. A high-point estimate for a state dummy coefficient indicates that, all else equal, plants in that state spend more on pollution abatement operating costs than do otherwise similar plants in the omitted state, New York.

Oddly, this plant-level index is not highly correlated with the more aggregate industry-adjusted index. The correlation between the plant-level index and the industry-adjusted index in 1988 is only 0.19. There are several possible explanations. The plant-level index is from a regression of 1988 PACE data on 1987 Census of Manufactures data for the same firms. (The PACE data were not collected in 1987). This mismatch may account for some of the discrepancy. Also, the plant-level index controls for plant vintage with a new-plant indicator, to account for the age bias already discussed, although its coefficient is small and statistically insignificant.

Table 4.3 presents correlations among the two cost indexes and the five conventional indexes. Although they were compiled at different times with widely different sets of components, the five conventional indexes are highly positively correlated. Except for the LCV index, the conventional indexes are all fairly ad hoc. Each is based on a list of component measures, with no objective guide as to what criteria are included or excluded from the index. Furthermore, each index either adds up the unweighted ranks of the separate components or weights the separate scores according to the subjective judgment of the index's authors. Nevertheless, there is remarkable consistency across the indexes.

On the other hand, the two cost-based indexes are negatively correlated

---

13. Implicit in this specification is a Cobb-Douglas production function in which output (Y) is estimated as a function of capital (K), labor (L), and pollution (P), with dummy variables for new plants, industries, and states: $Y = A \cdot K^{\beta_1} \cdot L^{\beta_2} \cdot P^{\beta_3}$. This estimation substitutes pollution abatement, which is observable, for pollution, takes the logarithm of both sides, and inverts the function to estimate abatement as a function of the other variables.

Table 4.3 Correlations among Indexes of State Environmental Effort

	Unadjusted Average Cost (S_s) (1)	Adjusted Average Cost (S_s^*) (2)	Conservation Foundation (3)	FREE Index (4)	Green Index (5)	Southern Studies (6)	LCV Index (7)
Unadjusted cost	1.00	0.75	−0.20	−0.45	−0.43	−0.53	−0.29
Adjusted cost		1.00	−0.19	−0.37	−0.27	−0.31	−0.29
Conservation Foundation			1.00	0.65	0.67	0.46	0.67
FREE index				1.00	0.58	0.35	0.63
Green index					1.00	0.89	0.72
Southern Studies						1.00	0.62
LCV index							1.00

with the conventional indexes. While the adjusted and unadjusted indexes are correlated with each other, they are both negatively correlated with each of the conventional indexes. While the conventional indexes may measure something systematic about states, it is not correlated with industrial pollution abatement expenditures.

There are several reasons for the negative correlation between the compliance-cost-based measures and the environmental organizations' indexes. The environmental organizations' indexes often include the quality of the environment in each state as part of their measure. The Green and Southern Studies indexes include measures of ambient air and water quality, and of pollution emitted. In many cases, environmental quality is inversely associated with compliance costs because plants in the dirtiest states are required to spend more effort cleaning up. Los Angeles, for example, has both the most polluted air and the toughest emissions standards in the United States (Berman and Bui 1997).

Another explanation has to do with the fact that the LCV index, which is itself included in the Conservation Foundation index, ranks states according to their congressional delegations' voting records on national legislation, rather than state legislation. Furthermore, U.S. senators and representatives appear to vote for stricter regulations when they are imposed on other states (Pashigian 1985). Finally, many of the indexes contain elements unrelated to manufacturers' pollution abatement costs, such as curbside recycling programs, spending on public parks, and automobile inspection programs. While these state characteristics may indicate something about the overall environmental sentiment in a given state, they are not necessarily related to the compliance costs faced by manufacturers.

In general, the two groups of indexes measure different concepts. The compliance cost index measures how much it costs to locate a manufacturing facility in any one state, relative to others, in terms of pollution abatement costs. The subjective indexes combine many different measures, including the quality of the environment, national delegations' voting records, and environmental effort unrelated to the manufacturing industry.

4.4   An Application: The Effect of Regulations on Foreign Direct Investment

As an example of the type of work that this index facilitates, in table 4.4 I present regressions of foreign direct investment (FDI) on characteristics of U.S. states, including their industry-adjusted indexes of environmental regulatory stringency, S^*.[14] Several studies have examined the effects of environmental regulations on FDI. However, all of the existing studies have either used a cross section of data, some unadjusted measure of regu-

---

14. This work is taken from Keller and Levinson (1999), where more details are provided.

Table 4.4   An Application: Foreign Direct Investment to U.S. States as a Function of Abatement Costs, 1977–94

	Mean (SD) (1)	Pooled OLS		Dynamic Panel Data Model: GMM, First Differences	
		Manufacturing (2)	Chemicals[a] (3)	Manufacturing (4)	Chemicals[a] (5)
Industry-adjusted index of abatement costs, S^*		500* (237)	267 (186)	2.4 (92.6)	−338* (100)
Lagged FDI				0.90* (0.02)	0.89* (0.03)
Market proximity	6,631 (8,220)	0.207* (0.019)	0.098* (0.015)	0.104* (0.018)	0.041* (0.014)
Population (thousands)	4,940 (5,134)	0.175* (0.033)	−0.016 (0.023)	−0.043 (0.051)	−0.003 (0.054)
Unemployment rate	6.61 (2.09)	122* (43)	86.0* (29.1)	−67.5* (15.7)	−56.6* (14.0)
Unionization rate	16.6 (6.7)	−108* (20)	−84.6* (13.9)	32.7 (21.4)	59.8* (20.0)
Wages	9.10 (2.24)	179* (87)	32.9 (66.7)	−135.7 (76.7)	5.8 (60.1)
Road mileage (thousands)	80.5 (48.4)	12.3* (2.6)	10.8* (1.8)	−0.37 (6.25)	−4.20 (5.48)
Land prices (per acre)	887 (775)	0.52* (0.12)	0.62* (0.10)	0.21* (0.10)	0.26* (0.08)
Energy prices	5.51 (1.70)	−288* (56)	−144* (41)	54.6* (27.7)	58.1* (24.4)
Tax effort	96.1* (16.1)	−31.0* (5.9)	−11.4* (4.1)	18.4* (4.9)	16.6* (4.6)
Year		166* (41)	32.4 (33.4)		
Constant		−11,602* (3,072)	−1,525 (2,516)	60.4 (25.9)	12.2 (21.5)
Number of observations	816	811	563	761	496
Number censored		5	109	7	272
R^2		0.70	0.47	0.10	0.15

Source: Keller and Levinson (1999). The dependent variable is property, plant, and equipment investment by foreign-owned manufacturers, from the BEA; see appendix.

Note: Standard errors are in parentheses. 1987 is dropped because no PACE data were collected that year.

[a] The chemical industry investment data are only available for 1977–91.

*Statistically significant at the 5 percent level.

latory stringency such as S in equation (1), or both (List and Co 2000; Friedman, Gerlowski, and Silberman 1992). Table 4.4 examines property, plant, and equipment investment by foreign-owned manufacturers, from the Bureau of Economic Analysis (BEA), as a measure of FDI. It presents regressions of FDI on state characteristics using a time series of data and the industry-adjusted index of environmental regulatory stringency.

The first column of table 4.4 presents means and standard deviations of the regressors. As a benchmark to compare to the previous literature, columns (2) and (3) contain pooled, ordinary least squares (OLS) regressions of FDI in the manufacturing sector and the chemical industry, respectively, on the industry-adjusted index of environmental stringency and other covariates. Controlling for other state characteristics, FDI appears to be positively correlated with stringency. In column (3), I examine FDI for the chemical industry (SIC 28). This is one of the only relatively pollution-intensive, two-digit SIC codes for which this measure of FDI is reported consistently by the BEA. Here, the coefficient on S^* remains positive, although it is smaller and statistically insignificant. These results, however, are based largely on the cross-sectional comparison of states. (Most of the variation in S^* is across states rather than within states over time.) These cross-sectional results are probably biased if states have unobserved characteristics correlated with both FDI and regulatory stringency.

To control for those characteristics, and to exploit the panel of data, consider a dynamic model. Suppose that a reduced-form relationship for FDI can be characterized by the following equation.[15]

$$(4) \qquad FDI_{st} = \delta FDI_{s,t-1} + \gamma S^*_{st} + X'_{st}\beta + u_{st},$$

where X_{st} is a vector of characteristics of state s at time t, and u_{st} is an error term composed of two parts, $u_{st} = \mu_s + v_{st}$. Equation (4) states that FDI is a function of current state characteristics and lagged values of FDI. Both FDI_{st} and $FDI_{s,t-1}$ are functions of μ_s and therefore OLS estimates of (4) will be biased because $FDI_{s,t-1}$, a regressor, is correlated with the error term.

Arellano and Bond (1991) suggest a generalized method of moments (GMM) estimate of equation (4) that uses lagged values of $FDI_{s,t-1}$ as instruments and first differences to eliminate the fixed state effects μ_s. First, take first differences of (4):

$$(5) \qquad \Delta FDI_{st} = \delta \Delta FDI_{s,t-1} + \gamma \Delta S^*_{st} + \Delta X'_{st}\beta + \Delta v_{st},$$

where Δ symbolizes first differences. Since $FDI_{s,t-2}$ is correlated with $\Delta FDI_{s,t-1}$, but not correlated with ΔFDI_{st}, it is a valid instrument. In fact, all past values $FDI_{s,t-3}$, $FDI_{s,t-4}$, and so on are valid instruments for $\Delta FDI_{s,t-1}$.

Columns (4) and (5) of table 4.4 present GMM estimates of (5) using Doornik, Arellano, and Bond (1999).[16] When equation (5) is estimated

---

15. This discussion is based on Baltagi (1995) and Arellano and Bond (1991).

16. Doornik, Arellano, and Bond's GMM estimation is written for the computer package Ox, and may be downloaded from http://www.nuff.ox.ac.uk/Users/Doornik/. See Doornik (1998) and Doornik, Arellano, and Bond (1999).

using all manufacturing FDI, in column (4), the large spurious positive coefficient on S^* disappears. Instead, the coefficient (2.4) is tiny and statistically insignificant, although still positive. Turning to the chemical industry, in column (5), the coefficient (-338) is negative and statistically significant. This suggests that the positive coefficients found in the cross-sectional evidence in this study and others are spurious, and are based on unobserved characteristics correlated with both environmental regulations and economic activity.

To interpret the size of these coefficients, consider the following. The fixed-effects coefficient in column (5) suggests that a 1-unit increase in the stringency index is associated with a decline in chemical industry FDI of $338 million. The standard deviation of this index within states over time ranges from 0.04 for Wisconsin to 0.56 for Colorado, and averages 0.18. So the coefficient suggests that a 1 standard deviation increase in the index, for the average state, is associated with a decline of investment by foreign-owned chemical manufacturers of $61 million. This amounts to approximately 6 percent of the annual average chemical industry FDI investment of $1,017 million per state.

The industry-adjusted index plays two important roles in the regressions in table 4.4. First, because the index spans 18 years, it is possible to use changes in the variables from year to year to control for unobserved fixed state characteristics that may be correlated with both stringency and FDI. The stringency coefficients in these first-differenced specifications are negative, while those in the pooled specification are positive, suggesting that these unobserved state characteristics are extremely important. Second, by adjusting for industrial composition, the index avoids merely measuring concentrations of polluting industries and instead assesses average abatement costs, holding industrial composition constant.

While these results are meant only as an example, they do suggest that an index such as the one described has considerable advantages over empirical approaches taken thus far.

4.5   Conclusion

The research described here creates an industry-adjusted index of state environmental compliance costs from 1977 to 1994. The index supports several conclusions. First, industry composition plays an important role in determining spending on environmental compliance costs for different states. Rank orderings of states by pollution abatement spending look very different once their industrial compositions have been controlled for. Second, differences among states are exaggerated by differences in their industrial compositions. The coefficient of variation of the unadjusted index is 0.65, while the coefficient of variation of the industry-adjusted index is 0.37. Third, once industrial composition has been accounted for, states appear to be converging in their environmental standard stringency. Fourth, when compared to conventional indexes of state environmental regulations, these cost-based indexes have opposite implications for the relative stringency of states. The two types of indexes are negatively correlated across states. Finally, when used in an analysis of the effect of regulatory stringency on economic activity (FDI in this case), time-series analyses using the industry-adjusted index have more sensible results than cross-sectional analyses or analyses using the unadjusted index. Together, these results imply that using conventional indexes or unadjusted cost indexes to analyze state environmental policies can lead to misleading conclusions about the effects of those policies on economic activity.

Appendix

Pollution Abatement Costs and Expenditures (PACE) Data

PACE data come from U.S. Department of Commerce, Bureau of the Census. The data are published in Current Industrial Reports: Pollution Abatement Costs and Expenditures, MA-200, 1974–94. The variable of interest from this source was the pollution abatement gross annual cost (GAC) total across all media types. Starting in 1977, the Census Bureau collected data only for establishments with 20 or more employees. Although PACE data were collected from all establishments for the years 1973–79, in order to lessen the administrative burden on small businesses, they were dropped from the survey starting in 1980. The PACE Survey was not collected in 1987. There are some censored observations for the state totals, and in those cases values were interpolated.

Gross State Product Data

All GSP data were acquired via the Regional Economic Information System CD, 1969–94, published by the U.S. Department of Commerce, Bureau of Economic Analysis, Regional Economic Measurement Division.

League of Conservation Voters (LCV) Index

This index is the unweighted average of the House and Senate environmental voting records. Each record is the average voting record for each member of the state's delegation. Thus, for states with more House members than Senate members, the Senate votes are weighted more heavily (and vice versa). The bills that are used to construct the index have been chosen by the LCV. See http://www.lcv.org. For the Senate votes, the years 1977–78, 1979–80, 1983–84, 1985–86, and 1987–88 each had only one scorecard. Therefore, the voting records for these years were entered separately for each year. For the House votes, the years 1987–88 had only one scorecard. Here also the same value was entered for both years. Also, the House had a scorecard for 1985 and for 1985–86. The information from the 1985 scorecard was used to disaggregate the 1985–86 scorecard by a weighted average.

Property, Plant, and Equipment Investment by Foreign-Owned Manufacturers

Bureau of Economic Analysis (BEA), Department of Commerce, *Foreign Direct Investment in the U.S.*, 1980, 1987, 1992, 1999.

Table 4A.1   Industry-Adjusted Index of State Environmental Compliance Costs

State	1977	1978	1979	1980	1981	1982	1983	1984	1985	1986	1988	1989	1990	1991	1992	1993	1994
AL	1.36	1.26	1.28	1.19	1.26	1.30	1.94	1.12	1.08	1.01	0.90	1.01	1.04	1.18	1.07	1.13	1.12
AR	2.59	2.43	1.70	1.33	1.47	1.12	0.89	1.72	1.64	1.29	1.46	1.06	1.11	1.02	1.02	0.87	0.86
AR	1.14	1.17	1.19	1.24	1.13	1.16	1.08	1.21	1.30	1.21	1.13	1.26	1.05	1.07	1.22	1.21	1.20
CA	0.93	0.96	0.83	0.83	0.75	0.67	0.79	0.82	0.90	1.03	1.00	1.01	1.11	1.00	0.90	0.86	0.96
CA	1.06	0.96	0.77	0.83	0.88	0.61	0.64	0.58	0.72	0.91	0.72	1.87	2.42	2.07	0.64	0.60	0.82
CT	0.49	0.55	0.60	0.60	0.59	0.51	0.58	0.63	0.63	0.62	0.76	0.92	0.92	0.80	0.67	0.68	0.80
DE	1.16	1.47	1.56	1.53	1.61	1.35	1.20	1.17	1.33	1.35	1.58	1.44	1.00	1.10	1.05	1.08	1.12
FL	1.70	1.55	1.35	1.24	1.15	1.35	1.19	1.17	1.22	1.19	1.18	1.19	1.04	0.92	1.00	0.97	1.10
GA	0.92	0.83	0.77	0.80	0.81	0.87	0.89	0.90	0.86	0.84	0.88	1.18	1.05	1.05	0.85	0.93	0.96
ID	1.59	1.64	1.92	1.93	1.98	2.29	2.11	2.02	1.63	1.92	1.43	1.16	1.25	0.94	1.44	1.41	1.54
IL	0.82	0.76	0.86	0.83	0.83	0.85	0.93	0.96	0.97	0.96	0.95	0.96	1.02	0.95	1.02	0.97	0.89
IN	1.15	1.14	1.15	1.28	1.30	1.44	1.28	1.18	1.16	1.08	1.13	1.04	0.97	1.02	1.01	1.02	0.99
IA	1.00	1.02	1.00	1.01	1.05	0.94	0.93	1.18	1.04	0.91	0.88	0.91	0.77	0.78	0.91	0.89	1.05
KS	0.68	0.64	1.30	0.72	0.68	0.69	0.62	0.84	0.65	0.75	0.62	0.85	0.90	0.79	0.67	0.81	0.77
KY	0.88	1.05	0.98	0.98	0.99	0.87	1.06	1.10	0.95	0.92	0.93	1.05	1.02	1.07	0.97	0.99	1.01
LA	1.26	1.27	1.52	1.34	1.62	1.59	1.79	1.92	1.97	1.75	1.32	1.21	1.17	1.18	1.56	1.56	1.57
ME	1.28	1.43	1.39	1.44	1.54	1.68	1.68	1.66	1.46	1.48	1.34	1.61	1.45	1.79	1.71	1.64	1.69
MD	1.08	1.23	1.25	1.24	1.10	1.06	1.11	1.17	1.25	1.25	1.32	1.10	1.04	1.19	1.17	1.25	1.03
MS	0.59	0.58	0.61	0.54	0.58	0.48	0.77	0.67	0.63	0.65	0.75	0.83	0.76	0.75	0.75	0.75	0.78
MI	0.98	0.96	1.02	1.11	1.12	1.05	1.00	1.11	1.04	0.94	1.07	0.99	0.99	0.96	0.97	0.95	0.88
MN	0.58	0.61	0.66	0.69	0.61	0.65	0.70	0.64	0.59	0.58	0.60	0.67	0.63	0.71	0.68	0.84	0.84
MS	1.72	1.85	1.54	1.56	1.18	1.35	1.52	1.64	1.52	1.42	1.29	1.16	1.32	1.37	1.62	1.60	1.39
MO	0.73	0.82	0.71	0.78	0.77	0.71	0.79	0.74	0.77	0.87	0.80	0.94	0.95	0.71	0.73	0.84	0.71
MT	1.19	1.23	1.36	1.35	1.53	1.85	1.69	1.69	1.85	1.58	1.93	1.88	1.67	1.56	0.76	0.63	1.65
NE	0.76	0.80	0.75	0.77	0.82	0.65	0.81	1.16	1.13	0.94	0.97	0.63	0.64	0.65	0.95	0.84	0.85
NV	0.42	0.59	0.58	0.61	0.57	0.62	0.61	0.86	0.69	0.68	0.93	0.63	0.48	0.60	0.72	0.60	0.45
NH	1.05	0.90	0.73	0.64	0.60	0.64	0.61	0.55	0.60	0.73	0.70	0.90	0.87	0.81	0.69	0.78	0.97

(continued)

Table 4A.1   (continued)

State	1977	1978	1979	1980	1981	1982	1983	1984	1985	1986	1988	1989	1990	1991	1992	1993	1994
NJ	0.95	0.80	0.93	0.99	0.79	0.81	0.84	0.77	0.72	0.73	0.82	0.86	0.86	0.87	0.85	0.69	0.61
NM	2.46	2.49	2.12	2.11	1.70	1.15	1.27	1.15	1.32	0.97	2.49	1.67	1.77	1.09	1.50	1.64	1.05
NY	0.71	0.71	0.79	0.78	0.80	0.83	0.73	0.65	0.66	0.80	0.84	0.84	0.80	0.85	0.82	0.84	0.71
NC	0.95	0.80	0.81	0.81	0.78	0.93	0.84	0.86	0.81	0.77	0.76	0.78	0.77	0.73	0.88	0.81	0.81
ND	0.77	0.69	0.53	0.47	0.40	0.36	0.48	0.79	0.84	0.88	1.09	0.51	0.68	0.74	1.74	1.55	0.64
OH	0.79	0.82	0.82	0.78	0.80	0.79	0.71	0.72	0.74	0.81	0.94	0.91	0.89	0.86	0.84	0.82	0.95
OK	0.77	0.68	0.63	0.69	0.48	0.65	0.78	0.51	0.43	0.54	0.59	0.49	0.49	0.44	0.55	0.57	0.60
OR	1.33	1.33	1.32	1.16	1.15	1.39	1.45	1.19	1.09	1.12	1.38	0.96	1.07	1.25	1.14	1.14	1.21
PA	0.93	0.95	0.87	0.92	0.98	1.02	0.93	0.75	0.97	0.96	0.96	0.98	0.97	0.80	0.81	0.78	0.86
RI	0.47	0.48	0.37	0.35	0.39	0.36	0.43	0.52	0.66	0.69	1.59	1.40	1.20	0.73	0.68	0.62	1.24
SC	1.12	1.05	1.07	0.78	0.83	0.91	0.83	0.92	0.97	0.97	1.02	1.15	1.16	1.08	0.99	1.08	0.98
SD	0.39	0.60	0.58	0.65	0.64	0.79	0.56	0.68	0.64	0.60	0.87	0.45	0.51	0.66	1.18	0.84	0.94
TN	1.14	1.03	1.05	1.02	1.00	0.97	1.05	1.30	1.31	1.28	1.19	1.01	1.10	1.05	1.06	1.08	1.11
TX	1.29	1.34	1.29	1.42	1.40	1.51	1.37	1.59	1.53	1.44	1.17	1.11	1.12	1.38	1.53	1.69	1.52
UT	0.80	0.98	0.93	1.04	0.96	0.89	0.82	0.77	0.75	0.96	0.76	1.08	1.05	0.84	1.18	1.16	0.88
VT	0.65	0.61	0.57	0.54	0.57	0.66	0.67	0.66	1.11	0.98	0.81	0.50	0.60	0.51	0.63	0.56	0.68
VA	0.99	0.99	0.93	0.90	0.93	0.95	1.05	1.08	1.04	0.97	1.03	0.88	0.82	0.78	0.98	1.01	0.96
WA	1.54	1.41	1.37	1.28	1.47	1.42	1.71	1.54	1.65	1.18	1.21	1.10	1.08	1.41	1.32	1.31	1.34
WV	1.65	1.87	1.94	1.83	2.04	1.82	1.25	1.09	1.02	1.20	1.03	1.31	1.40	2.32	1.89	1.78	1.38
WI	0.88	0.86	0.81	0.84	0.83	0.88	0.93	0.92	0.92	0.91	0.90	0.93	0.88	0.85	0.89	0.93	0.94
WY	0.38	0.38	0.37	0.36	0.35	0.23	0.34	0.29	1.10	1.25	1.72	0.75	0.92	0.77	1.37	1.00	0.69

Table 4A.2   Unadjusted Index of State Environmental Compliance Costs

State	1977	1978	1979	1980	1981	1982	1983	1984	1985	1986	1988	1989	1990	1991	1992	1993	1994
AL	0.0211	0.0205	0.0219	0.0215	0.0225	0.0208	0.0333	0.0193	0.0208	0.0192	0.0161	0.0191	0.0217	0.0250	0.0228	0.0232	0.0239
AZ	0.0237	0.0222	0.0177	0.0145	0.0166	0.0113	0.0101	0.0160	0.0177	0.0143	0.0156	0.0122	0.0134	0.0121	0.0126	0.0105	0.0105
AR	0.0125	0.0135	0.0150	0.0168	0.0153	0.0145	0.0148	0.0163	0.0177	0.0161	0.0169	0.0196	0.0172	0.0179	0.0205	0.0203	0.0201
CA	0.0104	0.0106	0.0098	0.0107	0.0101	0.0090	0.0106	0.0112	0.0123	0.0128	0.0131	0.0139	0.0160	0.0151	0.0134	0.0125	0.0139
CO	0.0100	0.0092	0.0079	0.0088	0.0093	0.0056	0.0067	0.0063	0.0081	0.0102	0.0077	0.0218	0.0294	0.0262	0.0077	0.0073	0.0091
CT	0.0041	0.0050	0.0060	0.0064	0.0063	0.0051	0.0069	0.0065	0.0069	0.0072	0.0090	0.0121	0.0128	0.0111	0.0090	0.0090	0.0102
DE	0.0252	0.0344	0.0386	0.0418	0.0420	0.0362	0.0319	0.0316	0.0377	0.0391	0.0392	0.0375	0.0292	0.0322	0.0304	0.0290	0.0294
FL	0.0164	0.0157	0.0145	0.0137	0.0126	0.0131	0.0132	0.0129	0.0118	0.0143	0.0145	0.0151	0.0138	0.0124	0.0128	0.0120	0.0135
GA	0.0099	0.0101	0.0099	0.0110	0.0109	0.0104	0.0113	0.0120	0.0121	0.0119	0.0128	0.0180	0.0167	0.0167	0.0135	0.0142	0.0145
ID	0.0152	0.0153	0.0196	0.0222	0.0232	0.0218	0.0201	0.0209	0.0180	0.0217	0.0161	0.0131	0.0152	0.0128	0.0185	0.0161	0.0174
IL	0.0099	0.0095	0.0115	0.0116	0.0109	0.0109	0.0132	0.0129	0.0136	0.0144	0.0140	0.0147	0.0162	0.0153	0.0168	0.0154	0.0140
IN	0.0144	0.0155	0.0173	0.0211	0.0216	0.0225	0.0225	0.0200	0.0212	0.0208	0.0191	0.0204	0.0195	0.0208	0.0202	0.0186	0.0176
IA	0.0083	0.0091	0.0095	0.0098	0.0100	0.0087	0.0102	0.0123	0.0121	0.0115	0.0103	0.0114	0.0100	0.0109	0.0120	0.0114	0.0134
KS	0.0086	0.0088	0.0179	0.0110	0.0102	0.0089	0.0105	0.0107	0.0102	0.0108	0.0101	0.0136	0.0153	0.0143	0.0109	0.0122	0.0119
KY	0.0092	0.0109	0.0114	0.0125	0.0127	0.0110	0.0152	0.0164	0.0142	0.0144	0.0146	0.0170	0.0184	0.0191	0.0177	0.0162	0.0169
LA	0.0408	0.0424	0.0465	0.0471	0.0576	0.0568	0.0616	0.0677	0.0682	0.0649	0.0430	0.0423	0.0461	0.0518	0.0606	0.0589	0.0580
ME	0.0163	0.0199	0.0197	0.0207	0.0217	0.0204	0.0227	0.0228	0.0220	0.0230	0.0206	0.0262	0.0263	0.0316	0.0309	0.0293	0.0288
MD	0.0141	0.0173	0.0193	0.0204	0.0186	0.0166	0.0183	0.0184	0.0190	0.0192	0.0205	0.0178	0.0174	0.0196	0.0198	0.0209	0.0168
MA	0.0044	0.0045	0.0051	0.0046	0.0050	0.0038	0.0072	0.0064	0.0061	0.0069	0.0073	0.0090	0.0087	0.0088	0.0089	0.0087	0.0091
MI	0.0086	0.0090	0.0107	0.0133	0.0127	0.0106	0.0111	0.0121	0.0120	0.0113	0.0134	0.0133	0.0145	0.0145	0.0144	0.0132	0.0110
MN	0.0065	0.0069	0.0078	0.0084	0.0076	0.0077	0.0094	0.0091	0.0089	0.0091	0.0080	0.0095	0.0098	0.0119	0.0114	0.0122	0.0119
MS	0.0179	0.0202	0.0180	0.0202	0.0157	0.0171	0.0207	0.0230	0.0224	0.0216	0.0198	0.0194	0.0241	0.0253	0.0285	0.0260	0.0226
MO	0.0070	0.0083	0.0081	0.0099	0.0097	0.0081	0.0101	0.0085	0.0097	0.0114	0.0110	0.0138	0.0155	0.0114	0.0118	0.0125	0.0104
MT	0.0221	0.0216	0.0247	0.0297	0.0375	0.0476	0.0396	0.0425	0.0454	0.0339	0.0334	0.0358	0.0373	0.0451	0.0217	0.0168	0.0448
NE	0.0058	0.0067	0.0069	0.0075	0.0080	0.0057	0.0081	0.0114	0.0122	0.0109	0.0110	0.0075	0.0078	0.0083	0.0116	0.0100	0.0101
NV	0.0047	0.0057	0.0061	0.0069	0.0070	0.0071	0.0069	0.0094	0.0083	0.0081	0.0109	0.0084	0.0062	0.0073	0.0083	0.0064	0.0049

(continued)

Table 4A.2   (continued)

State	1977	1978	1979	1980	1981	1982	1983	1984	1985	1986	1988	1989	1990	1991	1992	1993	1994
NH	0.0077	0.0063	0.0057	0.0053	0.0050	0.0050	0.0059	0.0049	0.0058	0.0073	0.0070	0.0096	0.0100	0.0094	0.0080	0.0087	0.0105
NJ	0.0141	0.0132	0.0155	0.0174	0.0138	0.0142	0.0154	0.0133	0.0136	0.0150	0.0153	0.0175	0.0192	0.0193	0.0195	0.0174	0.0155
NM	0.0413	0.0433	0.0392	0.0418	0.0380	0.0264	0.0284	0.0317	0.0313	0.0273	0.0424	0.0300	0.0337	0.0155	0.0193	0.0188	0.0112
NY	0.0067	0.0070	0.0081	0.0083	0.0085	0.0082	0.0083	0.0072	0.0089	0.0096	0.0089	0.0100	0.0097	0.0105	0.0103	0.0102	0.0085
NC	0.0074	0.0065	0.0072	0.0078	0.0075	0.0083	0.0084	0.0087	0.0089	0.0085	0.0087	0.0095	0.0099	0.0096	0.0115	0.0105	0.0112
ND	0.0083	0.0076	0.0061	0.0060	0.0052	0.0046	0.0066	0.0117	0.0122	0.0131	0.0154	0.0074	0.0099	0.0108	0.0243	0.0207	0.0087
OH	0.0101	0.0108	0.0119	0.0127	0.0136	0.0129	0.0132	0.0119	0.0132	0.0136	0.0155	0.0157	0.0173	0.0166	0.0158	0.0146	0.0165
OK	0.0115	0.0088	0.0088	0.0105	0.0078	0.0115	0.0140	0.0094	0.0095	0.0114	0.0100	0.0090	0.0087	0.0108	0.0112	0.0108	0.0122
OR	0.0117	0.0115	0.0128	0.0128	0.0134	0.0141	0.0157	0.0130	0.0129	0.0136	0.0170	0.0124	0.0146	0.0183	0.0148	0.0131	0.0142
PA	0.0150	0.0155	0.0150	0.0167	0.0177	0.0169	0.0181	0.0151	0.0179	0.0180	0.0168	0.0177	0.0185	0.0174	0.0175	0.0159	0.0178
RI	0.0037	0.0039	0.0035	0.0035	0.0038	0.0032	0.0044	0.0051	0.0069	0.0076	0.0152	0.0149	0.0147	0.0081	0.0074	0.0068	0.0146
SC	0.0128	0.0128	0.0143	0.0114	0.0120	0.0129	0.0125	0.0142	0.0160	0.0164	0.0161	0.0206	0.0218	0.0205	0.0198	0.0202	0.0181
SD	0.0030	0.0037	0.0040	0.0046	0.0048	0.0053	0.0047	0.0053	0.0058	0.0056	0.0080	0.0043	0.0049	0.0066	0.0105	0.0063	0.0079
TN	0.0148	0.0141	0.0152	0.0156	0.0147	0.0135	0.0155	0.0171	0.0188	0.0184	0.0174	0.0162	0.0186	0.0184	0.0175	0.0172	0.0173
TX	0.0282	0.0287	0.0265	0.0306	0.0294	0.0312	0.0337	0.0355	0.0341	0.0346	0.0260	0.0264	0.0296	0.0346	0.0343	0.0331	0.0317
UT	0.0117	0.0149	0.0145	0.0168	0.0161	0.0135	0.0131	0.0120	0.0124	0.0133	0.0126	0.0199	0.0207	0.0182	0.0246	0.0251	0.0191
VT	0.0047	0.0044	0.0045	0.0044	0.0046	0.0051	0.0063	0.0060	0.0109	0.0099	0.0084	0.0060	0.0070	0.0063	0.0077	0.0064	0.0079
VA	0.0097	0.0106	0.0107	0.0109	0.0111	0.0105	0.0122	0.0123	0.0125	0.0119	0.0128	0.0117	0.0116	0.0113	0.0136	0.0136	0.0130
WA	0.0177	0.0166	0.0178	0.0190	0.0221	0.0188	0.0230	0.0224	0.0229	0.0188	0.0183	0.0175	0.0192	0.0210	0.0194	0.0186	0.0203
WV	0.0370	0.0446	0.0487	0.0478	0.0536	0.0473	0.0347	0.0310	0.0325	0.0348	0.0272	0.0363	0.0416	0.0697	0.0577	0.0517	0.0401
WI	0.0080	0.0084	0.0086	0.0094	0.0093	0.0094	0.0113	0.0107	0.0117	0.0120	0.0114	0.0125	0.0124	0.0126	0.0130	0.0133	0.0131
WY	0.0174	0.0161	0.0124	0.0144	0.0141	0.0100	0.0137	0.0120	0.0375	0.0396	0.0595	0.0265	0.0310	0.0310	0.0489	0.0331	0.0235

References

Alexander Grant & Co. 1985. *General manufacturing climates of the forty-eight contiguous states.* Chicago: Alexander Grant & Co.

Arellano, M., and S. R. Bond. 1991. Some tests of specification for panel data: Monte Carlo evidence and an application to employment equations. *Review of Economic Studies* 58:277–97.

Baltagi, B. H. 1995. *Econometric analysis of panel data.* New York: Wiley.

Becker, R., and V. Henderson. 1997. Effects of air quality regulation on decisions of firms in polluting industries. NBER Working Paper no. 6160. Cambridge, Mass.: National Bureau of Economic Research.

Berman, E., and L. Bui. 1997. Environmental regulation and labor demand: Evidence from the south coast air basin. NBER Working Paper no. 6299. Cambridge, Mass.: National Bureau of Economic Research.

Crandall, R. W. 1993. *Manufacturing on the move.* Washington, D.C.: Brookings Institution.

Doornik, J. A. 1998. *Object-oriented matrix programming using Ox 2.0.* London: Timberlake Consultants Ltd.

Doornik, J. A., M. Arellano, and S. Bond. 1999. Panel data estimation using DPD for Ox. Nuffield College, Oxford University. Mimeo.

Duerkson, C. J. 1983. *Environmental regulation of industrial plant siting.* Washington, D.C. Conservation Foundation.

FREE. 1987. *The state of the states 1987.* Washington, D.C.: Fund for Renewable Energy and the Environment.

Friedman, J., D. A. Gerlowski, and J. Silberman. 1992. What attracts foreign multinational corporations? Evidence from branch plant location in the United States. *Journal of Regional Science* 32 (4): 403–418.

Gray, W. B. 1997. Manufacturing plant location: Does state pollution regulation matter? NBER Working Paper no. 5880. Cambridge, Mass.: National Bureau of Economic Research.

Greenstone, M. 1998. The impacts of environmental regulations on industrial activity: Evidence from the 1970 and 1977 Clean Air Act Amendments and the Census of Manufactures. University of California at Berkeley. Mimeo.

Hall, B., and M. L. Kerr. 1991. *Green index: A state-by-state guide to the nation's environmental health.* Washington, D.C.: Island Press.

Institute for Southern Studies. 1994. *Green and gold.* Durham, N.C.: Institute for Southern Studies.

Jaffe, A. B., S. R. Peterson, P. R. Portney, and R. N. Stavins. 1995. Environmental regulations and the competitiveness of U.S. manufacturing: What does the evidence tell us? *Journal of Economic Literature* 33 (1): 132–63.

Keller, W., and A. Levinson. 1999. Environmental compliance costs and foreign direct investment inflows to U.S. states. NBER Working Paper no. 7369. Cambridge, Mass.: National Bureau of Economic Research.

League of Conservation Voters (LCV). 1977–94. *National environmental scorecard.* Washington, D.C.: League of Conservation Voters.

Levinson, A. 1996. Environmental regulations and manufacturers' location choices: Evidence from the Census of Manufactures. *Journal of Public Economics* 61 (1): 5–29.

List, J., and C. Co. 2000. The effects of environmental regulations on foreign direct investment. *Journal of Environmental Economics and Management* 40 (1): 1–20.

Pashigian, P. 1985. Environmental regulation: Whose self interests are being protected? *Economic Inquiry* 23 (4): 551–84.

Comment   Domenico Siniscalco

The new analysis carried out in this paper on the effects of environmental compliance costs on economic activity stems from the unsatisfactory performance of the existing indexes of environmental effort in U.S. states. The author calculates a new industry-adjusted index of U.S. states' environmental regulatory efforts, taking into account environmental compliance costs and the differences in industrial composition across states and within states over time. This approach differs from subjective indexes, which are based on perceptions of the states' efforts at one time, and conventional cost-based measures, which do not control for industrial composition.

There are three key motivations for creating this index. First, there is poor enforcement of environmental federal standards in many U.S. states. An industry-adjusted index is thus needed to analyze the variation of environmental compliance costs across states. Second, the decentralization of the responsibility for monitoring and enforcing environmental regulation from the federal government to local regulators may cause the states' stringency to converge over time. Thus, an index is required to obtain data on the level of such stringency. Third, it is difficult to assess the effects of environmental regulation on economic activity because existing indexes do not take into account industrial composition and this reduces the link between compliance costs and environmental regulation.

Levinson's index provides new data on historical trends in U.S. states' regulatory differences. It differs from conventional wisdom regarding states' relative environmental efforts and provides a useful tool for researchers and policymakers exploring and testing the effects of compliance costs on economic activity.

The most striking finding is that when states' regulatory stringency is ranked according to the industry-adjusted index, results are quite different from the ranking according to cost-based unadjusted indexes. Moreover, the dynamic evolution of the index shows that only few states' rankings change over time. It is, however, worth noticing that some results, especially for small U.S. states, might be heavily affected by noisiness in the data. Nevertheless, there is an evident downward trend in the variability of regulatory stringency among states, although their industrial composition has become much more dissimilar.

This paper provides a new and important tool for both regulators and policymakers because it allows us to conduct meaningful analyses of states' environmental efforts, taking into account variations among states' industrial composition. When policymakers set compliance standards they need a complete picture of environmental regulation stringency and en-

Domenico Siniscalco is professor of economics at the University of Torino and director of the Fondazione Eni Enrico Mattei, Milan.

forcement over time, compared with other U.S. states, to guarantee high environmental standards and to avoid restriction of local firms' competitiveness. An effective comparison of the environmental effort across states implies the definition of adjusted indexes that discriminate according to the characteristics of each state area. An index that is not industry adjusted can lead to misleading conclusions about the actual effects of environmental regulation on economic activities; that is, highly polluting industries have higher compliance costs regardless of regulation strictness and enforcement. Even if Levinson's index is still not a measure of regulatory stringency alone, when it is used together with other state-specific information, such as age of manufacturers, average dimension of firms, and average wages, it allows us to answer several often-raised questions about environmental law and enforcement.

The information provided by the index may also help firms in deciding the location of new sites and the extent of investment in existing sites.

Considering the whole set of findings and the interesting implication for regulators and firms' decision making, this paper certainly indicates a promising research area.

Costs of Air Quality Regulation

Randy A. Becker and J. Vernon Henderson

5.1 Introduction

An ongoing debate in the United States concerns the costs environmental regulations impose on industry. In this paper, we explore some of the costs associated with air quality regulation. In particular, we focus on regulation pertaining to ground-level ozone (O_3) and its effects on two industries sensitive to such regulation—industrial organic chemicals (Standard Industrial Classification [SIC] 2865–9) and miscellaneous plastic products (SIC 308). Both of these industries are major emitters of volatile organic compounds (VOCs) and nitrogen oxides (NO_x), the chemical precursors to ozone. Using plant-level data from the U.S. Census Bureau's Longitudinal Research Database (LRD), we examine the effects this type of regulation has had on the timing and magnitudes of investments by firms in these industries and on the impact it has had on their operating costs. As an alternative way to assess costs, we also employ plant-level data from the U.S. Census Bureau's Pollution Abatement Costs and Expenditures (PACE) survey.

Our prior work has found a variety of effects on industry behavior at-

Randy A. Becker is an economist at the Center for Economic Studies of the U.S. Bureau of the Census. J. Vernon Henderson is professor of economics at Brown University and a research associate of the National Bureau of Economic Research.

The authors thank the National Science Foundation for support, as well as the National Bureau of Economic Research and the Fondazione Eni Enrico Mattei. This work has benefited from the comments of conference participants as well as those of seminar participants at the Center for Economic Studies (U.S. Bureau of the Census). The opinions and conclusions expressed herein are those of the authors and do not necessarily represent the views of the Bureau of the Census. All papers are screened to ensure that they do not disclose any confidential information.

tributable to environmental costs (Becker and Henderson 2000). Here we attempt to quantify some of these costs. To identify effects, we use spatial variation in regulatory stringency as well as temporal differences arising from the introduction of heightened regulation. Our previous research has shown that plant age and plant size are important determinants of who gets regulated when and how intensely, so we incorporate these elements into our analysis here as well. Our models also control for location-specific fixed effects, which is critical in this type of work. Here, we find that regulation indeed significantly increases production costs, especially for young plants, with estimates that (arguably) are higher than the expenditure data from the PACE survey suggest. Our results also show that regulation may lead plants to restrict their size, and the reasons why this might be the case are discussed. We also find that, in at least one of these two industries, investment profiles are significantly altered for plants subject to regulation, with relatively more up-front investment and less phasing-in.

In section 5.2, we offer a general overview of air quality regulation in the United States, introducing our key environmental variable and discussing some of the difficulties involved in identifying a control and a treatment group for empirical work. In section 5.3, we discuss the results from our prior research that led us to our current focus. We then turn to a description of our data in section 5.4. The three ensuing sections present results from our analyses of the size and timing of investments, regulatory costs using data from the LRD, and cost estimates using PACE data. The final section offers some concluding remarks.

5.2   The Nature of Air Quality Regulation

Each year (since 1978) each county in the United States is designated as either being in or out of attainment of the National Ambient Air Quality Standards (NAAQS) for ground-level ozone. Areas that are in *nonattainment* of this standard are, by law, required to bring themselves into *attainment* or face harsh federal sanctions. The primary way of achieving attainment is through the regulation of VOC- and NO_x-emitting sources within one's jurisdiction—particularly manufacturing plants in certain industries. As a result, these plants in nonattainment areas face much stricter environmental regulation than their counterparts in attainment areas.

For example, in nonattainment areas, plants with the potential to pollute are subject to more stringent and more costly technological requirements on their capital equipment. New plants wanting to locate in nonattainment counties (as well as existing plants undertaking major expansion and/or renewal) are subject to lowest achievable emission rates (LAER), requiring the installation of the "cleanest" available equipment without regard to cost. Existing plants in nonattainment counties, who are grandfathered from these strict requirements (at least until they update their equipment), are required to install reasonably available control technology (RACT)—usually some simple retrofitting—which is to take into account the economic burden it places on a firm. In contrast, in attainment counties, only new plants and only those with the potential to emit over (originally) 100 tons per year of a criteria pollutant are subject to regulation of their capital equipment, and the technological standard is a weaker one. Rather than LAER, large new plants in attainment counties are required to install best available control technology (BACT), which is negotiated on a case-by-case basis and is to be sensitive to the economic impact on a firm. Existing plants and small new plants in attainment areas face no specific requirements on their capital equipment.

In addition to more stringent technological requirements, nonattainment status also usually entails higher costs in other areas as well. Forced to "produce environmental quality," plants in nonattainment areas must purchase additional inputs. Additional labor is certainly required; however, "environmental production" may also call for more (and/or more expensive) materials and energy as well. Costly redesigns of production processes can also be involved. And any proposed expansion—either the construction of a new plant or the modification of an existing facility—must first be approved by environmental regulators. This permitting process can involve lengthy and costly negotiations over equipment specifications, emissions limits, and the like. The purchase of pollution offsets may also be required. Finally, plants in nonattainment areas face a greater likelihood of being inspected and fined than their counterparts in attainment areas.

As this discussion reveals, we have (at least in principle) a control and a treatment group with which to estimate the costs of regulation. In particular, given age (i.e., new vs. existing plants), we would expect capital costs, labor costs, operating costs, and so forth to be higher for plants in counties classified as being in nonattainment of the NAAQS for ozone than for plants located in counties classified as being in attainment. The reality of the situation, however, is a bit murkier than this neat dichotomy would suggest. First, within a county, regulatory scrutiny often varies by plant size. In attainment areas, large new plants are required to install BACT, while small new plants have no specific requirements. In nonattainment areas, differential treatment is de facto rather than by decree. Local regulators, who are generally resource constrained, focus their enforcement on larger (and hence more polluting) plants, while smaller plants have been slow to be classified as polluters, and once classified, may be inspected infrequently or not at all. Then, given plant age and size, regulatory treatment of otherwise similar polluters may differ from one nonattainment area to the next because of variation in state philosophies on how best to achieve attainment. Even within a state, nonattainment areas may face different degrees of regulation because they differ in the *extent* to which they are in nonattainment. Dissimilarities between attainment areas also exist: Some face a degree of regulation above what is normally required of them simply because they are in states with strong environmental agendas.

In the empirical specifications that follow, we are mindful of differences in regulatory treatment that are due to plant characteristics, such as age and size. The remaining differences, then, between attainment and nonattainment areas are "typical" differences—alert to the fact that each group itself may have some significant variation in regulatory intensity. We also note two potential qualifications that affect this interpretation of our results. First, there is the notion that plants in attainment areas may incur environmental costs "voluntarily," as opposed to being required to do so by regulators. Such plants, for example, may be reluctant to install "dirty" production equipment in this day and age for fear of protests and law suits, as well as inducing active regulation. Furthermore, for plants in many industries, dirty equipment that may still be permissible for use in attainment areas may no longer be available for purchase. Prior to the regulatory era, plants in polluting industries were mostly located in (what would become) nonattainment areas and a considerable proportion still remain there. These producers spur technological innovation and create a market for "green" production equipment, which have affected equipment choices for everyone. Therefore, plants in attainment areas may incur environmental costs that are not the result of regulation per se, but rather are the result of various other forces (social, political, technological, etc.). Our approach, therefore, of comparing plants in attainment and nonattainment areas, will not reveal the *full* costs of regulation (from comparing a world with regulation to one without regulation and these other forces), but should at least reveal a lower bound on such costs.

Our second qualification only serves to lower this lower bound even further. In particular, plants may self-select into attainment or nonattainment areas. For example, it may be the case that firms who choose to locate in nonattainment areas may, to some extent, be those who can best handle regulatory costs. Firms in attainment areas, on the other hand, may be ones for whom regulation would be particularly burdensome. This would suggest that our estimates of regulatory costs are for a select group—understating costs for the typical plant.[1] Both these qualifications should be kept in mind when interpreting the results.

---

1. In theory, one might control for self-selection by using plant fixed effects in modeling (rather than county fixed effects, which we use here). In practice, however, imposing plant fixed effects eliminates many young plants (since these fixed effects require each plant to appear in two censuses, at least 5 years apart), makes identification of age effects impossible, and greatly reduces sample sizes. We therefore resign ourselves to any selection bias that may be present, realizing that it will reduce our estimates of treatment effects.

5.3   Prior Findings and Current Motivation

In our previous work in this area (Becker and Henderson 2000), we investigate the effects ozone nonattainment (vs. attainment) status has had on the decisions of firms in polluting industries. In that study (described in more detail later) we focus on major VOC- (and NO_x-) emitting industries that (1) have had large numbers of plants and plant births (nationally), and (2) do not have (as) much other air pollution emissions. These industries are industrial organic chemicals (SIC 2865–9), miscellaneous plastic products (SIC 308), metal containers (SIC 3411–2), and wood household furniture (SIC 2511). In this current paper, we focus on just the first two of these. Industrial organic chemicals, as in turns out, is the heaviest polluter of all of these industries (it actually manufactures VOCs!) and has the largest average plant sizes. Miscellaneous plastic products uses VOCs in its production and has the convenient property of being the industry with the largest sample size. Plant-level data for both studies come from the 1963–92 Censuses of Manufactures.

Our current line of research expands on previous work by Henderson (1996). Prior to this, much of the literature found little effect of state or county differences in environmental regulation on firm behavior (e.g., Bartik 1988; McConnell and Schwab 1990; Gray 1997; Levinson 1996). Much of this work, however, has been based on cross-sectional data and/or methods, which has proven to be a critical limitation. In order to properly disentangle the inherent locational and productivity advantages typical of nonattainment areas from the adverse (regulatory) impacts of nonattainment status, panel data and methods are necessary, such as those used in Henderson (1996), Becker and Henderson (2000), and Kahn (1994). We, again, employ such data and methods here in this paper.

Our earlier research (Becker and Henderson 2000) yielded three key findings. First, plant births in these polluting industries (followed by the stocks of plants) have, with the advent of regulation, shifted over time from nonattainment to attainment areas, while general economic activity has not exhibited such a shift. Depending on the industry and time period one looks at, the expected number of new plants in these industries in ozone nonattainment areas dropped by 25–45 percent. The sectors targeted first and most intensely by regulators were those industries with the largest plants and, within industries, the "corporate" sector (with its larger plants) compared to the "nonaffiliate" (or single-plant firm) sector. This supports the notion that size matters in who is regulated when and how intensely.

Second, survival rates of plants in nonattainment areas, while originally the same as those in attainment areas, rose with the advent of regulation. Recall that existing plants are grandfathered from the strictest regulations (until they update or expand their operations) and are only subject to RACT requirements. New plants, on the other hand, are subject to costly LAER requirements. Existing plants, therefore, have a cost advantage over new entrants and reason to stay in business longer than they might have otherwise. Similarly, as regulations tighten over time, former new plants (with former LAER equipment) are exempt from the tightening. The net effect is better survival of existing plants in nonattainment areas and incentive to delay equipment renewal and changes in product composition. There is yet another explanation for this result. Older firms may get heavily involved in their states' regulatory process—working with regulators to formulate regulations, advocating for particular laws, and so forth. Even if the regulatory process remains without favoritism, these firms have insiders' knowledge of what their state regulators are most focused on. It may, therefore, be easier and less costly for them to meet the specifications and regulations issued by that particular state.

Third, it appears that plants in nonattainment areas, rather than phasing in investment over a 5- to 10-year period, do more up-front with less subsequent investment. In terms of sales and employment, we found that new plants in nonattainment areas started off anywhere from 25 to 70 percent larger, but after 10 years, no size differences remained. The permitting process for the construction of a new plant in a nonattainment area (as well as the proposed expansion of existing facilities) can require months of costly negotiations—involving the firm, its environmental consultants, state regulators, and the regional Environmental Protection Agency (EPA)—over equipment specifications, emissions limits, the purchase of pollution offsets, and the like. By investing all at once, these plants avoid incurring negotiation costs over and over again; moreover, they preserve their grandfathered status.

Our current paper expands on these findings in two ways. First, we revisit the issue of regulation's impact on the size and timing of plants' investments (the third key finding) by examining data on plants' capital stock formation, instead of using sales and employment data as we did before. The questions we ask here (and the methodology we employ) are similar to those in our previous paper. Namely, does nonattainment induce more up-front investment and less subsequent investment as a result of the costly negotiating and permitting process required for plant expansion under regulation? And, given that regulatory scrutiny seems to be closely related to plant size, is downsizing evident in nonattainment areas relative to attainment areas once the initial investment period of a new plant is past? Also, how does regulation impact the capital-to-labor use of plants in these industries?

The second (and major) focus of this paper is in actually quantifying regulatory costs. The birth model estimated in Becker and Henderson (2000) implies that the number of new plants in nonattainment areas drops because the net present value of profits in those areas falls. One view of the birth process is that, in any given year, there is a local supply of potential entrepreneurs to an industry in a county and a (demand) schedule of profit opportunities decreasing in the number of births. Nonattainment status shifts back the (demand) schedule of profit opportunities, moving the county down the supply curve and reducing births. The implied percentage drop in plant profits (which are unobserved) is unclear since both demand and supply elasticities are involved in a reduced-form specification of birth counts.[2]

In this paper, we look at this issue from the cost side. In particular, we ask what happens to a plant's operating costs if we move it from an attainment to a nonattainment area. We perform this experiment by comparing the production costs of plants in nonattainment counties (our treatment group) with those in attainment counties and those in existence before the advent of regulation (our control group). Since our prior work suggests that both plant size and plant age matter in regulation, we incorporate these factors into our analysis as well. And, as we mentioned earlier, this type of work suffers tremendously if inherent county characteristics are not controlled for, so we also employ county fixed effects in all our models. Given these fixed effects, the nonattainment effect is identified by differences between attainment and nonattainment counties arising from the imposition of regulation (in 1978), relative to any differences that might have existed in the preregulatory period (when there were no regulatory differences between these counties). Recall from our comments in section 5.2 that estimated cost differences between attainment and nonattainment areas are likely to represent a lower bound on *true* regulatory costs. We will see in the next section that there are additional reasons why we cannot estimate the full costs of regulation.

5.4 The Data

Our plant-level data come from the LRD, available through the Census Bureau's Center for Economic Studies. Here, we use only the quinquennial Census of Manufactures from the period 1972–92. And since we require (nonimputed) data on capital assets for both our examination of investment patterns and our estimation of cost functions, we use mainly those plants that are also in the Annual Survey of Manufactures (ASM) in those years.[3] We further eliminate any plants that are administrative record

2. Note that, even with regulation, nonattainment counties do have some births, given a local supply of entrepreneurs (with their own idiosyncracies) and local and regional demand forces.

3. Total assets (buildings and machinery together) was also asked of non-ASM plants in the 1987 and 1992 censuses. For our cost function exercises, since we are not interested in the separate components of capital stock, we also use these plants in our estimation.

cases, have their establishment impute flag set, or show signs of inactivity (i.e., have a zero value for any critical variable). We further restrict our attention to "corporate" (or "multiunit") plants. Controlling for age, these plants are much larger than single-plant firms and are therefore more likely to be regulated and exhibit regulatory effects. Their data may also be more accurate than those for single-plant firms, and they certainly account for most of an industry's output.[4] Finally, the inclusion of county fixed effects in our models requires plants to be in counties where at least one other plant-year is observed. The impact of this restriction on sample sizes is relatively slight; less than 5 percent of plants are lost as a result of this requirement. In the end, our samples contain 70–74 percent of all multiunit plants in industrial organic chemicals and 53–61 percent of all such plants in miscellaneous plastics.

In our investment regressions (in section 5.5) we use a plant's stock of real capital and its real capital-to-labor ratio as dependent variables. For these regressions we use, as our measure of real capital stock, end-of-year machinery and equipment assets (which are on an "original cost" basis) divided by capital-asset deflators constructed from Bureau of Economic Analysis (BEA) published data (see Becker 1998). Plant total employment serves as the denominator of our ratio. Here, and in our cost function regressions, plant age dummies are (generally) determined by the time elapsed since the plant's first appearance in the Census of Manufactures, regardless of its industry in that first appearance. In our empirical specifications below, we recognize three age categories: 0–4 years, 5–9 years, and 10+ years.[5]

For our cost-function regressions, a plant's total costs are defined as the sum of its salaries and wages; its costs of materials, fuels, electricity, and contract work; and the cost of "capital services." The last is calculated by multiplying beginning-of-year total assets (machinery, equipment, structures, and buildings) by an appropriate "user cost factor."[6] Note that these data from the Census of Manufactures do not subsume all of the (previously noted) costs associated with environmental regulation (e.g., fines,

---

4. In industrial organic chemicals, corporate plants account for about 97 percent of the industry's output. In miscellaneous plastic products, they account for about 72 percent.

5. For example, a plant in the 1972 census is 0–4 years of age if it is making its first census appearance in 1972. It is 5–9 years of age if it made its first census appearance in 1967 and 10+ years of age if it made its first census appearance in the 1963 census. The recognition of any additional age categories is not practical. Since the LRD does not contain any of the censuses prior to 1963, we are not able to distinguish between 1972 plants that are 10–14 and 15+ years of age. Excluding 1972 plants from the analyses (and using just 1977–92 plants) avoids this problem but unfortunately eliminates an important (control) group of preregulatory plants that help us identify the effects of regulation. On the opposite end, fewer age categories would not buy us any additional data. In principle, two age categories would allow us to use plants in the 1967 census as well; however, capital-asset data were not collected from these plants and therefore they are of no use to us for the types of analyses we wish to conduct here. Using three age categories, therefore, is ideal.

6. The difficulty here is that the asset information collected by the Census Bureau is on an original cost basis. It reflects the book value of assets (of various vintages and quality, etc.),

pollution offsets, and environmental consultants), which obviously affects plants in nonattainment areas more than those in attainment areas. This is yet another reason to view any estimated cost differences as a lower bound on the true cost of regulation. What will be captured here is regulation's impact on the use of labor, capital, and some of the other inputs, as well as any impact it may have on production (output). Here, total output of a plant in a given year is measured by its total value of shipments in that year, with appropriate adjustments for changes to inventory. This value of output is then divided by the industry's (national) output price index to yield a real measure of plant output. This price index, along with the industry-specific materials price index (referred to later), is taken from the NBER-CES Manufacturing Industry Database (Bartelsman, Becker, and Gray 1998).

In addition to these plant-level data, we also have information on county characteristics. In particular, we have LRD-derived, county-level measures of average manufacturing wages and total manufacturing employment, exclusive of the industry being analyzed. We also have county ozone nonattainment status, as recorded annually in the Code of Federal Regulations (Title 40, Part 81, Subsection C). Given that 1978 was the first year in which counties were designated as being in or out of attainment, the plants in the 1982 Census of Manufactures would have been the first ones to directly feel the effects of the 1977 amendments to the Clean Air Act. And since the previous year's attainment status determines the current year's regulation, we use 1981 nonattainment status for 1982 plants, 1986 nonattainment status for 1987 plants, and 1991 nonattainment status for 1992 plants.

5.5   The Size and Timing of Investments

The questions we pose in this section are (1) does regulation induce more up-front investment by plants (versus phasing-in) as a result of the (fixed)

---

but not necessarily their true economic value. Given the highly imperfect nature of these data, multiplying them by a *proper* user cost of capital series seems somewhat incongruous. What we have done instead is derive "user cost factors" such that capital's share of total costs in our samples (by industry and year) equals capital's share of total output (for the corresponding year and two-digit industry) in Dale Jorgenson's 35KLEM.DAT (available at http://www.economics.harvard.edu/faculty/jorgenson/ and described in Jorgenson 1990). For SIC 28 (chemicals), capital's share of total output in Jorgenson's data ranged from 15.2 percent (1982) to 21.3 percent (1987) for the 5 census years used in our study. To replicate these shares in our data for industrial organic chemicals required user cost factors ranging from 0.1495 (1972) to 0.2136 (1987). For SIC 30 (rubber and miscellaneous plastics), capital's share of total output in Jorgenson's data ranged from 4.1 percent (1982) to 6.6 percent (1972). To replicate these shares in our miscellaneous plastic products sample required user cost factors ranging from 0.0689 (1982) to 0.0979 (1972). We note that in the initial phases of this study we experimented with time-invariant user cost factors of 0.17 (e.g., a 10 percent interest rate plus a 7 percent depreciation rate) and 0.10, with results that are remarkably similar to the ones obtained using the factors computed here. We do not believe, therefore, that our results are sensitive to our treatment of the capital data.

costs involved in capacity expansion, and (2) does regulation ultimately lead to reduced plant sizes as plants seek to avoid regulatory scrutiny? Regarding the latter, plants could also downsize to reduce investment risk at any one location, in the face of uncertainty over local regulatory costs. In Becker and Henderson (1997), we explore general downsizing issues in these industries. In the miscellaneous plastics industry, we find that plants of the same age are of roughly comparable size across the generations. In the industrial organic chemicals industry, on the other hand, plants built prior to 1968 (i.e., before the 1970 amendments to the Clean Air Act) are found to be distinctly larger, at every age, than plants built after this point (i.e., those who made their first census appearance in 1972 or later). The size profiles of successive birth cohorts in the regulatory period, however, did not continue to decline. This suggests, of course, that technological rather than regulatory changes may have led to this "one-time" change in average plant size. But the issue of determining overall trends in plant sizes in this industry is complicated by the fact that there is a great deal of switching of plants into and out of the industry (in comparison with other industries). We find that, after 1972, the number of plants switching out of the industry and the average size of such plants rises quite dramatically, while the sizes of those switching into the industry actually diminishes somewhat. It is difficult, therefore, to come to any firm conclusions.

Here, rather than try to assess the general effects of regulation on plant size, we focus on the differences between attainment and nonattainment counties. Our main interest is in the effects that nonattainment status has had on plants' accumulation of real capital stock (machinery and equipment assets in particular), but we also consider possible impacts it may have had on real capital-to-labor use. We hypothesize that these two items are functions of county characteristics—wages (a cost factor), employment (a scale/demand factor), fixed effects, and ozone nonattainment status—and plant characteristics. In particular, we allow our dependent variables to be functions of plant age (which will allow us to gauge investment patterns over time) as well as age interacted with nonattainment status (which allows us to measure the differential impact of regulation). Results from these regressions are in table 5.1.

We clearly see that capital assets rise with plant age. In the industrial organic chemicals industry, relative to the base group (new plants in attainment counties), plants 5–9 years of age are 67 percent larger, those 10+ years are 97 percent larger, and those built prior to 1968 are 176 percent $(0.968 + 0.793 = 1.761)$ larger. In miscellaneous plastics, these percentages are 45 percent, 81 percent, and 122 percent, respectively. The final percentage in each of these trios reinforces the notion (discussed previously) that plants built before 1968 (and the 1970 amendments to the Clean Air Act) are simply larger than those constructed later.

What effect does regulation have on these patterns? In industrial organic

Table 5.1   **Capital Stocks under Regulation**

	Industrial Organic Chemicals		Miscellaneous Plastic Products	
	ln (K)	K/L	ln (K)	K/L
Age 5–9 years	0.668**	−49.05	0.455**	−7.124
	(0.188)	(35.63)	(0.057)	(7.366)
Age 10+ years	0.968**	−103.23**	0.809**	−7.740
	(0.229)	(43.36)	(0.071)	(9.207)
Plant built before 1968	0.793**	8.57	0.406**	7.643
	(0.202)	(38.30)	(0.069)	(8.891)
Nonattainment	0.794**	82.05	0.062	4.665
	(0.267)	(50.52)	(0.088)	(11.355)
× Age 5–9 years	−0.587*	−36.07	0.089	−0.182
	(0.324)	(61.40)	(0.098)	(12.696)
× Age 10+ years	−1.046**	−79.10	−0.086	−3.055
	(0.317)	(60.03)	(0.099)	(12.781)
× Plant built before 1968	0.555**	13.72	0.120	−2.493
	(0.241)	(45.64)	(0.092)	(11.852)
County wages and employment	Yes	Yes	Yes	Yes
Year and county effects	Yes	Yes	Yes	Yes
N (counties)	1,730 (220)	1,730 (220)	7,745 (820)	7,745 (820)
Adjusted R²	0.545	0.369	0.290	0

Note: Standard errors are in parentheses.
**Significant at the 5 percent level.
*Significant at the 10 percent level.

chemicals, new plants in nonattainment counties are 79 percent larger than new plants in attainment counties. Plants 10+ years of age, however, are actually 13 percent smaller in nonattainment counties than similar plants in attainment counties.[7] These results support our hypotheses: Regulation induces greater up-front investment in nonattainment counties, but tempers the size of mature plants.

The story is different, however, for plants in nonattainment areas built prior to 1968. In industrial organic chemicals, these plants are actually 11 percent larger than similarly old plants in attainment counties.[8] This suggests an intriguing possibility. These old plants in nonattainment areas have various competitive advantages over new entrants—aspects of their operations are grandfathered; they are experienced players in the local regulatory process, learning long ago how to work with regulators and how to coexist with their neighbors; and so forth. These plants, therefore, may

7. [(1 + 0.968 + 0.794 − 1.046) − (1 + 0.968)] ÷ (1 + 0.968) = −0.1280, or roughly 13 percent.

8. [(1 + 0.968 + 0.793 + 0.794 − 1.046 + 0.555) − (1 + 0.968 + 0.793)] ÷ (1 + 0.968 + 0.793) = 0.1097, or 11 percent.

be in a better position to exploit the scale economies inherent in production (see section 5.6), and, given grandfathering and an exodus of competitors, they may have access to relatively large regional demands, compared to similar plants in attainment areas. As such, it may be profitable for them to operate on a scale larger than that of their attainment-area counterparts, who face a substantial number of new entrants.

Turning to the miscellaneous plastic products industry, our hypotheses are really not borne out. In this industry, after controlling for plants built prior to 1968, real capital stocks are no different between plants in attainment and nonattainment areas at different ages. Since total capital investments in this industry are so much smaller than they are in industrial organic chemicals (in any given census, the average multiunit miscellaneous plastics plant has approximately 6–7 percent of the machinery and equipment assets of the average industrial organic chemicals multiunit plant) issues of phasing-in, downsizing, and so forth may be less relevant here.[9]

We also see no significant effects of nonattainment status on real capital-to-labor use in these industries. In fact, very few coefficients in either of these two regressions are actually significant. That we find no effect of regulation on capital intensity is somewhat at odds with our later findings with the PACE data, which show, at least for industrial organic chemicals, capital expenditures relatively more affected than labor costs.

5.6   Quantifying Regulatory Costs

In this section, we compare the average total costs of production for plants in nonattainment counties to those in attainment counties. We assume that, in any period, competitive plants face a constrained cost minimization problem. We could formalize regulatory constraints in various ways, but here we will specify a very general constraint. Suppose l, k, and m are inputs of labor, capital, and materials into production; l_R and k_R are inputs of labor and capital associated with regulation (i.e., pollution reduction); w, r, and p_m are the respective factor prices (which are exogenous to the firm), and X is plant output. A plant's constrained cost minimization problem (with respect to l, k, m, l_R, and k_R) could be written as

$$(1) \qquad w \cdot (l + l_R) + r \cdot (k + k_R) + p_m \cdot m$$
$$- \theta[X - X(l,k,m;\text{age})] - \lambda[R_h(l,k,m,l_R,k_R;\text{age})].$$

---

9. Having said that, we note that an identical regression (not reported here) on a sample that also includes single-unit firms (in addition to these multiunit plants) reveals some of the hypothesized effects. Namely, new plants in nonattainment areas were found to start with 20 percent more capital than their counterparts in attainment areas, but after 10 years, there was virtually no difference between the two groups. Why these effects might be found in the single-plant sector and not the multiplant corporate sector is puzzling.

Here, the $R_h(\cdot)$ function is the regulatory constraint, where h indexes the two possible regulatory states: attainment and nonattainment. Note that we have allowed plant age to affect both the technology of the plant as well as (and more critically) regulatory stringency. This minimization problem, with its choice of inputs, yields a reduced-form total cost that is a function of factor prices, output, and age. Dividing through by output and invoking linear homogeneity of cost functions, we are left with the following reduced-form average total cost function:

(2) $$ATC/p_m = f_h(w/p_m, r/p_m, X, \text{age}).$$

We have let our needs and interests dictate our empirical formulation of equation (2). Since we are not interested in estimating the elasticities of substitution between factors, a translog specification is too much. And a simple Cobb-Douglas cost function, which is linear in output, is also inappropriate for our purposes. We therefore choose a log-quadratic formulation, which allows for a classic U-shaped average total cost function:

(3) $$\ln\left(\frac{ATC_{ijt}}{p_{m_t}}\right) = \alpha_0 \cdot \ln\left(\frac{w_{jt}}{p_{m_t}}\right) + \sum_{s=0}^{5} D_{sit} \cdot [\beta_s + \gamma_s \cdot \ln(X_{it}) + \delta_s \cdot \ln(X_{it})^2]$$
$$+ d_t + C_j + \varepsilon_{ijt},$$

where

$D_{0it} = 1$ for all plants.
$D_{1it} = 1$ if plant i is 5–9 years old in year t; 0 otherwise.
$D_{2it} = 1$ if plant i is 10+ years old in year t; 0 otherwise.
$D_{3it} = 1$ if plant i is in a nonattainment county in year t; 0 otherwise.
$D_{4it} = 1$ if plant i is in a nonattainment county and 5–9 years old in year t; 0 otherwise.
$D_{5it} = 1$ if plant i is in a nonattainment county and 10+ years old in year t; 0 otherwise.

Note that the average total cost of plant i in county j at time t is a function of output, output squared, year effects (d_t), county fixed effects (C_j), and a contemporaneous error term (ε_{ijt}) that is independently and identically distributed. Wages (w_{jt}), as we discussed earlier, are average manufacturing wages in the county, exclusive of the industry being analyzed. Since we are not interested in factor price coefficients per se, we have taken some liberties with respect to the other two factor prices. We assume perfect capital markets, such that all plants in an industry in a given year face the same price of capital. This, then, is captured by our year effects (d_t). Our material prices (p_m), which come from the NBER-CES Manufacturing Industry Database, vary only over time (within an industry). We can either assume that the price of materials is the same for all plants within an industry within a year or, if we believe that there may be spatial variation in such prices, to the extent that these price differences are constant over time, spatial differences are captured by our county fixed effects (C_j). What we are most interested in here is the shape of the cost curve and how it changes with age and attainment status. To this end, we have included a series of dummy variables (D_{sit}), which are interacted with the intercept, output, and output squared. These terms allow the shape to differ for six categories of plants (three age categories times two states of regulation).

The results of these regressions are given in table 5.2. Note that all the coefficients on the U-shaped structure are statistically significant, with two exceptions in miscellaneous plastics where the coefficient on nonattainment $\times$ output has a t-statistic of 1.584 (significance level of approximately 11 percent) and the coefficient on nonattainment $\times$ output^2 has a t-statistic of 1.189 (significance level of approximately 23 percent). This poses a problem for evaluating the results for the plastics industry, which is why we focus mostly on industrial organic chemicals in our remaining discussion. In and of themselves, these regression coefficients are not very interesting. These point estimates, however, are necessary for the exercises that follow.

First, for each industry, we use the estimated coefficients to calculate the level of plant output that minimizes average total cost (ATC), for each of the six categories of plants. Every situation in each of the two industries happens to be characterized by a U-shaped ATC function (i.e., a negative linear term and a positive quadratic term). For young plants in attainment areas (the "all plants" category) in the industrial organic chemicals industry, for example, the minimum of the average cost curve occurs at $(1.192994 \div [0.0622169 \times 2])$, which equals 9.587. (Recall that output and costs are both measured in natural logs.) For industrial organic chemical plants that are 10+ years old and in nonattainment counties, minimum average total cost is achieved at $\{(1.192994 - 1.078772 - 1.030363 + 1.214576) \div [(0.0622169 - 0.0588917 - 0.0538479 + 0.0632474) \times 2]\}$, which equals 11.727. Table 5.3 contains the cost-minimizing level of output for all categories of plants in both industries. We will return to a discussion of these later.

Next, we take these cost-minimizing levels of output, plug them back into the estimated cost functions, and calculate cost differentials between comparable plants in attainment and nonattainment areas, operating at their respective minimum ATCs. For example, the cost differentials for young plants in the industrial organic chemicals industry is $\{-4.751786 - [(1.192994 - 1.030363) \times 9.7162 \ldots] + [(0.0622169 - 0.0538479) \times (9.7162 \ldots)^2]\} - \{-(1.192994 \times 9.5873 \ldots) + [0.0622169 \times (9.5873 \ldots)^2]\} = 0.1770$, or 17.7 percent (given costs and output are in natural logs). Table 5.4 contains the cost differentials computed for this and all

Table 5.2   **Average Total Cost Functions**

	Industrial Organic Chemicals	Miscellaneous Plastic Products
ln(Output)	−1.193**	−0.319**
	(0.122)	(0.051)
ln(Output)^2	0.062**	0.015**
	(0.007)	(0.003)
Age 5–9 years	−3.594**	−0.658*
	(1.116)	(0.338)
× ln(Output)	0.850**	0.165**
	(0.231)	(0.081)
× ln(Output)^2	−0.048**	−0.010**
	(0.012)	(0.005)
Age 10+ years	−4.802**	−0.931**
	(0.817)	(0.297)
× ln(Output)	1.079**	0.223**
	(0.168)	(0.069)
× ln(Output)^2	−0.059**	−0.012**
	(0.009)	(0.004)
Nonattainment	−4.752**	−0.734**
	(0.998)	(0.367)
× ln(Output)	1.030**	0.143
	(0.216)	(0.090)
× ln(Output)^2	−0.054**	−0.007
	(0.012)	(0.006)
× Age 5–9 years	4.908**	1.282**
	(2.263)	(0.604)
× ln(Output)	−1.144**	−0.308**
	(0.462)	(0.143)
× ln(Output)^2	0.064**	0.018**
	(0.023)	(0.008)
× Age 10+ years	5.625**	1.815**
	(1.252)	(0.487)
× ln(Output)	−1.215**	−0.391**
	(0.258)	(0.115)
× ln(Output)^2	0.063**	0.021**
	(0.013)	(0.007)
Wages, year, and county effects	Yes	Yes
N (counties)	1,847 (233)	8,878 (881)
Adjusted R^2	0.232	0.231

Note: Standard errors are in parentheses.
**Significant at the 5 percent level.
*Significant at the 10 percent level.

Table 5.3   Plant Output That Minimizes ATC (natural log)

	Attainment Areas	Nonattainment Areas
Industrial Organic Chemicals		
Young plants (0–4 years)	9.587	9.716
Plants 5–9 years old	12.459	9.600
Plants 10+ years old	17.175	11.727
Miscellaneous Plastic Products		
Young plants (0–4 years)	10.425	10.077
Plants 5–9 years old	13.719	9.516
Plants 10+ years old	16.166	10.163

Table 5.4   Cost Differentials between Plants in Nonattainment Areas and Attainment Areas Operating at Minimum ATCs

	Industrial Organic Chemicals	Miscellaneous Plastic Products
Young plants (0–4 years)	+17.7	+4.3
Plants 5–9 years old	+9.9	+8.6
Plants 10+ years old	+10.4	+11.2

Note: In percentage by which ATC in nonattainment areas exceeds that in attainment areas.

other comparisons. Here (and throughout) *differentials* will be defined as the percentage by which costs in nonattainment areas exceed those in attainment areas. These are, therefore, expected to be positive.

The results in table 5.4 indicate that costs are indeed higher for plants in nonattainment areas compared to those of similar age in attainment areas. In industrial organic chemicals, young plants in nonattainment areas experience costs 17.7 percent higher than their counterparts in attainment areas. The difference for older plants, although lower, is still quite considerable, at roughly 10 percent. This lower cost differential for older plants is consistent with the notion (discussed earlier) that regulatory requirements are stricter for new (as opposed to existing) plants. In the miscellaneous plastic products industry, production costs are also found to be more expensive for plants in nonattainment counties, but the pattern is the reverse. Young plants in nonattainment areas are found to have costs that are 4.3 percent higher than their counterparts in attainment areas, while plants 5–9 years old in nonattainment areas have 8.6 percent higher costs and plants 10+ years old have 11.2 percent higher costs. But again, the precision and accuracy of these estimates are compromised by the two statistically insignificant cost function coefficients used in their calculation. Nonetheless, all these results point in the same direction: Nonattainment status leads to higher operating costs for plants in these industries.

Table 5.5   **Industrial Organic Chemical Plant Output That Minimizes ATC,**
                        Average ln(Output), and ln(Average output)

	Attainment Areas	Nonattainment Areas
Young plants (0–4 years)	9.59, 9.16, 10.20	9.72, 9.28, 10.61
Plants 5–9 years old	12.46, 9.96, 10.94	9.60, 9.63, 10.57
Plants 10+ years old	17.18, 10.75, 11.62	11.73, 10.63, 11.77

A number of issues are raised by our analysis, and we focus on the industrial organic chemicals industry to explore them. First, one may ask why outputs that minimize ATCs might vary by age. As table 5.3 reveals, cost-minimizing outputs grow as plants age (although the growth is not always monotonic). Why do young plants minimize ATC at lower levels of output? It is probably not the case that young plants have technologies that dictate smaller plant sizes. Arguably, some sort of learning process is taking place. Young plants perhaps do best starting off small because they can only handle a simple organizational structure and a smaller scale of operation. As they gain experience, however, and learn more about their local factor (labor and material) markets, they expand. For plants in attainment areas (which show the largest growth in cost-minimizing output!), there is an additional reason for starting out small. Recall that if such plants start out too big, they may be subject to somewhat costly BACT requirements, whereas if they start out small they face no regulation. These small (initially) unregulated plants may then expand as they learn more about their local regulatory environment and, in particular, as they learn from other plants in the area how to best handle (or avoid) regulation. For plants in nonattainment areas, which exhibit smaller changes in cost-minimizing output, there are reasons (discussed previously) for not phasing in investments in this way.

Another issue, revealed in table 5.5, is that the output of the average plant can be far smaller than the level of output that minimizes ATC. There are a few reasons why this might be. It may be the case that regional goods markets are imperfectly competitive, leading firms to exercise some monopoly power (hence, production shy of cost-minimizing output). Risk avoidance behavior (to reduce exposure) may also lead firms to invest less than the amount necessary to minimize average total cost. Having said that, however, we note that the differences between actual and cost-minimizing output in attainment areas is absolutely enormous. For plants 5–9 years old in the industrial organic chemicals industry, the level of output that minimizes ATC (12.46) is approximately 1.76 standard deviations from the average ln(Output) of 9.96, and the gap for plants 10+ years of age is even larger! What is limiting the size of these plants? The obvious suggestion is regulation, or more specifically, the threat of regulation. If one believes these particular extrapolations out to the cost-minimizing levels of output, there are (virtually) decreasing average total costs throughout. Plants in attainment areas do not generally grow to these sizes because at some point they will attract attention from regulators, they will be sued by local interest groups, or they may even (single-handedly) pollute their counties into nonattainment. There are, therefore, regulation-related constraints even on these "unregulated" plants (that are not reflected in production costs). The plants that do grow to these sizes may be in lax states, where plants in attainment areas really are left alone—that is, areas that truly are devoid of *effective* regulation.

The oldest category of plants in attainment areas also contains two distinct groups: those built before the regulatory era (say, the 1970 amendments to the Clean Air Act) and those built after. Recall that we acknowledged this distinction (i.e., pre- and post-1968 plants) in our investment regressions. Attempts to control for this separate group of plants here in our cost functions result in coefficients insignificant at the 5 percent level. However, the coefficients (imprecise as they are) suggest that plants built before 1968 have much larger cost-minimizing levels of output than other plants 10+ years old. The estimates suggest that those pre-1968 plants could operate at much lower costs in attainment areas if they operated at a large scale—large enough to be regulated, but much less severely so than they would be in a nonattainment area. Post-1968 plants that are 10+ years old, on the other hand, operated at about the same costs in attainment and nonattainment areas. (Differentials for young plants and those 5–9 years old are unaffected by this reformulation.) All this might suggest that large pre-1968 plants in attainment areas, as grandfathered players with extensive experience, reap considerable advantages. Having said that, however, we note that it is still the case that very few of these plants operate at even a reasonable fraction of cost-minimizing output. We are, therefore, left with our same conclusion: Plants in attainment areas stay small to avoid triggering regulation.

How do differences between the *estimated* cost-minimizing levels of output and the *actual* levels of plant output affect the cost differentials computed in table 5.4? To see, we repeated the exercise using, instead, average ln(Output) and ln(Average output). The results of these (and our previous) computations for the industrial organic chemicals industry are contained in table 5.6. The cost differentials for young plants are fairly insensitive to the output measure chosen. Using average ln(Output), young plants are found to have costs 16.7 percent higher in nonattainment areas, compared to their counterparts in attainment areas. Using ln(Average output), this difference was found to be 16.0 percent. Originally, using ATC-minimizing output, we found a cost differential of 17.7 percent. For the older categories of plants, the results are less comparable across output measures. Using average ln(Output), cost differentials all but disappear for plants over 5 years old (+1.3 percent and −1.8 percent). With ln(Average output), plants 5–9 years old are found to have costs 9.0 percent higher in nonat-

Table 5.6   **Cost Differentials between Industrial Organic Chemical Plants in Nonattainment Areas and Attainment Areas**

	Minimum ATC	Average ln(Output)	ln(Average output)
Young plants (0–4 years)	+17.7	+16.7	+16.0
Plants 5–9 years old	+9.9	+1.3	+9.0
Plants 10+ years old	+10.4	−1.8	+0.2

Note: In percentage by which ATC in nonattainment areas exceeds that in attainment areas.

tainment areas (vs. 9.9 percent, using cost-minimizing output), but the differences virtually disappear (+0.2 percent) for plants 10+ years old. All of these estimates, however, recalling our earlier discussions, are likely to represent lower bounds on the true costs of regulation. If nothing else, they uniformly indicate that regulation is most burdensome for new (rather than existing) plants.

5.7   An Alternative Approach

Instead of quantifying the costs of regulation by inferring it indirectly from a plant's total costs (which we did in the previous section), we could also, in principle, examine directly the environmental costs incurred by the plant. The Census Bureau's PACE survey, for example, asks manufacturing plants about their capital expenditures and operating costs associated with various environmental efforts. This survey, however, has been criticized for potentially missing a large portion of environmental expenses (see Jaffe et al. 1995 for a discussion). It is generally the case that plants do not keep special track of their expenditures on environmental protection. These data therefore must be estimated. Capital expenditures of the *end-of-line* variety (e.g., scrubbers, filters, and precipitators) are rather straightforward to estimate, since these items are easily recognized and their sole purpose is pollution abatement. However, when capital expenditures are of the *production process enhancement* type (e.g., the installation of new equipment that both improves production efficiency and reduces air emissions) the task is much more difficult.

In these instances, survey respondents are asked to "estimate the pollution abatement portion [of such projects] as the extra cost of pollution abatement features in structures and equipment (i.e., your actual spending less what you would have spent without the pollution abatement features built-in)" (U.S. Census Bureau 1994, A-12). The Census Bureau acknowledges that "interviews with survey respondents indicate that estimating such an incremental cost is difficult in many instances," if not impossible (1994, 4). In 1992, the following "special instructions" were added to the survey form to help respondents in particularly difficult cases:

Do *not* include any of the project cost *unless* the primary purpose is environmental protection. If the primary purpose of the project is environmental protection, report the whole production process enhancement project expenditure. . . . *Caution:* A project with the primary purpose of improving production efficiency may include pollution abatement features added to meet legal requirements. Since the primary purpose of such a project is still not environmental protection, do not report *any* of the production process enhancement. (1994, A-12)

Given these guidelines, and the last two sentences in particular, it is not clear whether any of the costs of production equipment meeting strict LAER standards, for example, will be attributed to environmental protection and reported in PACE, especially in the absence of an obvious baseline.

Concerns also apply to operating expenses. The salaries and wages of a plant's environmental staff are rather easily accounted for, but what of a production team that spends a small but nonzero amount of time on various environmental tasks or of plant management who must also spend a fraction of its time and effort on environmental issues? Are these costs captured in PACE? Similarly, the cost of "materials, parts, and components that were used as operating supplies for pollution abatement, or used in repair or maintenance of pollution abatement capital assets" (U.S. Census Bureau 1994, A-10) might be easy to estimate, but what about the "incremental costs for consumption of environmentally preferable materials and fuels" or the "fuel and power costs for operating pollution abatement equipment" (A-9)? Surely these are not easy items to calculate, even for the most talented and organized (and patient) of plant staffs. Apart from the potential underreporting of capital expenditure and operating costs, there are certainly other potential costs that PACE makes no attempt to capture. For example, adverse impacts on plant output, either from the outright stoppage of production (e.g., to install pollution control devices) or through the loss of operational flexibility (to comply with certain regulatory requirements). All these factors argue for the approach we used in section 5.6, where environmental costs (and related effects) are subsumed by total plant costs (and output).

Nevertheless, we conducted some rudimentary analysis of our two industries using plant-level data from the 1992 PACE survey linked to 1992 Census of Manufactures (CM) data from the LRD. Only a relatively small sample of manufacturing plants are actually asked to complete the PACE survey in any given year (e.g., approximately 17,000 in 1992), focusing disproportionately on large (and hence older) plants and plants in polluting industries. After eliminating plants with imputed data (in either the PACE, the CM, or both), as well as other suspicious cases, we are left with approximately 15 percent of all plants in the industrial organic chemicals industry in 1992 and about 4.5 percent of all plants in the miscellaneous

Table 5.7   **Costs of Air Pollution Abatement Relative to Total Costs and Expenditures (%)**

	Industrial Organic Chemicals	Miscellaneous Plastic Products
Capital expenditures[a]	6.8 (6.9)	1.9 (1.6)
Labor costs	1.8	0.1
Operating costs	0.9	0.2

[a] Figures in parentheses are based on published totals.

plastic products industry. This is about one-third the industry coverage we had in our previous cost-function exercises. And young plants, a segment we found to be particularly affected by regulation in section 5.6, are underrepresented here in our PACE-LRD samples. In industrial organic chemicals, only 7 percent of our sample consists of young plants (compared to 23 percent of the 1992 population in this industry), and in miscellaneous plastics, 13 percent are young plants (compared to 35 percent in the 1992 population). In the previous section, using just multiunit plants from the period 1972–92, young plants accounted for 15 percent and 26 percent of our samples, respectively, compared to the 16 percent and 30 percent in the universes from which they were drawn. These differences in sample sizes and composition, as well as in the time period covered, should be kept in mind when comparing the results here to the ones presented before. In particular, an unfortunate consequence of the limited number of young plants we have here is that we are not able to properly distinguish separate age effects in the following.

Table 5.7 contains some basic statistics for our sample of plants. Specifically, we present the share of total plant capital expenditures, labor costs, and operating costs (in 1992) directly attributable to air pollution abatement activity. These shares are computed by comparing PACE and CM responses to questions on capital investments; salaries and wages; and the costs of labor, materials, energy (electricity plus fuels), and contract work. Note that operating costs as defined here (as opposed to the definition we used in the previous section) do not include the costs of capital services (essentially because we do not have data on the stock of pollution abatement capital equipment). Perhaps most striking here is that expenditures on air pollution abatement in these industries appear to be fairly low. Air pollution capital expenditures in industrial organic chemicals only account for approximately 6.8 percent of total capital expenditure in our sample of plants (6.9 percent based on published totals). In plastics, this number is less than 2 percent. The shares of plant labor costs and operating costs accounted for by air pollution concerns in industrial organic chemicals are 1.8 percent and 0.9 percent, respectively. In miscellaneous plastics, these shares are negligible.

Table 5.8   **Nonattainment Coefficients from Regressions of PACE-to-Total Ratio on ln(Output), ln(Age), Multiunit Dummy, and County Nonattainment Status**

	Industrial Organic Chemicals	Miscellaneous Plastic Products
Capital expenditures	0.038*	0.006
	(0.021)	(0.004)
Labor costs	−0.001	0.000
	(0.003)	(0.000)
Operating costs	−0.000	0.001
	(0.002)	(0.001)
N	135–141	571–586

Note: Standard errors are in parentheses.
**Significant at the 5 percent level.
*Significant at the 10 percent level.

While the impact of regulation generally appears to be much smaller here than we found before, a direct comparison is not possible given the aforementioned difference in the way operating costs are measured. We therefore instead turn to a comparison of costs between plants in attainment and nonattainment areas, using the three cost measures that we do have here. In particular, we run simple ordinary least squares (OLS) regressions, where our dependent variable is a plant's ratio of air pollution abatement expenditures (capital investment, labor costs, or operating costs) to total plant expenditures (in those same categories). Our explanatory variables include plant output, plant age, a multiunit dummy, and county ozone nonattainment status. The nonattainment coefficients from these regressions are reported in table 5.8. Only relative capital expenditure on air pollution abatement in industrial organic chemicals is significantly higher in nonattainment areas than it is in attainment areas, with a difference of almost 4 percent. All the other nonattainment coefficients are statistically insignificant and very close to 0.

These estimates obviously suggest much lower regulatory costs than we were finding with our cost-function approach in the previous section. This might be evidence of the long-held belief that PACE misses a substantial portion of environmental expenditures. The potential limitations of this survey (noted before) would obviously understate costs much more for plants in nonattainment areas than for those in attainment areas, narrowing the estimated gap between the two groups. Our earlier caveats, regarding possible self-selection as well as voluntary environmental expenditures, also apply here—serving to narrow this gap even more. And we note again that our results in table 5.8 do not (because we really cannot) distinguish regulatory effects by age of plant. Given our previous results, indicating that young plants are most affected by regulation, and given that the PACE sample is actually weighted toward older plants, differences in cost estimates may also (to some extent) be due to differences in sample composition. That this potentially heavily affected group is underrepresented in PACE obviously also has potential implications for the aggregate statistics published from this survey. The results here are suggestive, but much more work is needed in this area.

5.8   Conclusion

This paper examines the effects that air quality regulation has had on the size and timing of plant investment in two particular industries, and the cost such regulation imposes on firms in these industries. In the industry with high relative average capital assets, we find that new, regulated plants start out much larger than their unregulated counterparts, but then do not invest as much, such that after 10 years capital stocks of regulated plants are in fact smaller. This is consistent with our previous findings and highlights the substantial fixed costs involved in negotiating expansion permits, the benefits of preserving plants' grandfathered status, and the desire to keep plants small (or even downsize) in an environment where the amount of regulatory attention is often correlated with plant size. In terms of quantifying the costs of air quality regulation, our basic results show that heavily regulated plants indeed face higher production costs than their less-regulated counterparts. This is particularly true for younger plants, which is consistent with the notion that regulation is most burdensome for new (rather than for existing) plants. Unregulated plants, however, also appear to be affected by regulation (or at least the threat of regulation); we find that they produce at levels far short of the levels that minimize average total costs. This, again, demonstrates the role that plant size plays in regulatory efforts.

References

Bartelsman, Eric J., Randy A. Becker, and Wayne B. Gray. 1998. *NBER-CES manufacturing industry database.* Available at http://www.nber.org/nberces/.

Bartik, Timothy J. 1988. The effects of environmental regulation on business location in the United States. *Growth and Change* 19 (3): 22–44.

Becker, Randy A. 1998. *The effects of environmental regulation on firm behavior.* Ph.D. diss., Brown University.

Becker, Randy A., and J. Vernon Henderson. 1997. Effects of air quality regulation on decisions of firms in polluting industries. NBER Working Paper no. 6160. Cambridge, Mass.: National Bureau of Economic Research.

———. 2000. Effects of air quality regulation on polluting industries. *Journal of Political Economy* 108 (2): 379–421.

Gray, Wayne B. 1997. Manufacturing plant location: Does state pollution regulation matter? NBER Working Paper no. 5880. Cambridge, Mass.: National Bureau of Economic Research.

Henderson, J. Vernon. 1996. Effects of air quality regulation. *American Economic Review* 86 (4): 789–813.

Jaffe, Adam B., Steven R. Peterson, Paul R. Portney, and Robert N. Stavins. 1995. Environmental regulation and the competitiveness of U.S. manufacturing: What does the evidence tell us? *Journal of Economic Literature* 33 (1): 132–63.

Jorgenson, Dale W. 1990. Productivity and economic growth. In *Fifty years of economic measurement,* ed. Ernst R. Berndt and Jack E. Triplett, 19–118. NBER Studies in Income and Wealth, vol. 54. Chicago: University of Chicago Press.

Kahn, Matthew E. 1994. Regulation's impact on county pollution and manufacturing growth in the 1980s. Columbia University. Mimeo.

Levinson, Arik. 1996. Environmental regulation and manufacturers' location choices: Evidence from the Census of Manufactures. *Journal of Public Economics* 62 (1): 5–29.

McConnell, Virginia D., and Robert M. Schwab. 1990. The impact of environmental regulation on industry location decisions: The motor vehicle industry. *Land Economics* 66 (1): 67–81.

U.S. Census Bureau. 1994. *Pollution abatement costs and expenditures, 1992.* Washington, D.C.: U.S. Government Printing Office.

Comment   Aart de Zeeuw

Randy Becker and Vernon Henderson are producing a series of papers on the effects of air quality regulation. This is clearly an interesting topic and the empirical results on the basis of good data sets are an especially important contribution to the literature.

The first sentence of this paper reads "An ongoing debate in the United States concerns the costs environmental regulations impose on industry," and then the paper explores some of the costs associated with air quality regulation. It might be obvious, but we must not forget that the purpose of environmental regulation is to prevent further degradation of the natural environment. It is important to know the effects and the costs imposed on industry, but it is equally important to know the benefits of improving the environment in order to be able to consider the trade-offs between the two. This paper focuses on one side of the story, but we need the other side as well in order to be able to evaluate environmental policy.

On the other hand, even if one concentrates on one side of the story, the way environmental policy is implemented can be discussed and evaluated. This aspect is somewhat missing in this paper and will be the subject of one of the sections in these comments.

A second general comment is that costs imposed on industry by environ-

Aart de Zeeuw is professor of environmental economics at Tilburg University, The Netherlands, and director of graduate studies at the CentER for Economic Research.

mental regulation are often discussed in the context of the competitiveness of that industry in the world market. It is interesting to note that the distinction between counties that are in attainment and that are not in attainment in the United States resembles the distinction between countries that have a lax and that have a strict environmental policy, but the fundamental difference is, of course, that countries do not operate under a supranational government, while counties have to comply with federal regulations. What we can learn from this will be further elaborated in another section of these comments. A conclusion follows the discussion.

Policy Implementation

The aim of the federal government is to reduce ground-level ozone O_3 by reducing volatile organic compounds (VOCs) and nitrogen oxides (NO_X), the chemical precursors to ozone, by setting standards. Becker and Henderson describe in section 5.2 of their paper how the policy is implemented. Each year, each county in the United States is designated as either being in or out of attainment of the National Ambient Air Quality Standards. Counties that are in nonattainment are, by law, required to bring themselves into attainment. This leads to the following differences. In nonattainment counties, new plants and existing ones undertaking major expansion and/or renewal are subject to lowest achievable emission rates (LAER), requiring the installation of the cleanest available equipment without regard to cost. Other existing plants are required to install reasonably available control technology (RACT). In contrast, in attainment counties, only new plants and only those with the potential to emit over 100 tons per year are subject to any regulation. These firms are required to install best available control technology (BACT), which is negotiable and sensitive to the economic impact on the firm.

This way of implementing the policy gives certain incentives to the industry involved. In fact, it does not seem difficult to predict what will happen, and this type of analysis is somewhat missing from the paper, which immediately jumps to empirical conclusions. The authors probably had some hypotheses in mind before starting the empirical work, but it helps, in my view, to understand the results when the mechanisms have been discussed first. In nonattainment counties, the incentives of the regulation are to start big in order to prevent later expansions from being subject to the LAER standard of that time, and to extend life in order to take advantage of the RACT standard on existing plants. In attainment counties, the incentive is to start small in order to prevent regulation according to the BACT standard. Given the different regulation of new plants, it is also to be expected that plant birth shifts from nonattainment to attainment counties. This last effect will reinforce the extension of the life of firms in nonattainment counties because these firms will face less competition both on local markets and in their struggle with the regulators. The empirical conclusions of Becker and Henderson are exactly in line with this, and in the interpretation of their results they come up, of course, with the same arguments. They conclude that plant birth in this industry has shifted, that the survival rate has increased in nonattainment counties, and that plants in nonattainment counties do more up-front with less subsequent investment.

As stated before, the paper does not discuss the benefits of this regulation in terms of an improved natural environment. It is clear, however, that ground-level ozone is reduced in nonattainment counties, but probably increased in attainment counties. This is fine if the damage of ground-level ozone can be characterized as of the critical-load type, which means that concentrations below a certain level are harmless. If the increase in attainment counties does not exceed the critical load, no harm is done. Otherwise, the regulation may not have been as successful as will be concluded from focusing only on the nonattainment counties.

Counties and Countries

Environmental regulations impose costs on industry. This statement will be subscribed to in general, although it was challenged by Porter (1991). His main arguments are that firms might also be triggered to reconsider their operation and to move closer to their efficiency frontier, and that firms might get a first-mover advantage when regulations also get stricter in the countries of their competitors. This discussion took place in the international context where countries that start out to internalize environmental externalities are confronted with countries that lag behind in environmental regulation. Abstracting from Porter's arguments, firms in the latter countries have lower costs and, therefore, a competitive advantage over firms in the regulated countries, driving these firms out of the world market. Alternatively, firms relocate from more regulated countries to less regulated countries. In order to counter these effects, countries have an incentive to weaken environmental regulation and lower the costs imposed on their home industry, sometimes referred to as ecological dumping. In the absence of a supranational government, countries have to enter the very complex process of international environmental agreements in order to try to get closer to the first-best solution.

Counties in nonattainment and in attainment can be compared to countries with strict and lax environmental policies, respectively. Indeed, through the shift in plant birth, the regulated industry will gradually relocate from nonattainment to attainment counties. In this case, however, the U.S. federal government has less to worry about because the industry stays within its national borders. Of course, the process leads to adjustment costs, but the federal government has no incentive for ecological dumping. The local governments of counties in nonattainment have this incentive, but they are subject to federal law and face harsh federal sanctions if they do not try to come into attainment.

The reason I make this comparison is to stress the point that here costs imposed on industry by environmental regulation are a consequence of internalizing environmental externalities and are not a concern in the context of strategic motives for the government. It is important, of course, to estimate these costs, the main aim of the paper by Becker and Henderson, but only to be able to make a proper cost-benefit analysis for setting the proper standards. It is interesting to note that the argument put forward by Becker and Henderson to explain why plants in attainment counties install cleaner equipment than needed comes close to one of the arguments in the international discussion. They argue that nonattainment counties must have a large number of polluting plants. Because a considerable proportion of these plants remains there and have to comply with the stricter standard, new technology will be developed that becomes the standard in this industry. Plants in attainment counties therefore install the same equipment, driven by technological development and not by environmental regulation. This is exactly one of the arguments put forward by Jaffe et al. (1995) to explain why, in their survey of empirical results, the competitive advantage of laxer environmental regulation does not lead to as much international relocation as was generally expected.

Conclusion

These comments are intended as a discussion of the paper in a certain context, but not to criticize the paper. In fact, the paper is well written and contains important empirical results, using new and extensive data sets, and arguments to explain the results. It is one in a series of papers; this one is mainly concerned with the estimation of cost differences between counties in nonattainment and counties in attainment. The approach is to estimate quadratic cost curves and to compare the minimal costs. The result is that these costs are higher in nonattainment counties. The paper considers two industries, industrial organic chemicals and miscellaneous plastic products, and in all the analyses the results for the first industry are somewhat stronger than for the second one. The authors also compare their indirect method with a direct method using the PACE survey, which leads to lower estimates of costs of regulation. The authors argue that this might be evidence of the long-held belief that the PACE survey misses a substantial portion of environmental expenditures.

One thing remains a bit puzzling. Becker and Henderson note that the difference between actual and cost-minimizing output in attainment areas is absolutely enormous. They give as a reason that plants in attainment areas remain small in order not to attract attention from regulators. Earlier in the paper, however, the authors argue that plants in attainment areas only start small in order to prevent regulation, which only applies to new and big plants. It is not so clear now what the final conclusion is regarding the development of plants over time. By the same token, Becker and Henderson seem to attribute downsizing of firms in nonattainment areas to the fear of regulatory scrutiny, whereas here one would think that downsizing is mainly due to higher costs. Anyway, the research program is apparently not finished and a full picture might appear in a next paper in this interesting series. To conclude, Becker and Henderson can be complimented on a nice contribution to an important topic.

References

Jaffe, Adam B., Steven R. Peterson, Paul R. Portney, and Robert N. Stavins. 1995. Environmental regulations and the competitiveness of U.S. manufacturing: What does the evidence tell us? *Journal of Economic Literature* 33 (1): 132–63.
Porter, Michael. 1991. America's green strategy. *Scientific American* 264 (4): 96.

6

International Factor Movements, Environmental Policy, and Double Dividends

Michael Rauscher

6.1 The Issues

The main purpose of environmental taxes is "to get the prices right." Emissions taxes signal the scarcity of environmental resources to their users and thus help to internalize negative environmental externalities. The idea dates back to Pigou (1920) and was taken up by modern environmental economics in the late 1960s; since that time environmental economists have tried to persuade politicians to introduce this incentive-compatible instrument as an efficiency-enhancing substitute of the still predominant command and control approach. Apparently, success has been limited. Environmental taxes are still the exception rather than the rule in environmental regulation. This has led economists to search for other arguments in favor of the introduction of environmental taxes. Are there additional dividends to be gained from an environmental tax reform? Such gains may exist, for green taxes generate tax revenues that may be used to alleviate other problems of the economic system. Two problem areas come to mind. The first one is the tax system, which in most countries is distortional. Green tax revenues may be used to reduce other taxes that distort the allocation of factors of production. The second problem area is the imperfection of the labor market, where union bargaining and efficiency wages result in a downward rigidity of real wages and cause unemployment.

Michael Rauscher is professor of international economics at Rostock University.

The author is indebted to David Bradford, Lans Bovenberg, Don Fullerton, conference participants, and participants of the Rostock Economics Research Seminar. The usual disclaimer applies.

Thus, so the argument goes, green taxes can be used to alleviate the tax burden on labor and help to solve unemployment problems.

The efficiency of the tax system is a primary concern of the theory of public finance. The basic idea that green taxes may be used to create a more efficient tax system goes back to Tullock (1967). It has been reinforced by Pearce (1991) and Repetto (1992). Bovenberg and de Mooij (1994) challenge this view and show that the existence of other distortional taxes does not necessarily imply that green taxes should overinternalize environmental damages. On the contrary, it is more likely that an optimal environmental tax does not completely internalize the environmental damage. As Fullerton (1997) and Schöb (1997) show, this result decisively depends on the normalization of goods and taxes, and using other ways of normalizing one arrives at contradictory conclusions. For overviews summarizing the public finance aspects of the double dividend literature, see Goulder (1995, 1997). Fullerton and Metcalf (1998) extend the standard approach by allowing for rents that are appropriated by the private sector if the government chooses the command and control approach to environmental policy. In this case, less tax revenue is generated and the double dividend becomes even more unlikely.

In Europe, emphasis has been placed less on the efficiency of the tax system than on the reduction of unemployment. An early example is Binswanger et al. (1983). Recent contributions to this literature are Schneider (1997), Koskela, Schöb, and Sinn (1998), and Holmlund and Kolm (2000). The present paper looks at unemployment problems as well. I consider an open economy endowed with fixed quantities of labor and capital. Capital is mobile internationally, whereas labor is not. The environmental regulation uses the command and control approach, which generates rents that are appropriated by the private sector. This element of the model resembles the approach adopted by Fullerton and Metcalf (1998). However, in the present paper, this rent is linked to the mobile factor of production. This is an implicit subsidy, and it generates another distortion in the economy.

The paper is organized as follows. Section 6.2 presents the basic assumptions of the model. Then comparative static results will be derived and discussed. The next section is devoted to a revenue-neutral environmental tax reform. Afterward, I consider its welfare and employment effects. Then optimal environmental policies in first-best and second-best situations are discussed. Some final remarks conclude the paper.

6.2   A Basic Model of Interjurisdictional Competition

Consider a small open economy. It produces an aggregate good that can be used as an input and as a consumption good. There are one fixed and three variable factors of production. The fixed factor can be land and it is not considered explicitly in the model. Labor, L, is the immobile factor of production, which is tied to the jurisdiction under consideration. Labor supply is fixed at a level L_0. However, the wage rate is fixed as well, and I assume that there is unemployment. Capital, K, is the mobile factor of production and its domestic supply, K_0, is given and constant. The third factor of production is an environmental resource, E. Assume that E is used up during the production process. For example, clean water enters the production process and contaminated water is discharged into the ambient. Thus, E, which measures the use of environmental resources, can also be interpreted as being a measure of emissions and contamination. Let the production function, $F(\cdot, \cdot, \cdot)$, exhibit diminishing returns to scale in the three variable factors of production. Thus, output, Q, is

$$(1) \qquad\qquad Q = F(L, K, E).$$

Let the production function have the usual properties, that is, positive first derivatives, negative second derivatives, positive second principal minors, and a negative determinant of the Hessian determinant. Moreover, we assume that

$$(2) \quad F_{KL} \equiv \frac{\partial^2 F}{\partial K \partial L} > 0, \quad F_{KE} \equiv \frac{\partial^2 F}{\partial K \partial E} > 0, \quad F_{LE} \equiv \frac{\partial^2 F}{\partial L \partial E} > 0,$$

where the arguments of the production function have been omitted for convenience. This implies that the factors of production are technological substitutes. This kind of production function has become a standard tool for the modeling of international factor movements. See MacDougall (1960), Kemp (1964), and Ruffin (1984) for surveys. Applications to environmental policy can be found in Oates and Schwab (1988), Wang (1995), and Rauscher (1997, chap. 3). An example for such a production function is the Cobb-Douglas function,

$$(1') \qquad\qquad Q = L^\alpha K^\beta E^\gamma.$$

Reasonable guesses for these parameters are $\alpha = 0.6$, $\beta = 0.3$, and $\gamma < 0.1$.

Because there are three factors of production, there are three factor markets. Since there is only one good, there are no relative prices and the factors are remunerated in terms of this commodity. Assume that the demand sides of the markets are competitive. Then the factor is employed up to the point where its marginal productivity equals the factor price plus the rate of factor taxation.

The environmental resource market is governed by

$$(3a) \qquad\qquad F_E = t^E,$$

where t^E is the emissions tax rate measured in units of the aggregate good per unit of emissions. Note that this specification includes the possibility of a tradable-pollution-permits scheme or a command and control approach to environmental policy. In the case of a permits scheme, it is the market price of the right to discharge one unit of the pollutant. In the case of command and control, it is the implicit or shadow price of the environmental resource.

In the labor market, the net wage rate w is exogenously given and larger than the market-clearing wage rate. Let t^L be the labor income-tax rate. Labor demand is then determined by

$$(3b) \qquad\qquad F_L = w(1 + t^L).$$

Since the wage rate is fixed in this model, there may be unemployment. Of course, such an approach to modeling unemployment is simple—if not simplistic. A more realistic model would look at the process of wage bargaining between employers and the unions explicitly. In this case, the government would be in the strategic position to influence the game between the employers and the unions as a first mover. However, I wish to concentrate on the pure employment effects and to avoid complications stemming from solving multiple-stage games.

The capital market is also special in this model. There is an additional element of capital remuneration that is not present in the standard model of international factor movements. Consider a situation where command and control is the government's approach to environmental policy. Thus, firms are given the right to pollute up to an upper limit for free. In this case, the government has no environmental tax revenue, but the regulated private sector appropriates the rent, which is due to the scarcity of environmental resources. This rent equals the potential emissions tax revenue, $t^E E$. Who in the private sector is able to appropriate this rent? It is often assumed that it is redistributed to the factors of production in a lump-sum fashion. Then allocation would not be affected and market-based environmental policies and command and control would be equivalent. This approach is not followed here. Instead, we assume that the environmental scarcity rent accrues to those who employ capital as a factor of production. The underlying idea is that there is a kind of environmental regulation that fixes emissions per unit of investment. So, the right to pollute is tied to the capital stock. As an example, consider the case of electric power generation, where emissions are often linked to installed capacity, which in turn is closely related to the capital stock invested;[1] see Sinn (1994) and Wellisch (1995). Thus, for each unit of capital, a share $(t^E E)/K$ of this rent

---

1. An alternative way of justifying this approach is to extend the model to more than one good and to allow for international trade. If command and control is applied to the capital-intensive sector of the economy, then the owners of the enterprises that are active in this sector appropriate the resource scarcity rent. This has effects similar to an output subsidy. According to the Stolper-Samuelson theorem, the remuneration of the intensively used factor tends to rise.

can be appropriated. To make this a bit more general, we will consider a situation where the government distributes a share $(1 - s)$ of the emissions rights for free in a command and control framework and auctions s in a market for tradable pollution permits. The parameter s will be used for comparative statics in what follows. The market-clearing condition for the capital market is

$$(3c) \qquad F_K + (1 - s)\frac{t^E E}{K} = r(1 + t^K).$$

The net tax revenue of the government equals the sum of the revenues from all kinds of taxes minus the unemployment benefit. The unemployment benefit is determined by the benefit rate, u, and the level of unemployment, $L_0 - L$, where L_0 is the exogenous supply of labor. We only consider situations where $L_0 > L$. A wage rate lower than the equilibrium wage rate, where labor demand is constrained by labor supply, is ruled out.

$$(4) \qquad G = t^L wL + t^K rK + st^E E - u(L_0 - L).$$

In order to limit the number of possible scenarios in the remainder of the paper, I impose some additional assumptions. The first assumption is that the tax revenues from payroll taxes and from emissions taxes are both increasing in the tax rates. In other words, I consider situations to the left of the maximum of the Laffer curve:

$$(5) \qquad \frac{d(t^L wL)}{dt^L} > 0, \quad \frac{d(t^E E)}{dt^E} > 0.$$

This is the assumption underlying the unemployment version of the double dividend hypothesis as stated by Binswanger et al. (1983) and many others afterward. Green taxes generate additional tax revenue, and this tax revenue can be used to reduce labor taxes, which then leads to higher employment. The first part of this argument is probably correct. Currently, most emissions tax rates are so low that tax increases lead to less than proportional reductions in emissions and thus indeed generate additional tax revenue. This tax revenue can be used to reduce the tax revenue from labor taxes. Under the assumption made above, the reduction in tax revenue can be achieved by a reduction of labor taxes, which leads to less unemployment. If labor demand were very elastic, we would obtain increased tax revenues as a result of reduced labor taxes. This would be the third dividend of a green tax reform. However, tax revenue neutrality could only be achieved by higher taxes and less unemployment. I assume that this is not the case here.

The second assumption is made about the effect of changes in the capital stock on the environmental scarcity rent per unit of capital. Let this rent be denoted by R. Using equation (3a), we obtain

(6)
$$R = \frac{EF_E}{K},$$

and

(7)
$$R_K = \frac{EKF_{KE} - EF_E}{K^2} = \frac{R}{K}\left(\frac{KF_{KE}}{F_E} - 1\right) < 0.$$

R_K consists of two components. On the one hand, the increase in the capital stock raises the marginal productivity of emissions and, thus, the scarcity rent. On the other hand, this rent is distributed to a larger capital stock such that the rent per unit of capital is reduced. I assume that the second effect dominates the first. This assumption is satisfied in the case of a Cobb-Douglas production function since the elasticity, which is the first term in brackets on the right-hand side, is always less than one. This concludes the exposition of the model.

6.3   Comparative Statics

What are the effects of changes in environmental and tax policies on the employment of labor and on the capital stock in the economy? We eliminate t^E and rewrite equation (3c) such that we obtain the following system of factor market equations:

(3b)
$$F_L = w(1 + t^L),$$

(3c′)
$$F_K + (1 - s)\frac{F_E E}{K} = r(1 + t^K).$$

Total differentiation yields

(8)
$$\begin{pmatrix} F_{LL} & F_{LK} \\ F_{LK} + (1 - s)R_L & F_{KK} + (1 - s)R_K \end{pmatrix}\begin{pmatrix} dL \\ dK \end{pmatrix}$$

$$= \begin{pmatrix} w & 0 & -F_{LE} & 0 \\ 0 & r & -F_{KE} - (1 - s)R_E & R \end{pmatrix}\begin{pmatrix} dt^L \\ dt^K \\ dE \\ ds \end{pmatrix}.$$

The sign of the determinant of the matrix on the left-hand side is ambiguous. This is due to the term $(1 - s)R_L$ in the second-row first-column element of the matrix. All other terms in this matrix are such that the determinant is positive. In the case of a Cobb-Douglas function, these effects dominate and the determinant turns out to be

$$\Delta = (1 - \alpha - \beta)[\alpha\beta + (1 - s)\alpha\gamma]\frac{Q^2}{K^2 L^2} > 0.$$

Assume that this can be generalized. Then the following comparative static results can be derived:

- An increase in the payroll tax leads to less employment, since the gross wage that has to be paid by the employers is raised. This reduction in labor demand reduces the marginal productivity of capital. Capital leaves the country.
- An increase in the tax on mobile capital induces capital flight. With less capital, the marginal productivity is smaller and labor demand is reduced. Thus, unemployment is increased.
- The same effect shows up if s is increased, that is, if there is a move from command and control toward incentives in environmental policy. The reason is that the share of the resource rent going to the mobile factor of production is reduced. This means that the implicit subsidization of the employment of the mobile factor is diminished. A reduction of a subsidy must have the same effect as the increase of a factor tax.
- The effects of an emissions reduction (i.e., a tighter environmental policy) are ambiguous. There are two components of the effect. The first is the standard effect that is expected from stricter environmental standards: Employment is reduced and capital is driven out of the country, since the productivities of the other factors of production are reduced. Here, however, there is also a second effect. A tighter environmental policy implies a higher (implicit) environmental tax rate. Due to the Laffer-curve assumption, this has the effect that the (implicit) tax revenue (i.e., the environmental scarcity rent) is increased. Thus, the implicit subsidy going to capital is increased. Thus, demand for capital rises, and the marginal productivity of labor is increased. Employment rises.

6.4   A Tax-Revenue-Neutral Environmental Policy Reform

Let us consider an environmental policy reform that keeps the tax revenue constant. What do I mean by tax revenue neutrality here? Formally, tax revenue neutrality is defined as a constant level of G even though some tax rates are changed. Note that G is not the total tax revenue but the tax revenue minus unemployment benefits. So, the government aims at having a constant budget after the unemployment benefits have been paid. In what follows, it is assumed that the government uses G to provide public consumption goods. Tighter environmental standards lead to an increased emissions tax revenue and the additional revenue can be used to reduce distortional labor taxes. However, one should note that there are also some indirect effects on the tax revenue since environmental policies have an impact on factor demand and, thus, on other components of the tax revenue as well. This will be discussed later.

Another issue must be mentioned as well. In a recent paper, Kaplow (1996) emphasizes that income taxes can be used as a source of financing public goods without the generation of distortions. Applied to environmental taxation and green tax reforms, this implies that income (labor) taxation schemes are possible such that the optimal environmental tax is the Pigouvian tax, which equals marginal environmental damages. However, in this model, matters are different (as will soon be seen) since, first, the labor supply is fixed and, therefore, changes in labor taxes have no impact on the factor allocation via the supply side of the labor market; and, second, there are other distortions in this model that imply modifications of the Pigouvian rule for emissions tax rates.

The following changes in environmental policy are considered: (1) a move from command and control toward environmental taxes or tradable pollution permits; (2) a tighter environmental policy, that is, a reduction in total emissions; and (3) a combination of these instruments such that sE remains unchanged, that is, the stricter environmental policy is achieved by reducing the number of pollution permits that are given out for free, whereas the number of pollution permits auctioned by the governments remains unchanged. Two objectives will be considered: the reduction of unemployment and the increase of social welfare.

First consider the employment objective. A move from command and control has direct and indirect effects on the demand for labor. The direct effect is a reduction in labor demand, since labor productivity falls as a consequence of tighter environmental standards. There are three types of effects on tax revenue. There is a direct effect: A move from command and control toward pollution permits generates additional tax revenue. There are three first-order indirect effects. The first affects the tax revenue from capital taxation; since capital leaves the country when its indirect subsidization is reduced, the capital tax revenue is reduced. Second, if capital leaves the country, labor productivity declines, employment shrinks, and the revenue from payroll taxes is reduced. Third, due to the unemployment benefits that the government has to pay out of its budget, if employment is reduced, this exerts an additional pressure on the government budget.[2] Finally, there are second-order indirect effects. If employment and capital are reduced, then the marginal productivity of emissions declines. There is less demand for environmental resources and the emissions tax revenue is reduced.

The direct effect and the indirect effects can be compared easily. In order to do this, the tax revenue is differentiated with respect to s. This yields

---

2. In the case of an earmarked tax on labor, an instrument that is used in several countries to finance unemployment benefits, the tax rate must increase since fewer employees have to finance more of the unemployed. This generates additional costs of employing people and additional workers tend to be dismissed.

$$(9) \quad \frac{dG}{ds} = F_E E + \left[t^K r \frac{dK}{ds} + (t^L + u)w \frac{dL}{ds} \right] + sE \left[F_{KE} \frac{dK}{ds} + F_{LE} \frac{dL}{ds} \right],$$

where dK/ds and dL/ds follow from equation (8). The first term in brackets on the right-hand side contains the first-order indirect effect. It can be seen that the negative effects dominate if the tax rates are large in the initial situation. This is not particularly surprising. If the tax rates are large, then the impact of a decline in factor demand on the tax revenue is substantial. The second term in brackets contains the second-order indirect effects on the emissions tax revenue.

In a second step, consider a reduction in emissions at a constant level of s. The ratio of pollution permits allocated by command and control to the permits allocated via markets stays constant. The total effect on tax revenue is not clear. Again, this can be decomposed into direct and indirect effects. The direct effect of the tighter environmental policy is an increase in the tax revenue due to the assumption that the economy is on the up-ward-sloping part of the Laffer curve. There are two possible first-order indirect effects. On the one hand, tighter environmental policies reduce the productivities of the other factors; this implies declining factor demands and declining tax revenues. On the other hand, the increase in the tax revenue is equivalent to an increase in the scarcity rent and, therefore, in the hidden subsidy; this implies increasing factor demands and increasing tax revenues. The second-order indirect effects amplify the first-order effects. If unemployment increases and capital leaves the country, then the negative effect on the tax revenue is aggravated by the decline in the emissions tax rate.

Algebraically, we obtain

$$(10) \quad \frac{dG}{dE} = s \frac{(F_E E)}{dE} + \left[t^K r \frac{dK}{dE} + (t^L + u)w \frac{dL}{dE} \right]$$
$$+ sE \left[F_{KE} \frac{dK}{dE} - F_{LE} \frac{dL}{dE} \right].$$

The importance of the first-order indirect effects again depends on the initial tax rates as in the case of a movement from command and control toward market instruments.

An interesting side issue should be mentioned. Even if there is pure command and control regulation, tighter environmental policies may have positive effects on tax revenues. If the resource-rent effect of the strict policy dominates, it is possible that capital moves into the country, that as a consequence employment rises, and that, therefore, the tax base is expanded. Thus, it becomes possible to reduce labor taxes and increase employment even further.

The combination of a reduction in emissions and an increase in s yields ambiguous results again. The direct effect on the tax revenue is positive since the scarcity of the environmental resource is increased and, thus, so is the emissions tax revenue. Moreover, the move from command and control to market-oriented environmental policy makes producers pay taxes for a larger share of their emissions. The first-order indirect effects are ambiguous again, but they are less likely to be positive than in the case of a mere tightening of environmental regulation since the impact of the increase in s is unambiguously negative. The second-order indirect effects depend on the first-order effects and reinforce them.

The ambiguity of the results of this model has obvious and intuitive explanations. Unfortunately, despite its simplicity the model is already so complex that it is difficult to come up with interpretable algebraic conditions for the positive and negative effects of environmental tax reform. Comparing the results to those derived from closed-economy models, however, one can make the following statements. (1) In a standard model of an open economy without indirect subsidization of the mobile factor, a double dividend from tighter environmental policy or from the introduction of market instruments is less likely than in a closed economy. Because these policy measures induce capital flight, this reduces the tax base and the tax revenue. Moreover, there are repercussions on the labor market that increase unemployment. Finally, second-order effects on the emissions tax revenue must be taken into account and they amplify the first-order indirect effects. (2) If this model is extended by the introduction of an environmental scarcity rent associated with the mobile factor, these negative effects are mitigated or even reversed.

Now consider welfare effects of the environmental policy reform. For simplicity, I assume a social-welfare function that is quasi-linear and additively separable in its arguments. This causes no loss of generality here, since I am not interested in doing comparative statics with respect to welfare maxima, but just looking at welfare effects and, later on, at optima. For this purpose, interaction terms of different arguments of the utility function do not provide additional insights.

Let the welfare function be defined as

(11) $$W = Y + v(G) - d(E),$$

where Y denotes consumable income, $v(\cdot)$ is an increasing and strictly concave utility function evaluating the benefits accruing to the residents of the jurisdiction from the consumption of the public good supplied by the government, and $d(\cdot)$ is an increasing and convex function measuring the environmental damage. The net government revenue available for the generation of the public consumption good was defined in equation (4) as

$$G = t^L wL + t^K rK + st^E E - u(L_0 - L),$$

and the consumable income is output minus interest payments to foreigners minus the tax revenue needed for the generation of the public consumption good:

$$(12) \qquad Y = f(K, L, E) - r(K - K_0) - G.$$

Here the implicit assumption is that all residents have utility functions that are linear in income and that the total utility derived from income is the sum of all individual utilities. This implies, for instance, that the unemployed, who have lower incomes than those who have jobs, have the same marginal utility as the employed people.

Now consider a tax-revenue-neutral environmental policy reform. Revenue neutrality implies that the change in environmental policy is accompanied by a change in labor taxation such that the net tax revenue available for the provision of public goods remains unchanged. This change in labor taxation may be achieved by a change in general payroll taxes. The corrections that are necessary to achieve revenue neutrality have the same effects on the allocation of labor and capital. Algebraically, this follows from equations (8) and (11). The economic intuition is that tax money is fungible, and if the tax base is the same in both cases, then the allocation effects must be the same as well.

Let us now denote by (dK/ds) the total capital-stock effects and by (dL/ds) the labor-employment effects of the environmental tax reform taking the shape of a move away from command and control. That is, these terms include the adjustments that are necessary to achieve the balanced budget.[3] The welfare effect is

$$(13) \qquad \frac{dW}{ds} = [t^K - (1 - s)R]\frac{dK}{ds} + F_L\frac{dL}{ds}.$$

There are two components. The first one is due to the induced change in the capital stock. If the capital stock is increased, this is positive for the economy if the tax rate is larger than the rate of subsidization, that is, if capital is a tax base and if some additional tax revenue can be generated from attracting foreign capital. If the capital tax rate is low and the environmental resource rent is large, this effect may be turned around. It is beneficial to lose a factor that is subsidized. But under which circumstance does the country lose this factor? One effect of the environmental policy reform is that capital leaves the country because the implicit subsidy is reduced. However, if the reform generates enough additional tax revenue such that the labor tax rate can be reduced, employment rises and this helps the jurisdiction to attract foreign capital. Increasing the labor force is always positive from a welfare point of view. Unemployed labor is idle

3. Note that this implies that dK/ds and dL/ds differ from the values that follow from the simple comparative statics of equation (8).

and does not contribute to the general well-being. So, the marginal gain from increasing the employment by one unit is the marginal productivity of labor.[4]

Similar considerations apply if the environmental policy reform consists of an emissions reduction. The welfare effect is

$$(14) \qquad \frac{dW}{dE} = [t^K - (1 - s)R]\frac{dK}{dE} + F_L\frac{dL}{dE} + F_E - d'.$$

The interpretation of the first two terms corresponds to that of the previous case. There is a tax-base effect for capital and a benefit from reduced unemployment. The last two terms on the right-hand side of equation (14) denote the main dividend of environmental policy. They measure the difference between the marginal benefit of emissions (more output) and their marginal damage (environmental disruption). If the environmental policy is lax in the initial situation, then there is a substantial reduction in environmental damage and a relatively small reduction in output: Stricter environmental standards have a positive welfare effect. If the environmental policy is relatively strict already, then this term denotes a negative welfare effect of even tighter environmental policy.

In the case of a policy reform that combines a reduction in emissions with a move from command and control, the welfare effect is straightforward. Note that $ds/dE = s/E$ if sE is constant. This implies

$$(15) \qquad \left.\frac{dw}{dE}\right|_{sE=\text{const}} = [t^K - (1 - s)R]\left(\frac{dK}{dE} - \frac{s}{E}\frac{dK}{ds}\right) + F_L\left(\frac{dL}{dE} - \frac{s}{E}\frac{dL}{ds}\right)$$
$$+ (F_E - d').$$

Again, there are three effects: a tax-base effect, a labor-market effect, and the environmental dividend. The reduction in emissions and the increase in s may have opposing effects on capital and labor.

6.5   Optimal Taxation and Interjurisdictional Competition

Having derived the effects of a revenue-neutral tax reform, we can now proceed by asking the normative question of what an environmental policy should look like. This will be done in a first step for the first-best situation, when there are no restrictions on the policymaker's choice set; all tax instruments are available. Afterward, I consider restrictions on the choice set and second-best taxation. Also, issues of interjurisdictional competition are addressed. As in other tax competition models, we employ the

---

4. Note that this term vanishes if there is no unemployment. Algebraically, this can be seen if a nonnegativity constraint for the upper limit to employment is introduced. The corresponding Kuhn-Tucker multiplier is 0 in the case of unemployment and jumps to a positive value if labor demand hits labor supply.

helpful assumption that the jurisdictions are identical. This is done by Zodrow and Mieszkowski (1986) for a simple tax-competition model and by Rauscher (1999) for a variety of tax-competition models involving environmental taxes and different kinds of factor- and goods-market distortions. The advantage of the identity assumption is that situations with and without tax competition can be compared easily since the mobile factor does not move ex post. Ex ante, however, each jurisdiction has an incentive to influence the location of the mobile factor. This leads to fiscal and other externalities and to suboptimal taxation.

I will start with the first-best situation. All kinds of taxes are available to the policymaker including a lump-sum tax, Θ. Taking all restrictions into consideration, a benevolent policymaker maximizes

$$(16) \quad W = f(K, L, E) - r(K - K_0) - G + v(G) - d(E),$$

subject to

$$(17) \quad G = \Theta + t^L wL + t^K rK + st^E E - u(L_0 - L),$$

$$(18) \quad L_0 - L \geq 0.$$

Maximization with respect to the lump-sum tax rate yields the result that the marginal utility of the public good in the optimum equals unity. This is explained by the fact that the marginal cost of providing it is one unit of the aggregate good. Moreover, we have the optimal tax rates and other policy parameters:

$$(19) \quad t^K = (1 - s)R,$$

$$(20) \quad t^E = d',$$

$$(21) \quad t^L = F_L(K, L_0, E) - w.$$

The mobile factor should neither be taxed nor subsidized. Thus a combination of the capital tax rate and the command and control parameter must be chosen such that their impacts on the capital market just cancel out. This result corresponds to the well-known zero-taxation property in the international tax competition literature. The environmental tax rate equals the environmental damage. The labor tax rate is negative: labor is subsidized to the point where the labor demand equals the total labor force. Since there is no unemployment, the allocation is not affected by the design of the unemployment compensation. From the point of view of the environmental economist, it is important to note that the emissions tax rate equals the Pigouvian tax rate. There is no underinternalization of environmental externalities. The reason is the availability of an efficient tax instrument, the lump-sum tax. This is used to finance the public good, and the other taxes just correct the market failures in the markets for the environmental resource and for labor.

Let us now consider a second-best world, where lump-sum taxes are not available. In order to simplify the analysis a bit, I take into account the fact that capital taxes and the command and control parameter are perfect substitutes. Thus, I optimize with respect to the labor, capital, and emissions tax rates only. The system of first-order conditions is

$$(22a) \qquad (F_K - r)\frac{dK}{dt^K} + (F_L - \lambda)\frac{dL}{dt^K} = (1 - v')\frac{dG}{dt^K},$$

$$(22b) \qquad (F_K - r)\frac{dK}{dt^L} + (F_L - \lambda)\frac{dL}{dt^L} = (1 - v')\frac{dG}{dt^L},$$

$$(22c) \quad (F_K - r)\frac{dK}{dE} + (F_L - \lambda)\frac{dL}{dE} = (1 - v')\frac{dG}{dE} - (F_E - v'),$$

where λ is the Kuhn-Tucker multiplier associated with the condition that employment cannot exceed the labor force. This can be rewritten in matrix form:

$$(22d) \qquad \begin{pmatrix} \dfrac{dK}{dt^K} & \dfrac{dL}{dt^K} & 0 \\ \dfrac{dK}{dt^L} & \dfrac{dL}{dt^L} & 0 \\ \dfrac{dK}{dE} & \dfrac{dL}{dE} & 1 \end{pmatrix} \begin{pmatrix} F_K - r \\ F_L - \lambda \\ F_E - d' \end{pmatrix} = \begin{pmatrix} (1 - v')\dfrac{dG}{dt^K} \\ (1 - v')\dfrac{dG}{dt^L} \\ (1 - v')\dfrac{dG}{dE} \end{pmatrix}.$$

Substituting from equation (10) for dG/dE on the right-hand side of this equation and using similar results for the partial derivatives of G with respect to the labor and capital taxes, we have

$$(23) \quad \begin{pmatrix} \dfrac{dK}{dt^K} & \dfrac{dL}{dt^K} & 0 \\ \dfrac{dK}{dt^L} & \dfrac{dL}{dt^L} & 0 \\ \dfrac{dK}{dE} & \dfrac{dL}{dE} & 1 \end{pmatrix} \begin{pmatrix} F_K - r - (1 - v')(t^K r + sEF_{KE}) \\ F_L - \lambda - (1 - v')[(t^L + u)w + sEF_{LE}] \\ F_E - d' \end{pmatrix}$$

$$= \begin{pmatrix} (1 - v')rK \\ (1 - v')wL \\ (1 - v')\dfrac{d(sEF_E)}{dE} \end{pmatrix}.$$

For the case of constant returns to scale, full employment, and flexible wages, it has been shown by Lorz, Scholz, and Stähler (1999) that this system has the simple solution that environmental taxes are Pigouvian, capital taxes are 0, and the whole tax burden resulting from the necessity to provide the public good falls on labor. Formally the reason for this result is the constant-returns-to-scale assumption. Constant returns to scale imply that the factor-price frontier does not depend on output. That is, if two marginal factor productivities are given, the third one is determined. In the case of flexible wages, the labor tax leaves the gross wage (including the tax rate) that the producer has to pay unchanged and the whole tax burden is borne by the factor that is available in fixed supply. This is a kind of lump-sum result.

In the case of unemployment, matters are different. Higher labor taxes lead to less employment and higher unemployment benefits. Thus, it is not clear whether the state has to use taxes on other inputs to finance the supply of public goods. Assume for instance that the optimal supply of public goods is achieved, that is, $v' = 1$. Since the matrix on the left-hand side of equation (22) has full rank, it follows that $F_K = r$ and $F_E = d'$. The capital tax rate should be 0 and the emissions tax rate should equal marginal environmental damage. If there is unemployment, then the shadow price λ is 0. But this has the consequence that the marginal productivity of labor is 0. This is an obvious contradiction to the unemployment assumption that was just made. If there is full employment, the shadow price is positive, and a subsidy has to be paid to the employers to make them hire people whose marginal productivity is below the wage rate. This is in almost all imaginable cases incompatible with the tax revenue being generated by green taxes. Due to the complexity of the model, explicit optimal solutions can be computed only for calibrated versions of this model.

Similar considerations apply if we look at an optimal environmental tax reform where capital taxes are given. This makes matters simpler since there are only two first-order conditions, equations (22b) and (22c). Moreover, it is assumed that the government uses this possibility to reconsider its budget. Thus, the budget is flexible as well.

Now we discuss environmental tax reform where the government considers the possibilities of using tax revenues to reduce labor taxes and, thus, unemployment or to improve the supply of public goods. Two cases are imaginable. Let us start from a situation where there is Pigouvian taxation of emissions, zero taxation of labor, and a large demand for public goods. A marginal change in environmental regulation has an effect on the environment and on output that cancel out in the case of Pigouvian taxes. In the first case, emissions taxes generate additional tax revenue; then this tax revenue is used to reduce labor taxes or even to subsidize labor and to improve the supply of public goods. The tax rate tends to be higher than the Pigouvian tax rate. If higher environmental taxes happen to reduce total tax revenue (via their impact on the demand for the other factors), it is better to use lax environmental regulation to create more employment and to produce additional public goods. Then the emissions tax rate should be lower than the Pigouvian rate.

Finally consider the second-worst case, where the government has the environmental tax as its only policy instrument. Optimization of the welfare function with respect to emissions, E, yields the first-order condition

$$(24) \quad (F_K - r)\frac{dK}{dE} + (F_L - \lambda)\frac{dL}{dE} = (1 - v')\frac{dG}{dE} - (F_E - v'),$$

which can be rearranged to

$$t^E = v' + (F_K - r)\frac{dK}{dE} + (F_L - \lambda)\frac{dL}{dE} + (v' - 1)\frac{dG}{dE}.$$

The emissions tax consists of four components, one for each distortion in the economy:

1. Environmental externality: The larger marginal environmental damage, the larger is the (implicit) emissions tax.
2. Taxation of the mobile factor of production: If the tax dominates, the term in parentheses is positive; if the indirect subsidy is larger, then it is negative. The sign of dK/dE is indeterminate. So, if tighter environmental standards attract capital, we should use them if the capital is a tax base; and we should employ lax standards if capital is subsidized. The opposite conclusions should be drawn if tighter standards imply capital flight.
3. Labor market distortion: The tax rate should be adjusted to reduce unemployment. Depending on the sign of dL/dE, this can be achieved either by relatively lax or relatively tight environmental standards.
4. Absence of a lump-sum tax for financing the provision of the public good: If the public good is scarce, $v' > 1$, then the environmental policy should be designed such that it generates additional tax revenue. If the consumption good is scarce, then the environmental tax should be low such that consumption is raised at the expense of the supply of public good.

If in an interjurisdictional competition many jurisdictions interact, they all use the biased environmental policies, but they never manage to attract additional capital nor are they able to repel capital in the case of indirect subsidization through command and control. The outcome, thus, is an unfavorable tax competition, where everybody uses taxes that are either too high or too low without achieving their objectives. As in the tax competition and fiscal federalism literatures, there is a prisoner's dilemma that can only be solved by international policy coordination.

6.6   Summary and Conclusion

This paper has addressed environmental policy in a distorted small open economy. It is shown that the fact that the scarcity rent of the resource, which goes to the private sector in the case of command and control, is linked to capital has a number of interesting consequences. It may happen that tighter environmental standards attract rather than repel mobile capital from the jurisdiction imposing these standards. Whether or not this is relevant in practice is an empirical question. Given the specification of the model, it should be possible to determine the relevant parameters and to calculate some crude estimates for the comparative statics terms.

Another interesting result of this paper is that a move from command and control to more market-oriented environmental policies may do the economy some harm, since it can induce environmental capital flight if environmental scarcity rents are linked to the mobile factor of production.

One should expect, however, that all these effects are rather small. Environmental regulation has only a minor impact on the cost side of the manufacturing industry. This may explain the limited and often anecdotal, rather than statistically significant, evidence of capital relocation as a consequence of tighter environmental standards. Some authors argue that this lack of evidence may be due to methodological problems (e.g., Folmer and Jeppesen 1999); nonetheless, in most cases, only drastic changes in environmental policies will induce significant relocation. So, in spite of the fact that this paper seems to imply some deviations from standard environmental-policy recommendations, the simple rule that the emissions tax rate should equal marginal damage is still a good rule of the thumb. After 30 years of theoretical environmental economics, the main problem in practice is still to internalize externalities and to get the prices right.

References

Binswanger, H. C., H. Frisch, H. G. Nutzinger, B. Schefold, G. Scherhorn, U. E. Simonis, and B. Strümpel. 1983. *Arbeit ohne Umweltzerstörung: Strategien einer neuen Wirtschaftspolitik.* Frankfurt am Main: Fischer.

Bovenberg, A. L., and R. de Mooij. 1994. Environmental levies and distortionary taxation. *American Economic Review* 84:1085–89.

Folmer, H., and T. Jeppesen. 1999. Environmental policy and location behavior of firms: A synopsis of the micro and regional science literature. European Studies Discussion Paper no. 34. Odense: Syddansk Universitet.

Fullerton, D. 1997. Environmental levies and distortionary taxation: Comment. *American Economic Review* 87:245–51.

Fullerton, D., and G. E. Metcalf. 1998. Environmental controls, scarcity rents, and pre-existing distortions. University of Texas at Austin. Working paper.

Goulder, L. H. 1995. Environmental taxation and the "double dividend": A reader's guide. *International Tax and Public Finance* 2:157–83.

———. 1997. Environmental taxation in a second-best world. In *International yearbook of environmental economics 1997/1998,* ed. H. Folmer, and T. Tietenberg 28–54. Cheltenham, U.K.: Elgar.

Holmlund, B., and A. -S. Kolm. 2000. Environmental tax reform in a small open economy with structural unemployment. *International Tax and Public Finance,* 7 (3): 315–33.

Kaplow, L. 1996. The optimal supply of public goods and the distortionary cost of taxation. *National Tax Journal* 49:513–33.

Kemp, M. C. 1964. *The pure theory of international trade.* Englewood Cliffs, N.J.: Prentice Hall.

Koskela, E., R. Schöb, and H. -W. Sinn. 1998. Pollution, factor taxation and unemployment. *International Tax and Public Finance* 5:379–96.

Lorz, O., C. M. Scholz, and F. Stähler. 1999. Environmental taxes, public goods, and capital mobility. Institute of World Economics, Kiel. Mimeo.

MacDougall, G. D. A. 1960. The benefits and costs of investment from abroad, a theoretical approach. *Economic Record* 36:13–35.

Oates, W. E., and R. M. Schwab. 1988. Economic competition among jurisdictions: Efficiency enhancing or distortion inducing? *Journal of Public Economics* 35:333–54.

Pearce, D. 1991. The role of carbon taxes in adjusting to global warming. *Economic Journal* 101:938–48.

Pigou, A. C. 1920. *The economics of welfare.* New York: Macmillan.

Rauscher, M. 1997. *International trade, factor movements and the environment.* Oxford: Oxford University Press.

———. 1999. Interjurisdictional competition and environmental policy. In *International yearbook of environmental economics,* ed. H. Folmer and T. Tietenberg. Cheltenham, U.K.: Elgar, forthcoming.

Repetto, R. 1992. *Green fees: How a tax shift can work for the environment and the economy.* Washington, D.C.: World Resources Institute.

Ruffin, R. J. 1984. International factor movements. In *Handbook of international economics,* vol. 1, ed. R. W. Jones and P. B. Kenen, 237–88. Amsterdam: North-Holland.

Schneider, K. 1997. Involuntary unemployment and environmental policy: The double dividend hypothesis. *Scandinavian Journal of Economics* 99:45–59.

Schöb, R. 1997. Environmental taxation and pre-existing distortions: The normalization gap. *International Tax and Public Finance* 4:167–76.

Sinn, H. -W. 1994. How much Europe? Subsidiarity, centralization and fiscal competition. *Scottish Journal of Political Economy* 41:85–107.

Tullock, G. 1967. Excess benefit. *Water Resources Research* 3:643–44.

Wang, L. -J. 1995. Environmental capital flight and pollution tax. *Environmental and Resource Economics* 4:273–86.

Wellisch, D. 1995. Locational choice of firms and decentralized environmental policy with various instruments. *Journal of Urban Economics* 37:290–310.

Zodrow, G. R., and P. M. Mieszkowski. 1986. Pigou, Tiebout, property taxation, and the underprovision of public goods. *Journal of Urban Economics* 19:356–70.

Comment   David F. Bradford

Michael Rauscher's paper on the impact of alternative fiscal and environmental rules on the level of unemployment raises modeling and practical policy issues of general interest. These comments address three of them:

David F. Bradford is professor of economics and public affairs at Princeton University, adjunct professor of law at New York University, and a research associate of the National Bureau of Economic Research.

- The adequacy of the rigid-wage view of labor-market equilibrium as the basis for modeling the impact of policy, environmental or otherwise, on unemployment.
- The incentive impact of typical methods of regulating emissions, operating through rent creation.
- The consequences of modeling distributive instruments more explicitly.

With support from the literature, Rauscher sets the problem in a small open economy with an economic "imperfection" in the form of a fixed wage rate. The fixed wage rate is treated as the culprit in explaining an excess supply of labor (i.e., unemployment) in equilibrium. His paper investigates the consequences of various policies—including, importantly, environmental policies—on unemployment. In addition to the fixed wage, the model adduces two other instances of rigidity, in this case rigidity of regulatory and fiscal rules. The first is a requirement that a tax on labor be set to finance a particular fraction of unemployment benefits. The second is a requirement that the revenue from a tax on emissions be used to finance a subsidy to the employment of capital. Unlike the wage rigidity, variation in these fiscal rigidities is treated as available to policymakers or, at least, admissible as open to investigators.

Another fiscal rigidity lurks in the background. That is the infeasibility of lump-sum taxes, an assumption that is standard in second-best tax analysis. As has been emphasized by, for example, Stiglitz (1982), however, it is hard to motivate the nonavailability of lump-sum taxes without taking into account distributional concerns. Introducing and modeling distribution explicitly (e.g., endowing people with different labor productivities) can rather dramatically change second-best conclusions of the sort developed in the present paper.

Modeling Unemployment Equilibrium

Rauscher's model has a single uniform output (the good) and three uniform factors of production—labor, capital, and emissions—all treated as flows, and a further factor (perhaps land), described as fixed. The good is taken as numeraire and is treated as untaxed throughout. Suppliers of labor receive w per unit (which might be per hour or per year); suppliers of capital receive r per unit, and we can think of suppliers of emissions as receiving 0 per unit. Unless we are thinking about the productive sector of the economy as a single firm, it might be preferable to speak of the fourth factor as offered in perfectly inelastic supply, rather than as fixed (to particular firms). (In fact, I don't think it would change anything if the fourth factor were supplied with a general supply function.) The price of the fourth factor is not specifically discussed in the model. In order for the system to work, however, presumably the fourth factor is priced, let us say at p^L.

Purchasers of the four factors, the firms, pay these amounts gross of any taxes and net of any subsidies. The tax on emissions may take the form of an explicit price of allowances, t^E, or it may be implicit in a regulatory constraint. The firms are treated as price takers in all markets and they all operate with the same constant-returns-to-scale production function.

The wage, w, is treated as fixed by some political process in the background. The price of capital services, r, is fixed in the big world capital market. The price of emissions is set by policy. The price of land is set by market clearing. Once any other tax/subsidy rates have been specified (perhaps determined as a feature of equilibrium), the cost of factor inputs to the firms is determined. This means the factor proportions are determined. The scale of output is determined by the land market-clearing condition and this determines all the other factor demands in equilibrium. It is assumed that the demand for labor falls short of the supply, so that there is unemployment in equilibrium.

Having laid out this model, Rauscher shows us how variations in policies not specifically related to the labor market may, indirectly, perturb the equilibrium in the direction of greater employment. Specifically, a subsidy to the employment of capital, which might be indirectly implemented through environmental regulation, might have this effect.

As noted by Rauscher, a rigid-wage model along the lines described has been fairly widely employed in the European literature. Exercising the discussant's privilege to offer more or less unsupported opinion, I would express the view that, even if the politically fixed wage model were well supported empirically, adopting rather subtle combinations of policies apparently unrelated to the labor market seems to me a rather peculiar approach to the unemployment problem. If the policymakers cannot understand what the model takes for granted as obvious, will they be persuaded by much more sophisticated modeling features (and by conclusions that are contingent on parameter values as well as specification)?

I am also unconvinced of the empirical adequacy of the rigid-wage model. Here I am stepping even more over the bounds of areas of economics in which I can claim any expertise. But, again in the spirit of stimulating discussion, I offer my version of a minimum wage equilibrium as having perhaps some advantages, particularly in drawing attention to the distributive aspects of the policy problem.

An Alternative Specification

My suggestion is to deploy the model used in optimal income-tax analysis. Instead of assuming that workers are perfect substitutes, suppose they differ in their abilities to provide labor services. Specifically, assume that a worker of type (skill level) q who works for 1 unit of time provides q

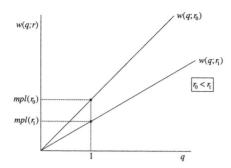

Fig. 6C.1   Two skill-premium schedules

units of standardized labor services. The value of a unit of standardized labor services in production will be denoted *mpl,* that is, the usual marginal product of labor.

Consider a two-factor version of the story. Under standard competitive conditions, firms will want to hire any worker whose hourly wage is equal to or less than the product of the worker's skill and *mpl.* (I am here assuming a worker's q is public knowledge.) Under competition for workers there will be an equilibrium skill-premium structure, which we may denote $w(q)$, relating the worker's skill level to the hourly wage received. Here the relationship is given by equation (1).

(1) $$w(q) = q \cdot mpl.$$

Any worker of type q willing to work for $w(q)$ will be employed.

With variable capital, firms will be willing to pay more for workers, the lower the going rental rate. So the marginal product of standardized labor in equilibrium will be a function $mpl(r)$ of the rental rate of capital. There will then be a skill premium structure, equation (2), that depends on the rental rate,

(2) $$w(q;r) = q \cdot mpl(r).$$

Figure 6C.1 illustrates this.

A possible model of unemployment caused by wage rigidity is now available. If the rigidity takes the form of a minimum wage per unit time, independent of skill, there would be a cutoff level of worker skill below which workers could not be legally employed. Since competitive firms will not pay a worker more than his or her marginal product, workers with subcritical skill levels will be unemployed. The extent of unemployment will be determined by the proportion of such workers in the labor force. With lower interest rate, ceteris paribus, more capital will be employed and the cutoff skill level will be lower, as illustrated in figure 6C.2, where $\bar{w}$ is the minimum wage allowed per unit of worker time. (Variations on the theme,

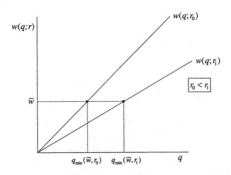

Fig. 6C.2   Interest rate and cutoff skill level

where the differential treatment relied on criteria other than simply a minimum wage, could be imagined that should result in models with similar properties.)

An Earnings Tax in the Alternative Specification

The effect of an earnings tax on wages and employment in this model would depend a little on the details of the tax. The wage-skill profile (which shows the amounts received by workers) will be shifted down by an earnings tax paid by employers. A flat percentage earnings tax, for example, would act like a proportional cut in each worker's skill level. If the minimum wage restricts what the employer is allowed to offer net of employer tax, the impact of the tax would be to raise the cutoff skill level and therefore raise the level of unemployment (fig. 6C.3). A tax paid by the workers, with a minimum expressed in terms of what the employer pays, would have a different impact. It is easy to spin out variations.

Subsidize Capital to Circumvent a Minimum Wage?

A subsidy to capital services would raise wages and reduce unemployment in this model. If the capital subsidy is financed by an earnings tax, there would be offsetting effects (and more complex effects once one recognizes the possibility of variable hours), but it is easy to imagine plausible stories that would produce a lower cutoff skill level and lower unemployment. The taxes on labor and subsidies to capital effected by the features of the tax-transfer and environmental tax-regulatory systems modeled by Rauscher would constitute instances of this phenomenon.

Modeling Environmental Regulation

I turn now to the question of the rents generated by environmental regulation. To investigate this subject, suppose capital and environmental services are uniform and that labor skills vary as in the model previously

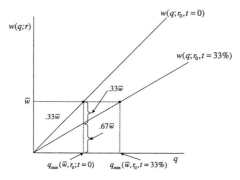

Fig. 6C.3   Effect of 33 percent earnings tax on cutoff skill

sketched out. Then, if we fix r, $\overline{w}$, and t^E there will, in general, be an equilibrium with some level of unemployment of the relatively unskilled, some level of capital services employed, and some level of environmental quality. Any taxes in the story will also affect the equilibrium; the unemployment rate will vary with variations in these parameters of the problem. (Reminder: The capital-labor and emissions-labor ratios are basically fixed by r and t^E, determining the equilibrium *mpl.* Total employment varies by changes in the skill required to top the minimum wage.)

I am here a little vague about the status of the price of emissions, t^E. If emissions services are sold by their owners, as are other factors of production, they should work into the general equilibrium system just the same way as do capital and labor services. A higher price of emissions will imply a lower equilibrium wage, given the price of capital services.

Emissions are, however, owned by the government in this story. The proceeds from the sale of emissions could be spent by the government on goods, but the interesting issues arise when the emissions proceeds are used to reduce other taxes. In particular, the double dividend sorts of questions come up where the revenues are used to lower distorting taxes on labor or capital.

Rauscher studies another form of rebate of the revenue from the sale of emissions that is supposed to describe the implicit effect of environmental regulation. A regulatory arrangement can typically be described as economically equivalent to a tax-cum-revenue-rebate scheme. The implicit tax on the regulated activity is the one that would induce the regulated entity to choose the regulated level voluntarily. Since there is no actual revenue generated by this shadow tax, it must be getting handed out implicitly to someone. In Rauscher's case of central interest, the revenue handout is in proportion to a firm's use of capital services and, hence, arguably acts as an implicit subsidy to such use. As a consequence, it has the same effect on equilibrium as would an explicit use of the revenue from an explicit emissions price to finance an explicit subsidy to the use of capital. As

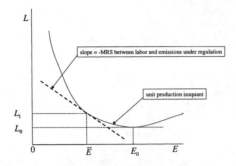

Fig. 6C.4   Regulation-constrained production

sketched out before, such a subsidy could well generate a reduction in the level of unemployment caused by a minimum wage.

What I would like to explore briefly is how the implicit revenue from an implicit tax of the type imposed by a typical regulatory program is implicitly distributed to implicit taxpayers.

A Representative Regulatory Scheme

Suppose the regulation takes the form of a prescribed maximum ratio of emissions to output. To get a feel for how this works, suppose there are just two inputs to production, labor and emissions, and forget about the minimum wage. Figure 6C.4 sketches the unit isoquant of a hypothetical production function using the two inputs. The illustrative isoquant is a little unusual in having a minimum; assuming the firm can choose its level of emissions, the upward-sloping part would never be observed. In the absence of regulation, the firm will treat emissions as free and use E_0 units of emissions per unit of output. Since labor is the only other input, under competitive conditions it must absorb the entire value of output, so the preregulation wage is $1/L_0$.

Regulation is presumed to set the prescribed maximum emissions rate per unit output, identified as $\overline{E}$ in the diagram. (I am here implicitly assuming that emissions constitute a consumption externality and do not affect production; it would be easy to change this assumption, provided only that the producer remains ignorant of the feedback between emissions and production costs.) Under the regulated conditions, the wage must fall to $1/L_1$. Suppose, for example, the initial wage is 1 and the new equilibrium is 0.75. With full employment of labor, regulation will generate a reduction in the aggregate emissions to $\overline{E}/E_0$ times their unregulated level. (If the regulation changes the level of labor used, there will be a corresponding further effect on emissions.)

The production technique chosen by the firms under regulation will be the same one they would choose if, instead of the regulation, emissions were priced at a level that resulted in the ratio of wage to emissions price equal to the marginal rate of substitution (MRS) between labor and emissions in the regulated equilibrium, as indicated in figure 6C.4.

Suppose, to put some numbers on things, the slope in figure 6C.4 were 2/3 (in absolute value). Then a wage of 0.45 and a price of emissions of 0.30 should do the trick. Assuming that, in general equilibrium, the seller of the emissions allowances is also a buyer of the output of the firm, this pair of prices should be an equilibrium combination—zero profits all around, and supplies equal demands. But a wage of 0.45 is obviously very different from a wage of 0.75, the wage in the regulated equilibrium. The difference is the implicit subsidy to the input of labor (equivalently, in this case at least, a subsidy to the output of the firm) of 0.30 per unit under the regulated equilibrium.

If the government-seller of emissions rights were to use the funds to finance a subsidy to the input of labor by the firm (or equivalently, in this case, to the output of the firm), the regulated- and the priced-emissions equilibria would be economically identical. If, instead, the government were to use the funds in some other way, for example, to provide a lump-sum tax rebate to somebody, the equilibria would be different, involving a lower wage in the priced-emissions equilibrium, with all that implies.

Three Factors

If we had three factors, labor, capital, and emissions, the story should be the same in general outline. But the supply conditions of capital have a notable impact on the results. In the regulated equilibrium there are an implicit tax on emissions and implicit subsidies on the use of the two other factors. We would like to identify these components.

The net payment to the other two factors, per unit of output, will, presumably, be lower than in an unregulated equilibrium. That is, there is a penalty in terms of the productivity of the two other factors imposed by the requirement to use less emissions. If the price of capital services is fixed in the world market, the brunt of the adjustment, relative to the unregulated equilibrium, will have to be in the wage. Suppose, for example, the prices of labor and capital are 0.60 and 0.15, respectively, whereas in the unregulated equilibrium the prices are 0.85 and 0.15, respectively.

Suppose, again, a price of 0.30 for allowances would do the trick in inducing the firms to satisfy the emissions limit voluntarily. If the price of capital services is fixed at 0.15, a price of labor of 0.30 would enable firms to break even at the regulated equilibrium factor proportions. But at this relative factor-price ratio between labor and capital, the firm will want to substitute toward labor input. The regulated equilibrium cannot be a priced-emissions equilibrium unless labor is subsidized at, in this case, 0.30 per unit. With that subsidy, the regulated equilibrium and the priced-emissions cum labor-subsidy equilibrium will coincide.

Rent to Be Acquired by Owners of Capital?

What this example suggests to me, and I have not had time to track down the details or work them out for myself, is that the implicit subsidy in the regulation regime (whereby the implicit revenue is expended) depends on the supply and demand details of the factor markets (as well as the details of the regulation). It does not seem likely, however, that the implicit subsidy would be to capital in the small open economy case, simply because the supply price of capital is completely determined in that case.

Is This Second-Best Framework the Right One?

Finally, I would like to raise the issue of whether the second-best model used in this and many related analyses is, empirically, the most appropriate one. The point, which has been made forcefully by Kaplow (1996), builds on Stiglitz's (1982) insight that we forgo lump-sum taxation in order to serve distributional objectives. The distributional objectives, in turn, are served in advanced economies by an earnings-based tax-transfer system. The option of adjusting the distributional consequences of policies by altering the parameters of the tax-transfer system is always, in principle, available; the other instruments may offer the possibility of serving efficiency ends, including effects derived from interaction with the earnings-based instruments.

Kaplow's Analysis

As Rauscher's paper makes clear, the basic issues here concern the effects of imposing environmental regulation on an economy with preexisting distortions. So, for example, a typical double dividend result is that interactions with distorting taxes imply that a Pigouvian tax should be set below the marginal social damage. Kaplow's (1996) treatment of the issue assumes that the preexisting distortion is due to a graduated income tax or, more precisely, a labor earnings tax. He concludes that for what one might think of as the central case, the Pigouvian tax should just equal the marginal social damage. (This does not imply that the earnings tax itself is nondistorting, by the way.)

To sketch the story, consider a model with just one consumption good and one public good (which may be an environmental-quality measure). The third object of individual preferences is labor supply. What I have called the central case assumption is that labor supply is separable in the preferences, which means that a person's trade-off between public good and consumption is independent of the amount of labor being supplied. Cobb-Douglas preferences, equation (3), for example, have this property,

$$(3) \qquad\qquad U(x,h,g) = x^{\alpha} h^{1-\alpha} g^{\beta},$$

where x is the quantity of the consumption good, h is hours of leisure (home use of time, i.e., time not spent at work in the labor market), and g is the level of the public good.

The detail that we need to use to get Kaplow's result—that the Samuelson condition is the right criterion for determining the level of a public good—is the separability of the ranking of (x,g) pairs from the amount of leisure obtained by the person. This separability is immediately apparent in the Cobb-Douglas case, since the ranking of (x,g) pairs is completely determined by $x^\alpha g^\beta$. Specifically, the marginal rate of substitution between the consumption good and the public good (the dollar value of an incremental unit of the public good) is given by

$$(4) \qquad \frac{U_g(x,h,g)}{U_x(x,h,g)} = \left(\frac{\beta}{\alpha}\right)\left(\frac{x}{g}\right),$$

independent of leisure consumption, h.

In the Kaplow story, as in the standard optimal income-tax analysis, everyone in the economy is assumed to have the same preferences, but people differ in their skills and therefore wage rates. The tax-transfer system that specifies the net tax paid by an individual as a function of that person's earnings is expressed by the function, $T(earnings)$. Earnings are the result of supplying labor, and we model each person as having an endowment of potential labor time, for example, 24 hours per day. Then the budget constraint for a worker of skill level q is given by

$$(5) \qquad x = w(q)(24 - h) - T[w(q)(24 - h)].$$

where, as before, the consumption good has been chosen as numeraire (so we can think of it as measured in dollars), T is the net tax paid (which could be negative—the person could be obtaining a net transfer), and $w(q)$ is the wage-skill relationship already discussed.

In general, when the separability property applies, the utility function can be written as a function of two arguments: a "subutility" function of the consumption good and public good, and the amount of leisure.

$$(6) \qquad U(x,h,g) = u(v(x,g),h).$$

In the Cobb-Douglas case,

$$(7) \qquad U(x,h,g) = v(x,g)h^{1-\alpha},$$

$$v(x,g) \equiv x^\alpha g^\beta.$$

Since the individual devotes whatever is left after earnings tax to the consumption of the numeraire good, there is a simple relationship between the amount earned and the subutility level achieved, given the tax system and the level of the public good:

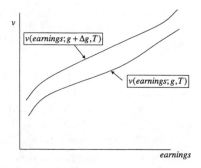

Fig. 6C.5   Earnings subutility schedules

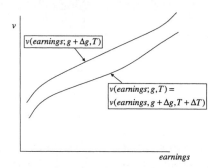

Fig. 6C.6   Fully offsetting tax system

(8) $v(earnings; g, T) = [earnings - T(earnings)]^{\alpha} g^{\beta}$.

Now suppose the public good is measured in units of private good forgone to produce it. If in (8) we increase the level of public good by some amount, Δg, then the new *v-earnings* relationship is given by

(9) $v(earnings; g + \Delta g, T) = [earnings - T(earnings)]^{\alpha}(g + \Delta g)^{\beta}$,

which lies above the original *v-earnings* relationship (fig. 6C.5).

Consider now the change in the tax system that will exactly zero out the impact of the increase in the public good (fig. 6C.6):

(10) $v(earnings; g + \Delta g, T + \Delta T) = v(earnings; g, T)$.

(Again, note that $g + \Delta g$ is a scalar and $T + \Delta T$ is a function.) Specifically,

(11) $[earnings - T(earnings)]^{\alpha} g^{\beta}$

$$= [earnings - (T + \Delta T)(earnings)]^{\alpha}(g + \Delta g)^{\beta}.$$

It should be clear that equation (11) defines a new tax-transfer function, called $(T + \Delta A)$, that does the trick. (Except that it would clutter the notation too much, we might indicate that it depends on the specific move from g to $g + \Delta g$.)

By the derivation of the tax-transfer function $T + \Delta T$, a person facing that tax-transfer system in a world with public-good level $g + \Delta g$ would have exactly the same *v-earnings* alternatives as a person facing tax-transfer system T in a world with public-good level g. So each person will make the same labor-supply choice in both cases. Suppose the amount of tax raised by the new system is more than enough to finance the increment to the public good. Then we know it is possible to make the change and have some money left over to make a Pareto improvement in the tax-transfer system. So we can always undertake a change that passes the Samuelson test, combined with a change in the earnings tax, so as to make everyone better off.

Notice that, as I have mentioned, the earning tax generates distortions before and after the change. The marginal tax rate before the change, as a function of earnings, is $T'(earnings)$; the marginal tax-rate schedule after the change is $(T + \Delta T)'(earnings)$. The original tax-transfer function is whatever it is and the new one is related to it by equation (10).

Long Preamble, Short Conclusion

With earnings- or income-based tax-transfer systems as elaborate as those of most developed countries, does it make sense to attempt to influence the level of unemployment by manipulating environmental policy?

References

Kaplow, Louis. 1996. The optimal supply of public goods and the distortionary cost of taxation. *National Tax Journal* 49/(4): 513–33.

Stiglitz, Joseph E. 1982. Self-selection and Pareto efficient taxation. *Journal of Public Economics* 17:213–40.

7

The Environmental Regime
in Developing Countries

Raghbendra Jha and John Whalley

7.1 Introduction

This paper explores the environmental regime in developing countries. By regime, we mean those environmental externalities that are commonly found in the developing world, along with the policy responses, if any, to them. Included are the direct effects of industrial emissions, air- and water-quality impacts of untreated waste (industrial and human), congestion effects of traffic, soil erosion, and open-access resource problems (including forests).

We note the many difficulties involved with adequately characterizing this regime, not the least of which is the heterogeneity across both environmental problems and policy responses in the developing world. Enforcement and compliance (which are typically lax in developing countries) also play a central role in defining this regime. In addition, we note the differences between the experiences of developed and developing countries more generally beyond the environmental area.

In the paper we make three main points. The first is that there is a tendency in much of the literature of the last few years to equate environmental problems in developing countries with pollutants (or emissions).

Raghbendra Jha is professor at Indira Gandhi Institute of Development Research, Bombay. John Whalley is professor at the University of Western Ontario and the University of Warwick, and a research associate of the National Bureau of Economic Research.

The authors are grateful to May Arunanondchai for research support and to their discussant Ed Barbier, Gib Metcalf, Carlo Cararro, Peter Newell, Ajit Ranade, Manoj Panda, Diana Tussie, and Ben Zissimos for comments and discussions. This paper draws on material from a project on a possible World Environmental Organization and the interests of the developing countries supported by the MacArthur Foundation, in which the two authors are jointly involved.

Such an approach has been partly influenced by data availability, including that collected by the Global Environmental Monitoring System (GEMS) supported by the United Nations Environmental Program (UNEP). This has yielded data on a range of environmental indicators including biological oxygen demand (BOD), airborne SO_2 concentrations, heavy metal levels,[1] untreated human waste, and other air- and water-quality indicators. This focus on pollutants has meant that in much of the literature there is less emphasis on what others have called degradation. This refers to the effects of uninternalized externalities seen in soil erosion, congestion, open-access resources, and other problems, where physical emissions are less of a problem. The paper argues that discussing environmental problems in developing countries (or comparing them with those in developed countries) without reference to these problems is incomplete; their effects are large and pervasive, and their severity and interaction with economic progress often differ sharply from the effects of pollutants.

The second point is in many ways an elaboration of the first. We have reviewed studies of the social costs associated with incomplete internalization of the externalities we list. The studies that are available are limited in their coverage of both countries and items and, in addition, do not always use consistent methodologies; but the picture they paint is that such costs seem large (perhaps in excess of 10 percent of GDP on an annual basis in some countries), and that these costs are dominated by degradation rather than by pollutant effects (accounting for perhaps three-quarters of the total effect). One implication we draw is that with large cost estimates of inaction, environmental policy in developing countries should perhaps have a higher ranking than it has currently, especially if these cost estimates substantially exceed those of inaction with regard to more conventional policy reform such as tax or trade policy. The other is that if the balance of costs is skewed more to degradation than to the effects of pollutants, degradation should perhaps receive more attention in the literature.

Our third point concerns the relationship among growth, policy reform, and environmental quality; and comparisons of the environmental situation either across economies or over time in light of our characterization of the environmental regime in developing countries. To the extent that recent literature focuses on differences in outcomes across countries or over time in terms of levels of various environmental indicators, the issue is whether degradation effects can give a different picture. We argue that degradation effects could well behave differently from pollutants effects; soil erosion problems, for instance, seem to progressively recede as income per capita rises, since the population in agriculture falls and plot sizes rise; while outward-oriented trade policies draw labor into urban areas from

1. Lead, arsenic, mercury, and cadmium.

rural areas, adding to congestion. We discuss the literature on the environmental Kuznets curve (Shafik and Bandopadhyay 1992; Grossman and Krueger 1995; Andreoni and Levinson 1998) and recent literature on trade and environment (Copeland and Taylor 1994, 1995; Antweiler, Copeland, and Taylor 1998). While the authors contributing to these literatures are clear in labeling their analyses as primarily of pollutant levels, users of this research naturally tend to think of the results as giving guidance on the wider environmental situation in the countries discussed; and without explicit reference to degradation effects, the picture once again can be incomplete.

In the final section, the paper argues that the welfare gains from moving to full internalization would seem to be the more appropriate comparative measure of the severity of environmental problems across countries (or changes over time). The studies referred to seem to suggest that internalization gains relative to GDP are significant for developing countries (and probably larger than for developed countries), raising the issue of why a higher degree of internalization has not occurred. We discuss briefly whether this outcome reflects income elasticities of demand for environmental quality above 1; or whether it reflects technology and capital intensity of environmental management and policy enforcement, so that abatement costs in developing countries are the barrier. We also touch on the role of the political structure in these countries, and on whether a key problem is also in defining and enforcing property rights. In the process we discuss the links between poverty and degradation taken up by Maler (1998).

In concluding, the paper discusses the implications of our characterization of the environmental regime in developing countries for environmental policy in these countries. Can the policy regimes in developed countries be simply transferred, or are there special features that need to be taken into account? Degradation, property rights, and compliance issues seem to be more prominent for developing than for developed countries.

7.2 The Environmental Regime in Developing Countries

We interpret the term "environmental regime" as applied to the developing countries to mean the set of externality-related problems often characterized as environmental, as well as the policy response they induce. Individually, these cover soil erosion, open-access resources (forests and fisheries), congestion (traffic), household emissions (fuel burning), industrial emissions, ground- and surface-water resources (shared aquifers and water-table problems), untreated human and nonhuman waste, and other problems. Property rights and their lack of clear definition, and compliance with environmental controls are two factors closely connected with these problems. Policy responses include regulation (command and control), local actions (village level concerning soil erosion), resource-management policies (forests), and infrastructure development (urban congestion).

For the purpose of our later discussion, we classify these externalities into two broad headings: pollutants, covering industrial and household emissions of various forms and untreated waste; and degradation, covering soil erosion, congestion, and open-access resources. For both of these problem areas, we identify the classical externality literature that applies: A Pigouvian tax will internalize the externality, and the Coasian issues of the assignment of property rights and whether partial internalization can take place through bi- (or multi-)lateral deals once property rights are established also arise.

We could group these externalities in other ways, such as agriculture and rural externality problems, urban externality problems, and environmental problems associated with varying forms of industrial waste. The reasons for grouping these environmental problems in the way we do here relate primarily to measurement issues. They do not reflect any major analytical distinction in terms of the economics, even though, for instance, open-access externality problems for renewable resources have a complex analytical literature characterizing both how replacement of the stock occurs and what constitutes optimal policy across sustainable harvests. Pollutants capture emissions and contaminants of various forms, which can be monitored by such efforts as GEMS. Degradation captures environmental effects for which emissions and contaminants are not the central issue, and for which direct monitoring is more problematic.

We note in passing that the developing countries in which these regimes occur are far from a homogenous group of countries. They vary by per capita income, GDP growth rates, size, the volume and pattern of their international trade, their degrees of urbanization, and many other characteristics. They also vary in the form their environmental problems take; some countries are heavily endowed with environmental assets such as tropical forests,[2] while others are arid and desert; some are mountainous, while others are low lying and flood prone. Generalizing across all developing countries and categorizing the environmental regimes they each face is thus difficult. A few generalizations seem to hold, however; for instance, lower-income countries have proportionately more significant agricultural and rural sectors.

7.2.1 Elements of the Regime

Notwithstanding these problems, in table 7.1 we have set out what we see as the main elements in our characterization of the environmental re-

---

2. Schatan (1998), for instance, notes that Latin America and the Caribbean account for 50 percent of the world's tropical forests, and five of the ten countries richest in biodiversity worldwide are in the region (Brazil, Colombia, Ecuador, Mexico, and Peru).

Table 7.1   **A Pollutant and Degradation Classification Scheme for Environmental Externalities in Developing Countries**

Pollutants	
Toxic contaminants	Organochlorines, dioxins, pesticides, grease and oil, acid, and caustic metals (mainly discharges from mines, chemical producers, pulp and paper plants, and leather and tanning factories), which cause health and other problems
Untreated fluid waste	Untreated sewage discharges into rivers, streams, open ditches, which causes waterborne disease
Domestic solid waste	Poorly managed solid waste, which spreads infectious disease and blocks urban drainage channels, with risk of flooding and waterborne disease
Smoke and burning	Burning dung, wood, coal, and crop residues; vehicle exhaust; and smoke, which cause respiratory damage, heart and lung disease, and cancer
Degradation	
Soil erosion	Sedimentary transfer of topsoil to neighboring plots, river estuaries, hydro dams, which causes silting, accompanied by leaching of soil
Soil quality	Pesticide residues, which affect production of neighboring plots
Open-access resources	Ill-defined property rights, which lead to overexploitation of resources (firewood/forests, fisheries; and shared aquifers and water tables)
Congestion and traffic	Poorly regulated traffic, which causes time loss, elevated accident risk, and lowered air quality in urban areas

gime in developing countries, using the broad categories of pollutants and degradation already discussed.

Pollutants in the form of toxic contaminants cover effluents of various types, which come largely from mines, chemical production, pulp and paper plants, and leather and tanning factories. They include organic chlorines, dioxins, pesticides, grease and oil, acid, and caustic metals. These generate health and other problems. The U.N. *Human Development Report* (HDR) (U.N. 1998) estimates that Asia's rivers, on average, contain lead levels 20 times in excess of those in European and North American countries, and claims, by way of example, that in China most toxic solid waste is disposed of in municipal waste streams without treatment.

A second category of pollutant-based externality problems consists of those associated with water quality and untreated fluid waste. It is common in many countries for there to be untreated sewage discharges into rivers, streams, and open ditches. The 1998 HDR suggests that as much as 50 percent of all discharges into waterways in developing countries are untreated. These, in turn, generate significant health problems, including waterborne diseases, which in some countries are rife. The HDR estimates that diarrhea and dysentery account for an estimated 20 percent of the total burden of disease in developing countries; that polluted water generates nearly 2 billion cases of diarrhea annually in the developing world; and that diarrhea-related diseases cause the deaths of some 5 million people annually, including 3 million children. They also estimate that contaminated water leads to 900 million cases of intestinal worms and 200 million cases of schistosomiasis, and that Asian rivers carry 50 times as many bacteria from human excrement as rivers in European and North American countries.[3] The high level of arsenic linked to phosphoric fertilizers in groundwater, which kill some of the people who drink such water, is a further problem in a number of countries.

Another component of the pollutant category is domestic solid waste. In most developing countries, there are only limited solid-waste disposal systems and the result is the spread of infectious diseases. The 1998 HDR estimates that between 20 and 50 percent of domestic solid waste in these countries remains uncollected, even with up to one-half of local government spending in some countries going to waste collection. In some areas, given the lack of sanitation, waste becomes mixed with excrement, further contributing to the spread of infectious disease. Uncollected domestic waste is the most common cause of blocked urban drainage channels in Asian cities, which in turn increases the risk of flooding and waterborne disease. Poorer households in these countries tend to live near waste-disposal sites.

Health-related problems (which include respiratory damage, heart and lung disease, and cancer) due to smoke from burning and to vehicle exhaust in both urban and rural areas reflect another pollutant-based element of the environmental regime. In lower-income countries, these problems come from burning dung, wood, and crop residues. The 1998 HDR estimates that 90 percent of deaths globally due to air pollution are in the developing world and of those 80 percent are due to indoor pollution.

Of the elements of degradation that we identify as part of the environmental regime in developing countries, soil erosion is a major component; although to identify the externality-related component one has to differentiate between on-site and off-site effects. Erosion arises from a variety of causes. One is population growth that results in progressive division of plot sizes, with spillover of topsoil into neighboring plots, river estuaries, hydro dams, and, in the case of countries with more desert areas, wind-borne soil loss. The 1998 HDR estimates that in Burkina Faso and Mali, one person in six has been forced to leave his or her land because it has turned into desert and that desertification has a worldwide annual cost of $42 billion in lost income, $9 billion of which arises in Africa. Soil erosion reduces agricultural productivity and in some cases the availability of agricultural lands per capita. Soil erosion has also had the effect of reducing fodder available for cattle.

A recent survey paper on studies of the cost of soil erosion in developing countries (Barbier 1998) places the annual losses by country in a range from 1 to 15 percent of GDP. Alfsen et al. (1996), in a study of Nicaragua,

---

3. This is consistent with Hettige, Mani, and Wheeler's (1997) finding that the environmental Kuznets curve does not hold for waterborne pollutants.

estimate annual productivity losses due to soil erosion by crop—of coffee 1.26 percent, beans 2.52 percent, maize 2.41 percent, and sorghum 1.35 percent. Magrath and Arens (1989), in a study of soil-erosion losses in Java in 1985, estimate annual losses of approximately 4 percent of the value of crops harvested. Cruz, Francisco, and Conway (1988),[4] examining two watersheds in the Philippines and focusing only on additional sedimentary costs for hydro-power installations (reduced water-storage capacity for hydro power, reductions in the service life of the dam, and reduced hydro power), estimate annual costs of $27 per hectare of agricultural land in the watershed, a significant portion of the value of crop yield. Soil-quality problems arise from the leaching of pesticides to neighboring plots, contaminating neighbors' soil.

In addition to soil erosion and soil quality, other degradation-type externalities arise with open-access resources (resources for which the property rights are ill-defined or poorly enforced) and the overexploitation of these resources. These include deforestation associated with land clearing, slash-and-burn cultivation, squatting, and, in some countries, the collection of firewood. These problems are especially severe in Africa and in Central and Latin America; Schatan (1998), for instance, identifies land degradation as the most serious environmental problem facing Latin and Central America. For Ghana, one of the less severe cases, López (1997) estimated that overcultivation of land at the expense of forests runs to 25 percent of land use. Overexploitation of fisheries is a further major problem. Shared access to water through common aquifers and groundwater is yet a further manifestation of the problem; this results in reduced water tables, causing especially severe problems in the north China plain.

Finally, in this regime under the heading of degradation are urban congestion problems. Rapid growth in urban population and vehicle densities, especially in high-growth economies, leads to congestion. This lowers air quality; increases the spread of infectious disease; and generates significant time loss from traffic, high accident rates, and noise. A 1990 study by Japan's International Cooperation Agency[5] produced the estimate that road congestion in Thailand (one of the worst cases) reduces potential output in the Bangkok region by one-third.

In closing this discussion, we also note that the environmental regime in developing countries is characterized by policy measures that frequently exhibit lax enforcement. As in the developed world, the primary form that environmental policy toward industrial emissions takes in developing countries is the use of command and control instruments of various forms. These involve the setting of standards and monitoring (with penalties for

---

4. Cited in Barbier (1998).
5. Cited in *The Economist,* 5 September 1998, although we should note that the estimate is substantially in excess of those in other studies we mention later.

violators), but a common feature is the presence of only limited compliance due to weak enforcement. For household wastewater, soil erosion, and other nonindustrial environmental problems, there is little or no abatement of damage in many countries.

7.2.2   The Costs of Environmental Damage in Developing Countries

If this is the regime, what are its consequences? In table 7.2 we report some estimates of the costs of environmental damage for six countries, each associated with the elements of the regime we identify. Cost estimates of this form are relatively few and are scattered throughout the literature. The methods and data used to construct them are not always fully available, and there are variances in their findings. Most of these estimates do not directly refer to the welfare costs of the environmental damage, but instead use some other measure (such as the value of work time lost due to health impacts). We rely heavily on a synthesis of studies of environmental damage for a sample of Asian economies that has recently been drawn together by the Asian Development Bank (ADB) and reported in the 1998 HDR. These, together with results of a related study by the World Resources Institute (WRI), are presented in table 7.2.

In the case of China, the ADB studies suggest that annual productivity losses due to soil erosion, deforestation, and land degradation could be as high as 7 percent of GDP for the early 1990s. If the health and productivity losses from pollution in cities are added (in the region of 1.7–2.5 percent of GDP), combined annual cost estimates from environmental damage are in the region of 10 percent of GDP. Even this estimate excludes a number of key components of environmental damage, such as those due to conges-

Table 7.2	Some Estimates of Environmental Costs in Selected Asian Countries
China	Productivity losses due to soil erosion, deforestation and land degradation, water shortages, and destruction of wetlands in 1990 of $13.9–26.6 billion annually or 3.8–7.3 percent of GDP; health and productivity losses from pollution in cities in 1990 of $6.3–9.3 billion or 1.7–2.5 percent of GDP
India	Total environmental costs of $13.8 billion in 1992 or 6 percent of GDP; urban air pollution costs of $1.3 billion; health costs from water quality of $5.7 billion; soil erosion costs of $2.4 billion; deforestation costs of $214 million (traffic-related costs, pollution costs from toxic wastes, and biodiversity losses excluded)
Indonesia	Health costs of particulate and lead levels above WHO standards in Jakarta of $2.2 billion in 1989 or 2.0 percent of GDP
Pakistan	Health impacts of air and water pollution and productivity losses from deforestation and soil erosion of $1.7 billion in the early 1990s or 3.3 percent of GDP
Philippines	Health and productivity losses from air and water pollution in the Manila area of $0.3–0.4 billion in the early 1990s or 0.8–1.0 percent of GDP
Thailand	Health effects of particulate and lead levels in excess of WHO standards of $1.6 billion or 2 percent of GDP

Source: Agarwal (1996); ADB (1997); U.N. (1998).

tion from traffic-related problems. A further study of China by Smil (1992) based on 1988 data puts losses due to environmental degradation (farmland loss, nutrient loss, flooding, and timber loss) at around 10 percent of GDP, compared to losses from pollutants (waterborne pollutants that reduce crop yields, fish catches, industrial output; airborne pollution that results in higher morbidity, reduced plant growth, damage to materials; and soil pollution that reduces crop yields) of perhaps 2 percent of GDP.

Estimates of the cost of damage from a series of environmental sources in India in 1992 are approximately 6 percent of GDP in the ADB studies. The elements included urban air pollution, health costs from water quality, soil erosion, and deforestation, while the study excludes traffic-related costs, pollution costs from toxic wastes, and biodiversity losses.

The other studies included in table 7.2 are less complete in their coverage of environmental damage. Studies for Indonesia of the health costs of particulate and lead levels (gasoline related) set these levels above those laid down as standards by the World Health Organization (WHO), at approximately 2 percent of GDP in 1989. In Pakistan the health impacts of air and water pollution along with productivity losses from deforestation and soil erosion are estimated at approximately 3.5 percent of GDP in the early 1990s. The ADB studies of the Philippines concentrate on the Manila area alone and look at the effects of lowered air and water quality; the cost estimate for this component of damage is approximately 1 percent of GDP. In Thailand, the health effects of particulates and lead levels (gasoline related) in excess of WHO standards are put at 2 percent of GDP.

Table 7.3 reports estimated time-loss costs from traffic congestion for a sample of Asian cities. These are also cited in the 1998 HDR and are in addition to those costs listed in table 7.2. For Bangkok time-related costs from traffic are estimated at 2 percent of local product in 1994; these estimates are 0.4 percent for Seoul in the same year. Health-related costs of traffic are already included in the studies referred to in table 7.2.

Table 7.3	Estimates of Time Losses due to Traffic Congestion in Asian Cities, 1994	
City	Annual Cost of Time Delays (millions of dollars)	Cost (percentage of local citywide product)
Bangkok	272	2.1
Kuala Lumpur	68	1.8
Singapore	305	1.6
Jakarta	68	0.9
Manila	51	0.7
Hong Kong	293	0.6
Seoul	154	0.4

Source: WRI (1996); U.N. (1998).

What is striking about these two sets of studies is that, in the case of the two more comprehensive country studies (China and India), the estimates for the combined environmental damage are large, in the region of 10 percent of GDP in the case of China, neglecting damage from additional sources such as time loss in traffic. Given that model-based analyses of the gains from more conventional policy reform (such as tax or trade reform) in those countries often produce estimates that are lower (perhaps 1–3 percent of GDP), this suggests that environmental policy should perhaps receive a higher weighting in the overall policy stance in these countries than it does currently.

In addition, the composition of environmental damage costs in these countries is striking. The studies of China in the ADB compendium suggest that perhaps 70–80 percent of environmental damage occurs through degradation, largely in rural areas; a range echoed in Smil (1992). While the numbers for India are perhaps less dramatic, the high estimates of the costs of soil erosion outside Asia[6] seem to us to support our contention that the degradation of the environment rather than damage caused by pollutants may well be the more important environmental issue in developing countries.

7.2.3   Transborder Environmental Externalities and the Environmental Regime in Developing Countries

Developing countries both contribute to and are affected by a range of transborder and global externality problems. In table 7.4 we list some of the more major transborder and global environmental externalities involved, both those affecting and those contributed to by developing countries. These also form part of the typical environmental regime in developing countries, and, although we do not emphasize them here, we mention them nonetheless.

Global warming is perhaps the more major transborder environmental issue for the developing countries, with temperature rise and microclimate changes as the projected outcome, combined with increased frequency of extreme weather events. The possible impacts on developing countries are thought to be potentially more significant for low-terrain countries such as Bangladesh, as are the adjustment problems faced by smaller countries as microclimates change (such as in western Africa) and labor flows across borders.

Further transborder elements forming part of the environmental regime in these countries include the thinning of the ozone layer, which increases ultraviolet-light penetration of the atmosphere. These effects are more severe in the temperate climates of developed countries than in the developing countries, but the ability of the developing countries to abate damage of this type is more limited than that in the developed world, especially

6. See Barbier (1998) and Schatan (1998).

Table 7.4	Transborder and Global Environmental Externalities Affecting Developing Countries
Global warming	Temperature rise and microclimate change, combined with increased frequency of extreme weather events
Ozone depletion	Thinning of ozone layer increases ultraviolet-light penetration of the atmosphere; effect more severe in temperate climates
Biodiversity loss and deforestation	Loss of gene pool through forest and wildlife erosion (e.g., mangrove losses linked to shrimp farming); loss of forests affects local populations who use nontimber forest products, reduces carbon absorption by forests, and increases water runoff in flooding
Acid rain	Airborne acid depositions; high in areas such as south and east China, north and east India, Korea, and Thailand (e.g., wheat yields halved in areas of India close to SO_2 emissions)

because much of the population spends a larger fraction of their time outdoors.

We also include problems associated with loss of biodiversity and deforestation as a part of the transborder and global regime. For loss of biodiversity, the issue is loss from the gene pool through flora and fauna damage. The environmental effects of economic activities that affect resources with global existence value (including species and biodiversity) is one aspect. Shrimp farming, for instance, has grown in the last 2 decades from initially low levels in Thailand and other countries, and with it has come a significant loss of mangroves and a resulting loss of biodiversity. Many pharmaceutical products sold worldwide each year are generated from forest-related sources in developing countries. The global impacts of forest loss occur through many channels, including carbon-sink reduction and impacts on existence value abroad. But forest loss also affects local populations who use nontimber forest products can cause increased water runoff in the event of flooding.

Acid-rain problems include airborne acid deposits affecting buildings and agricultural yields; these problems are especially significant in such areas as south and east China, north and east India, Korea, and Thailand. The 1998 HDR reports that in areas in India that are close to SO_2 emissions sources (admittedly mostly originating in India) the wheat yields are estimated to have been halved due to these emissions. While these global and transborder externalities are also part of the environmental regime in developing countries, both their impact on individual countries and the contribution of countries to global damage because of them remain poorly quantified.

7.3 Growth, Policy Reform, and the Environmental Regime in Developing Countries

The discussion in section 7.2 emphasized the wide range of externalities that make up the environmental regime in developing countries, along with the seeming quantitative dominance of environmental problems associated with degradation over those associated with pollutants. But how does this regime change as countries grow? Does environmental quality improve or worsen, and in what dimension and for what reasons? And what policy measures contribute to the environmental situation, either positively or negatively?

7.3.1   The Environmental Kuznets Curve

One of the more prominent of the recent discussions on these issues focuses on the environmental Kuznets curve (EKC). The EKC refers to the relationship between environmental indicators of certain types and per capita incomes of countries; its origins lie in Kuznets's work in the 1950s on income inequality measures across developing countries, which documented a clear trend initially toward increased inequality as per capita income grows, with a subsequent fall. This work suggested an inverted U shape for a cross-country plot of an inequality measure such as a Gini coefficient against income per capita. The EKC hypothesis is that environmental indicator levels first rise (e.g., pollutant levels per capita rise) as per capita income rises; then the relationship reverses after some threshold level of income.

The implication drawn by some from EKC plots is that growth need not be inconsistent with the objective of improving environmental quality in the medium to longer run: Environmental concerns can be delinked from growth objectives. Indeed, some authors have gone further and argued that the best way to improve environmental quality is to follow policies that make countries rich in the shortest possible time, since in the long run there is no conflict between growth and environmental protection. Andreoni and Levinson (1998) and Jaegar (1999) have recently provided microfoundations for the EKC, arguing that the characteristics of cleanup technology are key to the EKC.

The first paper in this area, by Shafik and Bandopadhyay (1992) (a background study for the 1992 *World Development Report,* World Bank 1992 with results given prominent profile in the report itself), examines a range of environmental indicators. These include lack of clean water, lack of urban sanitation, ambient levels of suspended particulate matter (SPM), ambient sulfur oxides, change in forest area during the period 1961–86, the annual rate of deforestation between 1961 and 1986, dissolved oxygen in rivers, fecal coliforms in rivers, municipal waste per capita, and carbon emissions per capita. Their sample consists of observations of up to 149 countries for the period 1960–90, although their coverage is incomplete. Some of the dependent variables are observed for cities within countries, in other cases for countries as a whole. Only in the case of air pollutants is an EKC type relation found. Lack of clean water and lack of urban sanitation are found to decline uniformly, both with increasing income and over time. Deforestation seems to be unrelated to income. River quality tends to monotonically worsen with income.

Selden and Song (1994), following Shafik and Bandopadhyay (1992), focus exclusively on air pollutants in their examination of possible EKC relationships. They study emissions of SO_2, NO_x, SPM, and CO. Emissions are measured in kilograms per capita on a national basis with pooled cross-sectional and time-series data drawn from WRI. The data are averages for 1973–75, 1979–81, and 1982–84. There are 30 countries in their sample: 22 high-income, 6 middle-income, and 2 low-income countries. Their results indicate that emissions of CO are independent of income, whereas emissions of other pollutants follow an EKC pattern. However, the turning points occur at much higher levels of income than in the Shafik and Bandopadhyay (1992) study.

Grossman and Krueger (1995) have subsequently investigated EKC relationships using the GEMS cross-country data on air quality for the period 1977–88 and isolate a series of environmental indicators: SO_2 concentration in selected cities, smoke, dissolved oxygen in water, BOD, chemical oxygen demand (COD), nitrates, fecal coliform, total coliform, lead, cadmium, arsenic, mercury, and nickel. The data measure ambient air quality at two or three locations in each of a group of cities in a number of countries during the period 1977–88. The number of observations varies over time (52 cities in 32 countries in 1982, but only 27 cities in 14 countries in 1988). The authors claim that the data are representative of countries at varying levels of economic development and with different geographical conditions, and they find an EKC type relation for SO_2, smoke, dissolved oxygen, BOD, COD, nitrates, fecal contamination of rivers, and arsenic. The evidence is less compelling for total coliform and heavy metals.

Also in the literature is Panayotou (1993), which estimates EKC-type relationships for SO_2, NO_x, and SPM, and deforestation using cross-sectional data for 1985 and, as in Selden and Song (1994), pollutants measured in emissions per capita on a national basis. Panayotou finds EKC-type relations for SO_2, NO_x, and SPM. Turning points were at levels of income lower than those in Selden and Song (1994). Cooper and Griffiths (1994), in contrast, estimate three regional (Africa, Latin America, and Asia) EKCs for deforestation only, using pooled cross-sectional time-series data for each region for the period 1961–91 and for 64 countries. They find no EKC relationship.

These findings are such that it is now often argued that attempts to estimate EKC-type relationships should be confined to air pollutants alone, and, in particular, to SO_2 emissions. As a result, drawing conclusions from any EKC plot as to how overall environment damage behaves as income changes is thought to be fraught with problems.

But even for SO_2, the EKC does not appear to be a particularly robust description in the current literature of the behavior of environmental pollutants vis-à-vis income per capita. Kaufman et al. (1998) point out a number of econometric problems with EKC estimates, including violations of homoscedasticity, the nonuse of random- and fixed-effects methods in panel data, the improper definition of dependent and independent variables, and other problems. Kaufman et al. try to circumvent these difficulties in their attempt to identify an EKC-type relation in the case of SO_2, defining SO_2 concentrations as annual average concentrations in ground-level atmosphere at a particular location in a city. Using a panel of 23 countries (13 developed, 7 developing, and 3 centrally planned) during the period 1974–89, their analysis shows an EKC-type relation between emissions per capita and spatial intensity of economic activity, as well as between emissions per capita and GDP per capita. However, they also find evidence that still further increases in income per capita lead to a further increase in emissions per capita—an N-type rather than inverted-U-type relation between emissions per capita and GDP per capita.

Unruh and Moomaw (1998) evaluate whether the transition from a high emissions to a low emissions state occurs mechanically at a particular income level, as suggested by earlier papers. They identify some industrialized countries that seem to have gone through EKC-type transitions, discovering that these transitions span a broad range of income levels.[7] Furthermore, the transitions occur abruptly and cotemporally and do not appear to be the consequence of endogenous income growth. Rapid and cotemporal historical events, technological progress, and the need to react to external shocks seem to drive the EKC structure. Ekins (1997) argues that the pattern of emissions of selected air pollutants does not indicate the environmental impact of such emissions and examines an aggregate indicator of environmental impact developed by the Organization for Economic Cooperation and Development (OECD). Examining the relationship between this indicator and income per capita, Ekins finds no evidence in favor of an EKC.

Thus, even considered within its own confines, the relation between economic growth and environmental damage seems more complex than portrayed by the EKC (Barbier 1997). There appears to be nothing automatic about this relation, nor is any inference on causality necessarily justified. Once degradation effects are added in, drawing conclusions as to how overall environmental quality changes with income is even more treacherous. For instance, soil erosion problems, measured relative to aggregate income, would seem to recede as growth occurs and (in relative terms) the agricultural sector shrinks. But growth is accompanied by urbanization

---

7. In a recent paper, Torras and Boyce (1998) take the existence of the EKC at face value and ask whether it is merely the level of income or also its distribution that affects emissions per capita. They argue that a more even distribution of income, higher literacy rates, and other indicators of power lead to lower emissions per capita.

and congestion problems, which, relative to income, may recede after a transitional period when growth and new infrastructure are taking hold.

7.3.2   The Environmental Effects of Policy Reform (Trade and Environment)

A further strand of the recent literature attempts to assess how environmental quality changes with policy changes, including trade liberalization; in particular how various kinds of pollutant concentrations can be affected. Copeland and Taylor (1994), for instance, evaluate the role of trade where environmental quality is a local public good (i.e., damage from pollutants remain in the country). They consider a two-country single-period equilibrium where goods differ in pollution intensity in production. Countries differ in their endowment of a primary factor (human capital); environmental quality in both countries is a normal good in preferences, and, with assumed endogenous setting of pollution policy, the higher-income country has higher environmental standards. They find that free trade shifts pollution-intensive production toward countries scarce in human capital and raises world pollution levels.

Copeland and Taylor (1995) consider a different case where environmental quality is a pure public good to which all countries are exposed. Trade effects are different in this case, since relocation of pollution-intensive industries to countries with less stringent environmental protection can increase the exposure of residents in the home country and works against more conventional gains from trade. Since there are transborder externalities in this case, nationally based pollution regulation does not lead to Pareto optimality, and free trade need not raise welfare.

More recently, Antweiler, Copeland, and Taylor (1998) first generate and then test a series of propositions as to how economies behave in terms of their trade and environment linkages. They assume a small, open economy formulation: The economy has a number of agents, produces two final goods, and uses two primary factors. One product is labor intensive and involves no pollution, whereas the other is capital intensive and causes pollution. They assume producers have access to an abatement technology, which, for simplicity, only uses the polluting good as an input. They also assume that the government uses emissions taxes to reduce pollution, and, given the pollution tax rate, they generate a firm-level profit function.

The level of the tax actually used is assumed to be an increasing function of what an optimally set tax would be. This treatment allows government behavior to vary across countries and also allows for environmental policy to respond and differ by country: On the demand side, consumers maximize utility, taking pollution as given; they assume preferences over goods are homothetic, while there is constant marginal disutility of pollution.

The model allows them to decompose a total change in pollution levels into scale, composition, and technique effects. This, in turn, allows them to generate a number of theoretical propositions to test. Thus, if economies differ only with respect to their degree of openness to trade, and both countries export the polluting good, then pollution will be higher in the country that is less open. Where the world price is fixed, then for a given level of income and for certain settings of key model parameters, they show that the composition effect associated with trade liberalization in such countries is to increase pollution. These and other propositions as to how the links between trade and environment operate emerge from their analysis as they focus on emissions associated with trade-related polluting activity.

However, as our earlier discussion indicates, emissions are likely to constitute only a portion of the overall welfare cost of environmental externalities in liberalizing developing countries, and other environmental externalities may well have different interactions with trade. Thus, if with increased trade labor moves from rural to urban areas and if this generates increased congestion, these adverse consequences linked to trade can easily dominate the overall environmental impact compared to changes in emissions. Impacts on soil erosion from agricultural trade liberalization abroad can be adverse, while being beneficial at home. Liberalization in the manufactured sector can produce opposite implications for soil erosion. A wider view of the environmental regime in developing countries can thus also produce different conclusions as to what the key linkages between policy changes and the environment actually are.

7.4   Measuring the Degree of Environmental Failure in Developing Countries

Given the preceding discussion, if pollutant levels across economies do not provide a complete picture for the evaluation of comparative environmental performance across countries or over time, either in analytical or empirical work, what is a more appropriate way to proceed? Unfortunately, the problem is not only the incomplete coverage of environmental externalities in developing countries; one also needs estimates of damage functions, which allow the losses involved to be computed in terms of welfare. Thus, even if economies have high levels of emissions per capita, if the ability to abate differs across economies (such as health care capabilities to deal with adverse effects of emissions), then differences in emissions levels across countries do not necessarily map onto comparable differential welfare losses due to environmental failures. In the appendix to this chapter we show for the special case of a stock externality that, even if an EKC relationship is followed in emissions per capita, this need not map onto a comparable relationship in terms of welfare.

For these reasons, therefore, an alternative approach is needed to evaluate the significance of environmental failures across economies or over time, and hence to assess the impact of the environmental regime in devel-

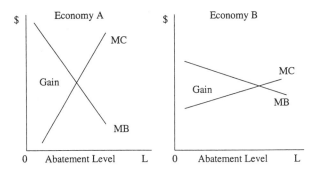

Fig. 7.1   Abatement levels and welfare gains from internalization in two economies

oping countries. The appropriate concept to us seems to be a distance measure reflecting the severity of the departures from Pareto optimality associated with externalities—how far away are economies from Pareto optimality in a welfare sense, and what would be the potential welfare gains from moving from the current allocation of resources with uninternalized or partially internalized externalities to complete internalization? The implied measure seems to be a money-metric measure (say a Hicksian measure) of the gain from internalization relative to a current noninternalized equilibrium. Income effects associated with different assignments of property rights would affect the precise fully internalized equilibrium, although we put these issues to one side for now. Such a measure of gain is implicit in the literature discussed in section 7.2, which produces estimates of the costs of various kinds of environmental failures in terms of GDP per capita; but much of this literature is not explicit about the precise welfare formulation used.

Such measures need not behave in any way that is necessarily collinear with levels of emissions or intensity of environmental failure. Figure 7.1 shows schematically how a comparison across two economies with differing levels of emissions may yield larger gains to the economy with smaller emissions. Here, we represent marginal benefit (MB) and marginal cost (MC) of abatement functions for two economies. Economy A has more steeply sloped functions, and in Pareto optimality has smaller abatement than B. But the gains from abatement (internalization) are larger in A than in B because of the more shallowly sloped functions in B. Comparing pollutant levels across economies need give no guide as to the relative size of the gains from internalization.

The seemingly large estimates we reported earlier of the gains from internalizing environmental externalities in developing countries also suggest the perhaps obvious question of why it is that if internalization gains are so large, more internalization has not occurred. It would be wrong to say that no internalization has occurred in these countries. At the village level, terracing and other schemes are designed to remedy some of the ills of soil erosion. National environmental regulation often approaches levels of stringency seen in regulation in developed countries, but is accompanied by problems of enforcement and compliance. In many developing countries, environmental nongovernmental organizations (NGOs) are also extremely active, generating a rising profile for environmental issues in local policy debate, even though large potential gains from internalization still seem to remain.

Various explanations abound for the presence of these seemingly large potential gains. One is that the technology of internalization is both capital intensive and high cost for low-income countries. Monitoring devices and administration of environmental fees and fines all require inputs on a scale not easily attained in low-income countries. Another is that if environmental quality is costly to provide, then models with traditional preferences and technology would naturally imply that abatement levels are lower in low-income countries. These effects, in turn, would be exacerbated by income elasticities of demand for environmental quality exceeding 1, as is often claimed.

Another direction explored in recent literature (see López 1998; Maler 1998) is that outside shocks to social systems are a significant compounding factor, either disrupting or delaying internalization and producing lowered environmental quality. Particularly important in this discussion is the observation that environmental management systems in developing countries commonly rely on informal social norms, which can partially or wholly break down under rapid population growth, technological innovation, or changes in market outcomes. Previously reasonably well-managed resources can become open-access, poorly managed resources, with worsened environmental quality the result. Dasgupta and Maler (1991) argue that, viewed in these terms, poverty and degradation can even be reinforcing. Thus, if deforestation moves the available firewood in forests progressively further from villages, families may have more children to offset the increased time required to collect firewood.[8] Population growth is higher and with it the demand for firewood, producing further degradation.

7.5   Comparing Policy Regimes in Developed and Developing Countries

We often tend to think of developing countries as following the developmental experience of developed countries with a form of compressed lag.

---

8. This hypothesis has been tested empirically by Filmer and Pritchett (1996) using household data for Pakistan for 1991–92. They conclude that households living in areas in which the distance from firewood is greater have more children.

OECD countries over some 200 years have grown and developed, transforming themselves first from agrarian societies to industrial economies based on heavy industry (steel and chemicals) to modern high-technology service-based economies. Developing countries are following this experience at varying speeds and in different ways, but the transition time is clearly shorter. Korea, for instance, has transformed itself from a country with lower income per capita than India in the mid-1950s to a lower-income OECD country today—a 40-year transition. Furthermore, unlike developed countries at the height of their industrial growth, developing countries today are under considerable pressure to reduce environmental stress. This pressure (sometimes backed by the threat of punitive action) comes from a number of sources, such as governments of developed countries, international funding agencies, academia, local and international NGOs, and the developing countries' own bodies of jurisprudence. Such pressures were unheard of during the days of rapid industrial growth of the currently developed countries.

It is only relatively recently, however, that developed countries have gained the environmental awareness they now have and have developed systems of environmental management that control emissions, treat waste, and otherwise abate environmental damage. At the height of the OECD countries' industrial revolutions, effectively no environmental controls were in place.

What then should developing countries do? The experience of developed countries would seem to indicate that they should adopt few environmental controls and that with income growth environmental quality will improve. Indeed, a great fear is that attempts to heighten environmental regulation will only serve to slow growth and, hence, slow eventual achievement of higher environmental quality through growth. On the other hand, because of problems of compliance one can argue that perhaps developing countries have no choice but to follow the older developed countries' industrial revolution experience of largely benign neglect.

There are, however, some key differences in the developing countries' experience in this area compared to the industrial revolution of old. First, the time periods involved are compacted, and hence the flow of environmental damage per year during industrialization is larger. Second, the shocks that hit the economies are also much more severe than was true of the old industrial revolutionizers. The industrial-revolution-era economies of the developed countries simply did not experience population growth rates of 3 percent or more per year, massive growth in urban vehicle densities, and other elements that contribute to today's environmental ills in the developing world. Not only is the process more compact, the time-adjusted severity of damage probably exceeds that experienced in the OECD 100 years ago. Third, even though weakly administered, there are abatement technologies that can be and are being employed; and even though there is political opposition, environmental management is taking root.

Thus, the large cost estimates we reported earlier and the scope of environmental problems in developing countries suggest to us that a much more activist environmental policy regime will continue to emerge in developing countries than was true in industrial countries some 100 years ago as they grew and industrialized. And, unlike the past, this regime will have an equal focus on degradation, if not a dominant focus on degradation over pollution.

7.6   Concluding Remarks

This paper discusses the environmental regime in developing countries, stressing both the complexity of the regime and the wide-ranging nature of environmental externalities that go beyond more conventional literature discussion of pollutant levels. It suggests that a full characterization of this regime needs to focus on externality problems such as soil erosion, open-access resources, and congestion problems in urban areas. The paper stresses that from available studies the gains from internalization of these externalities seem to be large, potentially exceeding numerical (model-based) estimates of gains from conventional policy reforms (such as trade or tax reform) by substantial orders of magnitude. Also, the majority of such gains seem to arise from internalizing externalities associated with degradation (soil erosion, open-access resources, and congestion) rather than pollution. We also stress how existing literature that discusses how the environmental situation changes with growth (the EKC) covers only part of the environmental situation; a point that also applies to other parts of the literature such as that on policy reforms (trade liberalization) and environmental quality.

Having developed this picture of the environmental regime in developing countries, the paper concludes by suggesting that a measure is needed of overall environmental performance in terms of departures from Pareto optimality so as to give a money-metric welfare measure of the gains of moving to complete internalization. It also discusses some of the reasons for the lack of internalization, citing recent literature that argues that social conventions defining implicit management regimes come under stress as rapid urbanization, rapid population growth, and other shocks to social systems occur. The overall theme of the paper, repeated throughout, is that in discussing the environmental situation in developing countries, a more comprehensive sense of what this regime comprises is needed.

Appendix

Internalization Gains and the Environmental Kuznets Curve

The EKC literature discussed in this chapter seemingly points to the conclusion that there is no clear evidence in favor of the EKC. Even though the EKC itself may be empirically dubious, its welfare interpretation also has to be highly qualified. Here, we develop a model where optimality is defined as internalization, and since such internalization is, in principle, independent of the level of emissions, the EKC even if it were to exist lacks any welfare content. We use an amended version of the growth with stock externalities model, showing that alternative technological assumptions can give us different (optimal) relations between emissions and income, and each such relation is consistent with perfect internalization. The emphasis in the model is on shadow pricing the external effect appropriately (Ko, Lapan, and Sandler 1992).

In the model, (1) labor is normalized to equal 1. (2) Output, y, depends upon capital, k, and emissions, e, $y = f(k, e)$. An important point here is the nature of the relationship between k and e. We assume that $f_{ke} > 0$, (i.e., capital and emissions are substitutes). We further assume that there exists a level of emissions $\bar{e}$ such that the marginal product of emissions for a given level of capital is 0. (3) Capital accumulates according to the equation

(A1) $$\dot{k} = f(k,e) - c - \delta k,$$

where c is consumption and δ is the rate of depreciation of capital. (4) Pollution accumulates according to the relation

(A2) $$\dot{S} = -bS + e,$$

where b is fixed.

The social planner's problem is to choose nonnegative consumption and emissions paths that solve the infinite horizon maximization problem,

(A3) $$\int_0^\infty e^{-\rho t} U(C,S) dt,$$

subject to equations (1) and (2). Here $U(\cdot)$ is the instantaneous utility of the representative consumer and ρ is the discount rate. The Hamiltonian for this problem is

(A4) $$H(k,S,c,e) = U(c,S) + \theta_1(t)[f(k,e) - c - \delta k] + \theta_2(t)(e - bS),$$

where θ_1 and θ_2 are costate variables. First-order conditions imply

(A5) $$U_c = \theta_1,$$

assuming we always have an interior solution; and

(A6) $$\theta_1(\partial f/\partial e) + \theta_2 \leq 0,$$

with equality if $e^*(t) > 0$. The canonical equations are

(A7) $$\dot{\theta}_1 = [\rho + \delta - (\partial f/\partial k)]\theta_1,$$

(A8) $$\dot{\theta}_2 = [(\rho + b)\theta_2] - \partial U/\partial S,$$

and transversality conditions apply:

(A9) $$\lim_{t \to \infty} e^{-\rho t}\theta_1(t) = \lim_{t \to \infty} e^{-\rho t}\theta_2(t) = \lim_{t \to \infty} e^{-\rho t} k(t) = \lim_{t \to \infty} e^{-\rho t} S(t) = 0,$$

which require that the present value of capital and pollution becomes negligible at infinity.

This welfare exercise refers to the optimal solution obtained in a command economy. From the first-order conditions we can solve for optimal consumption and optimal emissions as $c^* = c(k, S, \theta_1, \theta_2)$ and $e^* = e(k, S, \theta_1, \theta_2)$. If we assume that the production and the utility functions are strictly concave in this case, then for given values of parameters, c^* and e^*, the issue is how this may be expected to vary with c. If we use this result of θ_2, then it follows that, from a welfare point of view, the relationship between consumption and emissions is monotonically falling. Richer countries will have higher θ_2 and therefore, lower emissions ceteris paribus than poorer countries.

In a competitive market economy the representative consumer takes as given time paths $\{w(t), r(t), \pi(t)\}$ for $t\varepsilon [0,\infty)$, of wages, interest rates, and profits. The instantaneous utility of the consumer is defined by $U(c, S)$ as before. The consumer sells the fixed labor input (normalized to unity) to a representative firm at the market-determined wage rate, and rents out capital, $k(t)$, at the market rate of interest to the firm. The representative firm maximizes profits under competitive conditions. It generates emissions $e(t)$ per unit time and pays a tax $\lambda(t)$ on these emissions. Total tax proceeds collected by the government are redistributed to the consumer. The consumer maximizes utility and has perfect foresight about market wage rates and other variables.

The consumer maximizes

(CP) $$\int_0^\infty e^{-\rho t} U(c(t), S(t)) dt,$$

subject to

$$\dot{k} = \pi(t) + rk(t) + \lambda(t)e(t) - c(t) - \delta k(t),$$

and treats S as a parameter. The variable ρ is the consumer's discount rate and δ is the rate of depreciation of capital.

The firm takes as given (and has perfect foresight about) time paths of emissions taxes $\{\lambda(t), t\varepsilon, [0,\infty)\}$ along with the time paths of wage and interest. The firm can reduce its tax liabilities by reducing output. Output is produced according to a standard neoclassical production function so that the firm chooses $k(t)$ and $e(t)$ to solve the problem

(FP) $\max \pi(t) = f(k(t), e(t)) - r(t)k(t) - \lambda(t)e(t).$

Given that the consumer perfectly predicts the time paths of $\{w(t), r(t), \pi(t)\}$ and the firm perfectly predicts the time paths of $\{w(t), r(t), \pi(t)\}$, then the consumer will determine consumption demand (c^d) and capital supply (k^s), whereas the firm will determine consumption supply (c^s) and capital demand (k^d) and the emissions $e(t)$. The paths $\{w(t), r(t), \pi(t), \lambda(t)\}$ are a perfect foresight competitive equilibrium with emissions taxes if the solution $\{c^s(t), k^d(t), e(t)\}$ of equation (FP) is such that if profits are defined by $\pi(t) = f(k(t), e(t)) - r(t) k(t) - \lambda(t) e(t)$ for each t and if $\{c^d(t), k^s(t)\}$ solves equation (CP), then for all $t\varepsilon [0,\infty)$ we have (1) $c^d(t) = c^s(t)$ goods market or flow equilibrium; (2) $k^s(t) = k^d(t)$ capital market or stock equilibrium; (3) $e^c(t)$ is the competitive emissions; and (4) $S = e^c(t) - bS(t)$, $S(0) = S_0$ (evolution of pollution stock).

An examination of the planner's problem in equation (A1) immediately reveals that if the emissions tax is defined as $\lambda(t) = -\theta_2(t)/\theta_1(t)$, the competitive equilibrium solutions for equations (CP) and (FP) for the firm are identical to the solution of the social optimization problem. To see this, assume that equation (FP) has an interior solution; then we must have

(A10) $\partial f/\partial k = r,$

(A11) $\partial f/\partial e = \lambda.$

These determine the demand for capital and the competitive supply of emissions. Given this, then the consumer maximizes the following Hamiltonian:

(A12) $H = U(c,S) + \gamma(t)(\pi + rk + \lambda e - c - \delta k).$

The first-order conditions are

(A13) $\partial U/\partial c = \gamma,$

(A14) $\dot{\gamma} = (\rho + \delta - r)\gamma,$

(A15) $\dot{k} = \pi + rk + \lambda e - c - \delta k \quad \text{with } k(0) = k_0.$

The transversality conditions are

(A16) $$\lim_{t \to \infty} e^{-\rho t} \gamma(t) = \lim_{t \to \infty} e^{-\rho t} \gamma(t) k(t) = 0.$$

If we compare this solution to that for the planner's problem, it is clear that for $\gamma = \theta_1$ and $\lambda = -\theta_2/\theta_1$ the solutions to the two problems are identical. Hence, by solving the social optimization problem and using an optimal and flexible emissions duty, the planner can induce profit-maximizing firms to follow the socially desirable emissions policy.

An important implication of the solution to the market problem is that if we have incomplete internalization ($\lambda \neq -\theta_2/\theta_1$), then this carries a welfare cost. The EKC, even if it is observed, then does not give any indication of the welfare cost of noninternalization across countries.

References

ADB (Asian Development Bank). 1997. *Emerging Asia—Changes and challenges.* Manila: Asian Development Bank.

Agarwal, A. 1996. Pay-offs to progress. *Down to Earth* 5 (10): 31–39.

Alfsen, K. H., M. A. de Franco, S. Glomsrod, and T. Johnson. 1996. The cost of soil erosion in Nicaragua. *Ecological Economics* 16:129–45.

Andreoni, J., and A. Levinson. 1998. The simple analytics of the environmental Kuznets curve. NBER Working Paper no. 6739. Cambridge, Mass.: National Bureau of Economic Research.

Antweiler, W., B. R. Copeland, and M. S. Taylor. 1998. Is free trade good for the environment? NBER Working Paper no. 6707. Cambridge, Mass.: National Bureau of Economic Research .

Barbier, E. 1997. Introduction to the environmental Kuznets curve special issue. *Environment and Development Economics* 2:369–81.

———. 1998. The economics of soil erosion: Theory, methodology and examples. In *The Economics of environment and development,* ed. E. Barbier. London: Edward Elgar.

Cooper, M., and C. Griffiths. 1994. The interaction of population and growth and environmental quality. *American Economic Review* 84:890–97.

Copeland, B. R., and M. S. Taylor. 1994. North-south trade and the environment. *Quarterly Journal of Economics* 109:755–87.

———. 1995. Trade and transboundary pollution. *American Economic Review* 85:716–37.

Cruz, W., H. A. Francisco, and Z. T. Conway. 1988. The on-site and downstream costs of soil erosion in the Megat and Pantabangan watersheds. *Journal of Philippine Development* 15 (1): 85–111.

Dasgupta, P., and K. G. Maler. 1991. The environment and emerging development issues. In *Proceedings of the World Bank annual conference on development economics,* ed. S. Fischer, D. de Tray, and S. Shah, 101–52. Washington, D.C.: World Bank.

Ekins, P. 1997. The Kuznets curve for the environment and economic growth: Examining the evidence. *Environmental Planning* 29:805–30.

Filmer, D., and L. Pritchett. 1996. Environmental degradation and the demand for children. World Bank, Discussion paper. Washington, D.C.

Grossman, G., and A. Krueger. 1995. Economic growth and the environment. *Quarterly Journal of Economics* 110:353–77.

Hettige, H., M. Mani, and D. Wheeler. 1997. Industrial pollution in economic development. Washington, D.C.:World Bank. Development Research Group.

International Cooperation Agency. 1990. The second country study for Japan's official development assistance to the Kingdom of Thailand. Tokyo: International Cooperation Agency.

Jaegar, W. K. 1999. Economic growth and environmental resources. Department of Economics, Williams College. Mimeo.

Kaufman, R. K., B. Davidsdottir, S. Garnham, and P. Paul. 1998. The determinants of atmospheric SO_2 concentrations: Reconsidering the environmental Kuznets curve. *Ecological Economics* 25:209–20.

Ko, I. D., H. E. Lapan, and T. Sandler. 1992. Controlling stock externalities: Flexible versus inflexible Pigovian corrections. *European Economic Review* 36: 1263–76.

López, R. 1997. Evaluating economywide policies in the presence of agricultural environmental externalities: The case of Ghana. In *The greening of economic policy reform*. Vol. 2, *Case studies,* ed. W. Cruz, M. Munasinghe, and J. J. Warford 27–54. Washington, D.C.: World Bank Environmental Department and Economic Development Institute.

———. 1998. Where development can or cannot go: The role of poverty-environment linkages. In *Annual World Bank conference on development economics,* ed. B. Pleskovic and J. E. Stiglitz, 285–306. Washington, D.C.: World Bank.

Magrath, W., and P. Arens. 1989. The costs of soil erosion on Java: A natural resource accounting approach. Environment Working Paper 18. Washington, D.C.: World Bank.

Maler, K. 1998. Environmental poverty and economic growth. In *Annual World Bank conference on development Economics,* ed. B. Pleskovic and J. E. Stiglitz, 251–70. Washington, D.C.: World Bank.

Panayotou, T. 1993. Empirical tests and policy analysis of environmental degradation at different stages of economic development. Working Paper WP238. Geneva: Technology and Employment Program.

Schatan, C. 1998. Some crucial environmental Latin American issues: Precedents for a WEO discussion. Economic Commission for Latin America and the Caribbean, Mimeo.

Selden, T., and D. Song. 1994. Environmental quality and development: Is there a Kuznets curve for air pollution emissions? *Journal of Environmental Economics and Management* 27:147–62.

Shafik, N., and S. Bandopadhyay. 1992. Economic growth and environmental quality: Time series and cross-country evidence. Background paper for the World Bank *World Development Report 1992.* Washington, D.C.: Oxford University Press.

Smil, V. 1992. Environmental changes as a source of conflict and economic losses in China. Paper prepared for a workshop on Environmental Change, Economic Decline, and Civil Strife, Institute for Strategic and International Studies, Kuala Lumpur, Malaysia, 1991.

Torras, M., and J. Boyce. 1998. Income inequality and pollution: A reassessment of the environmental Kuznets curve. *Ecological Economics* 25:147–60.

United Nations (U.N.). 1998. *UN human development report.* New York: Oxford University Press.

Unruh, G. C., and W. R. Moomaw. 1998. An alternative analysis of apparent EKC-type transitions. *Ecological Economics* 25:221–29.

World Bank. 1992. *World development report, 1992.* Washington, D.C.: Oxford University Press.

WRI (World Resources Institute). 1996. *World resources 1996–97*. New York: Oxford University Press.

Comment   Edward B. Barbier

The paper by Jha and Whalley on the environmental regimes found in developing countries is a timely and cogently written analysis of the subject. The authors review a wide range of literature and, in doing so, cover many issues concerning the diverse environmental problems faced by developing countries.

The authors confine their analysis of this potentially broad topic to three main points. First, they argue that, although much attention has focused on the growing welfare implications of increased pollution levels in developing countries, problems of degradation (i.e., soil erosion, deforestation, overexploitation of fisheries, etc.) deserve much more attention. Second, available evidence on the economic costs to developing countries of environmental problems suggests that these costs are large, particularly with respect to the degradation problems. Finally, citing evidence from the emerging environmental Kuznets curve (EKC) and trade and environment literature, the authors point out that the relationship between growth, policy reform, and environmental quality may differ significantly depending on whether one is examining degradation or pollution problems. The authors conclude by examining a perplexing issue: If the welfare costs of many environmental externalities are so great in developing countries, why has more internalization of these costs not occurred?

I would like to make some general observations concerning these key points of the paper. First, I commend the authors for basing their paper on a distinction between conventional pollution problems and a wider, more pervasive problems of environmental degradation in developing countries. The need to make such a distinction is critical for analyzing environmental and resource issues in developing countries, because the fundamental economic and physical processes underlying degradation problems require a different approach to analyzing degradation as opposed to pollution problems (Barbier 1989; Dasgupta 1982). Unfortunately, we still need to improve our understanding of the economic aspects of environmental degradation in developing countries. As the discipline of environmental and resource economics has been developed largely in the richer or advanced industrialized countries, most of the analytical approaches are better suited to analyzing more conventional environmental

Edward B. Barbier is the John S. Bugas Distinguished Professor of Economics in the Department of Economics and Finance, University of Wyoming.

problems such as pollution, nonrenewable depletion, and standard timber- and fishery-harvesting issues. More complex environmental problems, such as poverty, land degradation, and deforestation linkages in developing countries, clearly require a different category of analysis and one focusing in particular on the incentives of poor rural households to manage their land and other resources at their disposal (Barbier 1997a).

However, I do have some issues to raise concerning the way in which the authors distinguish environmental pollution and degradation "externalities," as indicated in table 7.1 of the chapter. In turn, these issues lead to more substantive points detailing why environmental degradation is a fundamental development problem in low-income countries.

It is important to recognize that two types of environmental externalities can occur through environmental degradation and pollution: flow externalities and stock externalities. The former is the more common type of third-party externality that is associated with conventional pollution problems, or as listed in the table, soil erosion and pesticide runoff (incidentally, the latter is really an agrochemical pollution problem and not a "soil quality" issue as table 7.1 suggests). In contrast, stock externalities arise through nonoptimal depletion of a renewable resource stock over time; that is, they represent the forgone future income associated with excessive depletion or overexploitation of a resource today. The authors seem to imply that the open-access resources of forests, fisheries, and aquifers may suffer from overexploitation, hence causing a stock externality problem. However, suboptimal depletion, degradation, or overexploitation of any type of resource stock (renewable, semirenewable, or nonrenewable) can lead to stock externalities. For example, later in the paper, the authors quote numerous estimates of the on-site cost of soil erosion across developing countries, which is essentially the stock externality problem of the forgone crop income arising from excessive topsoil and soil-fertility depletion. Yet, surprisingly, this erosion cost is not included in table 7.1. An equally surprising omission is depletion of freshwater resources, which has been cited as a critical environmental issue in developing countries in coming decades by a number of international sources, including those cited by the authors (UNDP 1998; WRI 1996).

This point may seem trivial, but it is not. The authors suggest, correctly in my view, that in developing countries resource degradation problems may be economically more significant than pollution. However, it is also possible that the most significant economic costs associated with resource degradation in developing countries take the form of the forgone income associated with resource depletion and degradation stock externalities, rather than the conventional flow externalities associated with air and water pollution, off-site impacts of sedimentation, and agrochemical runoff. This in turn implies that developing countries should be more concerned about the stock externalities arising from resource degradation, because they translate into forgone income opportunities, thus undermining economic development efforts more directly. The authors are clearly acknowledging this point, at least implicitly, by stressing that resource degradation imposes disproportionately high welfare costs on these economies. This is certainly true. However, a more fundamental reason why this is the case is that many poor countries continue to remain economically dependent on natural resources for their current development efforts, and, for the foreseeable future, efficient and sustainable management of this resource base is critical to sustaining economic development. Hence, given the economic importance of many rural renewable resources in these economies, it is not surprising that degradation of these resources imposes significantly large welfare losses and economic costs.

There is also evidence emerging from the recent literature suggesting that developing countries endowed with abundant natural resources are wasting this potential "natural capital" wealth rather than efficiently exploiting it for sustainable economic development. For example, many low-income and lower-middle-income economies—especially those displaying low or stagnant growth rates—are highly resource dependent (Barbier 1994). Not only do these economies rely principally on direct exploitation of their resource bases through primary industries (e.g., agriculture, forestry, and fishing), but also over 50 percent or more of their export earnings come from a few primary commodities. These economies tend to be heavily indebted and are experiencing dramatic land-use changes—especially conversion of forest area to agriculture—as well as problems of low agricultural productivity, land degradation, and population-carrying-capacity constraints. A recent cross-country analysis by Sachs and Warner (1995) confirms that resource-abundant countries (i.e., countries with a high ratio of natural resource exports to GDP) have tended to grow less rapidly than countries that are relatively resource poor.

Explanations as to why resource dependence may be a factor in influencing economic growth point to a number of possible fundamental linkages among environment, innovation, trade, and long-term growth that are relevant to poor economies. For example, the limitations of resource-based development have been examined by Matsuyama (1992) and Sachs and Warner (1995). Matsuyama shows that trade liberalization in a land-intensive economy could actually slow economic growth by inducing the economy to shift resources away from manufacturing (which produces learning-induced growth) toward agriculture (which does not). Sachs and Warner extend the Matsuyama model to allow for the full "Dutch disease" influences of a mineral- or oil-based economy; that is, when an economy experiences a resource boom, the manufacturing sector tends to shrink and the non-traded-goods sector tends to expand. The authors' theoretical and empirical analyses support the view that a key factor influencing endogenous growth effects is the relative structural important of tradable manufacturing versus natural resource sectors in the economy.

Of course, such models do not include the effects of resource degradation or depletion per se. However, it is fairly straightforward to demonstrate some of the possible influences of environmental-asset depletion on innovation and growth in a resource-dependent economy, as well as the role of policy and institutional failures in this process (Barbier 1999; Barbier and Homer-Dixon 1999). In terms of policy implications, this suggests that low-income countries should be pursuing a two-pronged strategy for sustained economic development. On the one hand, correcting problems of chronic policy failures, social instabilities, and poor institutions that inhibit innovation and long-term growth prospects should also enhance the capacity of these economies to reinvest the rents from natural-resource exploitation into more dynamic and advanced sectors of the economy (Barbier 1999; Matsuyama 1992; Sachs and Warner 1995). However, focusing simply on policies and institutions to foster improved innovation in the advanced economic sectors of low-income economies may not be sufficient. Because these economies are highly dependent on their natural resource base for economic growth and development over the medium term, the take-off into higher growth rates and economic development will be directly related to their ability to manage natural resources efficiently and sustainably over the medium to long term. Once again, therefore, we are back to the main issue raised by the authors of this paper: The need for developing countries to recognize the economic consequences and welfare losses arising from pervasive rural resource degradation.

A major cause of environmental degradation in developing countries is the distortion in economic incentives caused by misguided policies. Curiously, Jha and Whalley do not discuss this aspect of the problem very much in their paper. Yet there is substantial evidence emerging that policy distortions and failures are a key factor in the economic disincentives for rural households to improve long-term, efficient management at their disposal (Barbier 1997a). There are two aspects of this disincentives problem that are routinely ignored by policymakers. First, empirical evidence suggests that poorer households in rural developing regions are more constrained in their access to credit, inputs, and research and extension services necessary for investments in improved land and resource management. Poverty, imperfect capital markets, and insecure land tenure may reinforce the tendency toward short-term time horizons in production decisions, which may bias land-use decisions against long-term management strategies. Second, poverty may severely constrain the ability of poor households to compete for resources, including high-quality, productive land. In periods of commodity booms and land speculation, wealthier households generally take advantage of their superior political and market power to ensure initial access to better-quality resources in order to capture a larger share of the resource rents. Poorer households are either confined to marginal land areas where resource rents are limited or only have access to higher-quality resources once they are degraded and any rents dissipated.

Economic and sectoral policies in developing countries usually reinforce these structural disincentives for improved land management rather than mitigating them. For example, in Colombia distortions in the land market prevent small farmers from attaining access to existing fertile land (Heath and Binswanger 1996). That is, because the market value of farm land is only partly based on its agricultural production potential, the market price of arable land in Colombia generally exceeds the capitalized value of farm profits. As a result, poorer smallholders and, of course, landless workers cannot afford to purchase land out of farm profits, nor do they have the nonfarm collateral to finance such purchases in the credit market. In contrast, large land holdings serve as a hedge against inflation for wealthier households, and land is a preferred form of collateral in credit markets. Hence, the speculative and nonfarming benefits of large land holdings further bid up the price of land, thus ensuring that only wealthier households can afford to purchase land, even though much of the land may be unproductively farmed or even idle.

Thus unless better policies are designed to correct such fundamental distortions in low-income economies, the disincentives for improved land management will remain. This in turn implies that economic growth in developing countries will continue to be accompanied by rapid land-use change and resource degradation.

As the authors imply, evidence that this may be a problem is emerging from the recent EKC literature. Recently I had the privilege of editing a special journal issue on the EKC. In my review of the literature, it became clear that perhaps the only significant resource-depletion indicator that has been examined for evidence of an EKC relationship has been deforestation (Barbier 1997b). However, as Jha and Whalley have also indicated, the evidence on this relationship is mixed. Some studies suggest that deforestation conforms to the EKC hypothesis; others have found it difficult to establish a relationship between any indicator of deforestation and income (Cropper and Griffiths 1994; Shafik 1994; Antle and Heidebrink 1995; Panayotou 1995). Perhaps most worrying is that, where an EKC relationship for deforestation has been established, the real per capita income levels of virtually all developing countries in the world are well to the left of the turning-point level of income on the curve, where deforestation starts to decline. The implications of this for medium-term global deforestation trends were illustrated when colleagues and I combined an estimated EKC deforestation relationship with aggregated forecasts of income and population levels for individual countries (Stern, Common, and Barbier 1996).

Our projections show that global forest cover declines from 40.4 million square kilometers in 1990 to a minimum of 37.2 million square kilometers in 2016, and then increases slightly to 37.6 million square kilometers in 2025. However, in stark contrast, over the same period tropical forests are nearly halved from 18.4 to 9.7 million square kilometers.

On a more positive note, recent studies also demonstrate that EKCs are highly susceptible to structural economic shifts and technological changes, which are in turn influenced by policy. For example, Komen, Gerking, and Folmer (1997) point to the key role of public investments for environmental improvements in reducing environmental degradation as income levels rise, which may explain the strong EKC and even decreasing relationships found for some pollution indicators in OECD countries. Panayotou (1997) finds that improved policies and institutions in the form of more secure property rights, better enforcement of contracts, and effective environmental regulations can help to flatten the EKC for SO_2 across countries.

There are also encouraging signs that reform of environmental policy is beginning to progress in developing countries. For example, a recent review by the World Bank (1997) identifies a vast range of such environmental policy innovations that have been implemented across the globe since the 1992 Rio Environment and Development Conference to improve resource management and control pollution. Of particular importance is that many of these policies are being adopted by developing countries and that they include market-based instruments as well as removal of major policy distortions (see Huber, Ruitenbeek, and Serôa da Motta 1998). In addition, some of these reforms have been targeted at improved land and forestry management. What is more, they are being implemented as part of more general economy-wide and sector-specific reforms in these economies. This is an exciting prospect because it suggests that market-based instruments, the removal of economic disincentives, and environmental policy improvements are being considered together as important instruments in improving the link between economy and environment, thus helping to reverse the chain of unsustainable development in poorer economies.

Finally, I endorse the general view expressed by the authors that the potential welfare gains from the internalization of environmental degradation externalities in developing countries are likely to be large. Further studies of these potential gains are therefore an important priority. In support of this view, the authors cite the few available studies that attempt such valuations, including estimates of the cost of soil erosion in developing countries contained in a recent paper of mine (which is now published as Barbier 1998). However, my paper also sounds a note of caution about such cost estimates. Virtually all of the studies of the on-site costs of soil erosion in developing countries that I have reviewed have involved a very flawed methodological approach for estimating this cost. In most cases, this has led to inaccurate estimates of the income losses associated with erosion. Although this has been inevitable given the data limitation and other constraints faced by many of the studies, as I outline in my paper, it is time that we begin employing more methodologically sound approaches and thus improve our estimations of the economic costs of land degradation in developing countries. I believe this view is shared by Jha and Whalley in their paper.

References

Antle, J. M., and G. Heidebrink. 1995. Environment and development: Theory and international evidence. *Economic Development and Cultural Change* 43 (3): 603–25.

Barbier, E. B. 1989. *Economics, natural resource scarcity and development: Conventional and alternative views.* London: Earthscan Publications.

———. 1994. Natural capital and the economics of environment and development. In *Investing in natural capital: The ecological economics approach to sustainability,* ed. A. M. Jansson, M. Hammer, C. Folke, and R. Costanza, 291–322. Washington D.C.: Island Press.

———. 1997a. The economic determinants of land degradation in developing countries. *Philosophical Transactions of the Royal Society* ser. B, 352:891–99.

———. 1997b. Introduction to the environmental Kuznets curve special issue. *Environment and Development Economics* 2 (4): 369–82.

———. 1998. The economics of soil erosion: Theory, methodology and examples. In *The economics of environment and development: Selected essays,* ed. E. B. Barbier. London: Edward Elgar.

———. 1999. Endogenous growth and natural resource scarcity. *Environmental and Resource Economics* 14 (1): 51–74.

Barbier, E. B., and T. Homer-Dixon. 1999. Resource scarcity and innovation: Can poor countries attain endogenous growth? *Ambio* 28 (2): 14–147.

Cropper, M., and C. Griffiths. 1994. The interaction of population growth and environmental quality. *American Economic Review* 84 (2): 250–54.

Dasgupta, P. 1982. *The control of resources.* Oxford: Basil Blackwell.

Heath, J., and H. Binswanger. 1996. Natural resource degradation effects of poverty and population growth are largely policy-induced: The case of Colombia. *Environment and Development Economics* 1 (1): 65–83.

Huber, R. M., J. Ruitenbeek, and R. Serôa da Motta. 1998. *Market-based instruments for environmental policymaking in Latin America and the Caribbean: Lessons from eleven countries.* World Bank Discussion Paper no. 381. Washington, D.C.: World Bank.

Komen, M. H. C., S. Gerking, and H. Folmer. 1997. Income and environmental R&D: Empirical evidence from OECD countries. *Environment and Development Economics* (Special issue on environmental Kuznets curves.) 2 (4): 505–15.

Matsuyama, K. 1992. Agricultural productivity, comparative advantage, and economic growth. *Journal of Economic Theory* 58:317–34.

Panayotou, T. 1995. Environmental degradation at different stages of economic development. In *Beyond Rio: The environmental crisis and sustainable livelihoods in the Third World,* ed. I. Ahmed and J. A. Doeleman, 171–87. London: Macmillan.

———. 1997. Demystifying the environmental Kuznets curve: Turning a black box into a policy tool. *Environment and Development Economics* (Special issue on environmental Kuznets curves.) 2 (4): 465–84.

Sachs, J. D., and A. W. Warner. 1995. Natural resource abundance and economic growth. *Development Discussion Paper no. 517a.* Cambridge, Mass.: Harvard Institute for International Development.

Shafik, N. 1994. Economic development and environmental quality: An econometric analysis. *Oxford Economic Papers* 46:757–73.

Stern, D., M. S. Common, and E. B. Barbier. 1996. Economic growth and environmental degradation: The environmental Kuznets curve and sustainable development. *World Development* 24 (7): 1151–60.

United Nations Development Program (UNDP). 1998. *Human development report 1998.* Oxford: Oxford University Press.

World Bank. 1997. *Five years after Rio: Innovations in environmental policy.* Washington, D.C.: World Bank.

World Resources Institute (WRI). 1996. *World resources 1996–97.* Oxford: Oxford University Press.

8

Environmental Information and Company Behavior

Domenico Siniscalco, Stefania Borghini,
Marcella Fantini, and Federica Ranghieri

8.1 Introduction

Environmental policy is traditionally based on two sets of tools: (1) command and control regulations; and (2) economic or market instruments, such as environmental taxes, emissions charges, and tradable permits. The two sets of instruments have been adopted in subsequent waves, partly in response to economic analysis that shows command and control environmental policies are not cost-effective or are incapable of achieving the desired objectives in many circumstances.

In the last few years, some policymakers, the business community, and the media have increasingly emphasized the role of information-based environmental instruments. Such instruments, which are typically voluntary, range from company environmental reports to environmental audit and management schemes, such as International Standards Organization (ISO) 14000, Eco Management Audit Scheme (EMAS), and related award and compensation systems.

Information-based environmental policies are the subject of a lively debate. Their supporters claim that environmental reports and environmental management schemes are fundamental instruments for achieving the

Domenico Siniscalco is professor of economics at the University of Torino and director of Fondazione Eni Enrico Mattei. Stefania Borghini is a researcher at Fondazione Eni Enrico Mattei. Marcella Fantini is a researcher at Fondazione Eni Enrico Mattei and a lecturer at the University of Bergamo. Federica Ranghieri is a researcher at Fondazione Eni Enrico Mattei.

The authors are grateful to Kevin A. Hassett and other conference participants for their helpful comments. Special thanks go to Carlo Carraro and Gilbert Metcalf for encouraging the publication of a work which is still in a seminal form. The usual disclaimers, more than ever, apply.

desired environmental quality. Their critics claim they are only "green-washing," basically ineffective and devoid of any real effect.

This paper tries to shed some light on companies' behavioral responses to information-based environmental policies, dwelling on two building blocks: an original database at the company level collected by Fondazione Eni Enrico Mattei (FEEM) since 1995, and some recent literature on information and incentive schemes in companies.

This paper is divided into seven sections. Sections 8.2 and 8.3 briefly describe the main information-based environmental management tools and recall the theoretical rationale for their adoption; section 8.4 describes the database and identifies a subset of homogeneous companies in three polluting industries: (1) oil and gas, (2) petrochemicals, and (3) power generation. Sections 8.5 and 8.6 present some empirical results on the relation between information-based environmental strategies, economic performance, and environmental performance at the company level. Section 8.7 contains some concluding remarks.

The paper presents preliminary work that needs refinement. Information-based environmental policies are still in their infancy and their history is too recent to allow for a sound econometric analysis. The existing data and the relevant theory, however, seem to support the hypothesis that information-based environmental policies are indeed an instrument for changing company behavior and implementing environmental policies and regulations.

8.2   The Theoretical Background

In the textbook institutional setting, governments set environmental standards and companies comply. In addition to this, companies try to follow sound environmental strategies in order to avoid litigation and the emergence of future environmental liabilities. In some industries, such a strategy may also establish a good environmental reputation, which can be a powerful tool in their relationship with consumers, communities, and environmentalists. In the two latter cases, far-sighted companies may even exceed environmental standards.

In the situation we have just described, information plays a crucial role. In a world with imperfect information, regulators, investors, consumers, and other stakeholders want to know the companies' environmental performance, the achieved results, and the remaining problems and the schedule to solve them. Companies, symmetrically, need to communicate their environmental strategy and performance, in order to deal with their shareholders, stakeholders, and regulators. Against this background, the communication aimed at the external stakeholders has been widely discussed in the recent literature (Musu and Siniscalco 1993; Tietenberg 1997; Lanoi, Laplant, and Roy 1997; Khanna and Damon 1999; McIntosh et al.

1998). The same flow of environmental information, moreover, can play a key role in reshaping company behavior, and this is the focus of our paper.

A useful starting point can be found in two papers, Brehn and Hamilton (1996) and Pfaff and Sanchirico (1999), which claim that the lack of internal information (i.e., ignorance) is often responsible for the noncompliance with environmental regulation by big companies and for their wrong assessment of environmental damage, hence, the need for information tools and self-audit. The issue, however, is more complex than this.

For many years, companies (as well as regulators and the general public) have somewhat neglected environmental issues, concentrating their efforts on economic and financial performance. But neglecting environmental standards, particularly in the traditional industries, has gradually created hidden liabilities that can seriously harm shareholders' value through various channels: trials and litigation about health, safety, and pollution; loss of reputation with clients and consumers; conflict with local communities and environmental groups; and so forth. Such new issues, which are well known to shareholders and companies' chief executive officers, require a change in company behavior that can be pursued using an information-based environmental strategy that aims at changing company behavior through appropriate flows of information, audit, and incentives (Sinclair-Desgagné and Gabel 1997; Pendergast 1999).

Given the nature and the objectives of information-based environmental strategies, governments and regulators too have a clear interest in promoting their standardization and wide adoption, sometimes proposing guidelines themselves to define such schemes and make them mandatory. In such cases, we can refer to information-based environmental policies.

8.3   Some Information-Based Environmental Management Tools

The best-known environmental management tool adopted by firms is the corporate environmental report (CER) published annually by companies to audit and communicate the most relevant environmental issues related to their operations (emissions, effluents, wastes, and expenditure and investment in the environmental area).

The number of companies publishing environmental reports has been rapidly growing from 1992 to 1998 (fig. 8.1). Data show that the release of environmental information, which actually began in 1990, was started by firms in highly polluting industries, such as chemicals and oil and gas. But environmental reporting quickly spread to other industries such as the automotive and transportation industries, telecommunications, electronic appliances, financial services, and consumer goods.

As previously mentioned, the quality of published environmental reports can vary substantially across companies and time. The earlier reports typically included many statements and very few data, typically referring

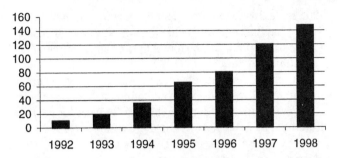

Fig. 8.1   Number of companies producing CERs worldwide, 1992–98
Source: ERM (1999).

to hot spots in the company operations, while recent reports include most comprehensive environmental data, together with environmental indicators and analyses that usually cover all the companies' activities.

In order to conduct a quality analysis, a specific rating system has been defined by FEEM within the Environmental Reporting Monitor (ERM) and published regularly since 1997 (see appendix C). If we adopt such system we can easily see that the quality of environmental reports has been constantly increasing (fig. 8.2).

Following the publication of CERs in the mid-1990s, companies began to introduce more sophisticated environmental audit systems aimed at promoting continuous improvements in the environmental performance of their operations. In order to facilitate and standardize the implementation of such audit systems, in 1993 the European Commission adopted the EMAS regulation. This scheme recommends voluntary participation by companies and gives them guidelines, with the objective of promoting better environmental performance at the site level. Similarly, worldwide the International Standards Organization launched the ISO 14000 scheme for the certification of corporate environmental management[1] at the company level.

Since the mid-1990s, the number of companies that certify their environmental management systems for EMAS and ISO 14000 has been constantly increasing. Since 1996, the year of the publication of the first five ISO 14000 standards, 10,439 companies have been certified. Since 1993, more than 2,790 sites have been certified for EMAS. A similar growth can be seen in our sample (fig. 8.3).

In addition to CERs and auditing schemes, other management tools, such as compensation programs and award schemes, have been gradually introduced by many big companies in order to link environmental performance to economic incentives. In this respect, the adoption of award and compensation programs related to environmental results can be viewed as

1. It should be remembered that the first national standard on environmental management was the BS 7750, developed in the United Kingdom in the early 1990s.

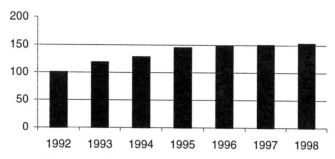

Fig. 8.2   **Average quality of CERs (1992 = 100) worldwide, 1992–98**
Source: ERM (1999).

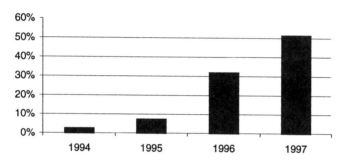

Fig. 8.3   **Percent of EMAS or ISO 14000-certified companies worldwide, 1994–97**
Source: ERM (1999).

an incentive-compatible strategy for integrating environmental issues into the company's management. Compensation and award schemes quickly spread and now they are common practice for almost all big companies. In our sample, the percentage of companies that implemented a compensation program increased from 32 percent to 73 percent from 1994 to 1997 (fig. 8.4).

From the theoretical point of view, environmental reports, audit schemes, and compensation mechanisms can be viewed as components of an integrated information-based environmental strategy aimed at changing company behavior. Let us see how these instruments work by analyzing an appropriate database at the company level.

8.4   The Database

Our database covers 476 CERs published worldwide from 1993 to 1997. To carry out a meaningful empirical analysis, we selected a sample that includes 39 big firms, based in 16 countries, belonging to three highly polluting industries that produce comparable emissions (such as NO_x and

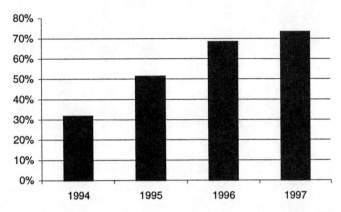

Fig. 8.4   Percent of companies adopting compensation programs and award schemes, worldwide, 1994–97
Source: ERM (1999).

SO_x) using similar feedstocks: (1) petrochemicals, (2) oil and gas, and (3) electric power generation (see appendix A) for the period 1993–97.

In addition to the CERs we gathered information on whether the companies adopted an environmental management system (i.e., ISO 14000 and/ or EMAS), adopted environmental compensation and award schemes, and collected data on the main economic variables at the company level (this was extracted from the standard annual reports).

Starting from this data, we built a panel that includes several variables: (1) a standardized index of pollution; (2) measures of the size and economic performance of the company; (3) an indicator of the quality, comprehensiveness, and transparency of the environmental information; (4) an indicator of the adoption of one or more environmental audit, compensation, or award schemes; and (5) several control variables at the company, industry, and country levels (see appendix A).

The panel is obviously affected by a sample selection bias because it includes only companies that voluntarily decided to publish CERs. Our analysis, however, focuses on the effects of more detailed instruments in this population of relatively caring industries.[2] In this case the sample-selection critique does not apply because publishing a CER does not imply the adoption of the environmental management tools we are considering.

The environmental performance variable (LPOLL) is defined on an annual basis as SO_x plus NO_x emissions per unit of output.[3] The indicators

2. Although in this case, our sample has no sample selection bias, it is worth noticing that, while data for environmental information accuracy are available for the whole period under consideration (1993–97), for more recent tools such as the environmental audit, compensation, and award systems the sample includes many zeros in early years.
3. The output is expressed in tons of oil equivalent (TOE), which seems to be a suitable measure in the three industries under review.

have been chosen on the basis of their impact on the environment and on data availability. SO_x is a main indicator used by the regulators as a base for the environmental taxation system, and NO_x plays a major role in land acidification. At this stage, we cannot consider data on waste and water discharges because classification across countries and regulations on waste have significantly changed over the last 5 years, and the currently available data do not account for the damage associated with different discharged pollutants (a firm emitting a large quantity of a relatively harmless substance would be ranked as a larger polluter than another firm emitting a small quantity of a very toxic substance).

The size of the company (WORK) is proxied by the number of employees, which also indicates the complexity of the agency problems in the organization, while the economic performance (OPERATING INCOME) is measured by the operating income in current U.S. dollars.[4]

The quality of the information disclosed in the environmental reports (INFO) is measured by a scoring system, developed by the ERM at FEEM (see appendix B). The system evaluates the descriptive information contained in the report (i.e., mission, objectives, strategy, organization, and programs), the quality of environmental variables and indicators (e.g., some reports contain data on emissions but omit economic data, such as defensive and environmental expenditure, while others include indicators but do not publish raw data for emissions, effluents, and wastes), and the thoroughness of the report (e.g., many reports cover a subset of sites or ignore some foreign countries where the company operates).

Information-based environmental management is measured by a 0–3 index (environmental audit, award, and compensation; EAC), which is the sum of three dummy variables: the first (E) records the adoption of EMAS and/or ISO 14000;[5] the second (A) records the existence of an environment-related award system, which does not give immediate benefits but directly influences the future career of the managers and the employees; and the third (C) records the adoption of an environment-related compensation scheme.[6] A more detailed description of the variables we use in our analysis can be found in table 8.4, later.

We are well aware that both company variables and indicators are rather

4. We used companies' annual reports to collect data on their operating income. Unfortunately, most financial statements are expressed only in local currency. In order to make them comparable we decided to convert all financial variables into current U.S. dollars by using the nominal exchange rate of the local currency against the dollar.

5. Data on companies' environmental management certification were obtained from the EMAS official register and ISO 14000-competent body in each country.

6. To gain information about environmental compensation programs and award schemes we relied on CERs and annual financial reports and, for U.S. listed companies only, also on official disclosure required by the U.S. Securities and Exchange Commission (reports such as 10K for American companies quoted on the New York Stock Exchange and 20F for non-American companies quoted on the New York Stock Exchange). If this information was not available in corporate publications, we directly interviewed companies' environmental managers and external relation managers.

raw and must be improved, but CERs have not been published for very long and the data we can collect are quite limited. In addition to company data, some control and regulation data have been collected at the country level.

8.5   A First Look at the Data

Do information-based environmental policies work? How do they influence company behavior? Some preliminary answers to such questions can be found by broadly comparing companies that adopted some information-based environmental strategies (henceforth EAC companies) between 1993 and 1997 with companies that did not adopt such schemes.

At a first glance, we observe that on average the companies that have implemented compensation and award schemes and have certified environmental management systems present better environmental performances. A t-test performed on the mean shows that although the difference is not significant at the standard 5 percent or 10 percent levels (except for 1997), EAC companies do perform better (table 8.1). Table 8.2 reports the average pollution growth rates for EAC companies versus the whole sample year by year. Once again, EAC companies are observed to perform better, as their average pollution rate is lower than the one for the whole sample for two out of the three years. When we consider average pollution rates (see fig. 8.5), we observe that EAC companies pollute much less than the total sample; and they reduce pollution throughout the time span we considered, while for the whole sample pollution drops from 1994 to 1996 but has an upward trend between 1996 and 1997.

To clarify this point we select three important case studies from among the companies in our sample: BP Chemicals (petrochemicals); ELF (oil

Table 8.1   **Average Pollution Rates**

	1994	1995	1996	1997
EAC = 0	0.0038	0.0028	0.0036	0.0052
EAC > 0	0.0014	0.0011	0.0009	0.0007
t-test	0.78	0.81	1.34	1.87

Table 8.2   **Average Pollution Growth Rates**

	1994–95	1995–96	1996–97	Average 1994–97
Whole sample	−0.39688	−0.086121	0.068504	−0.1618
EAC	−0.20206	−0.21496	−0.14913	−0.18921

Source: See appendix D.

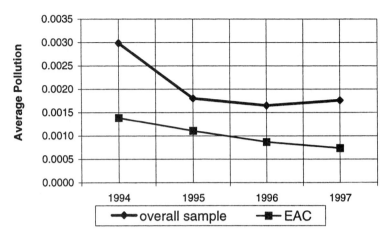

Fig. 8.5   The impact of information-based environmental policies; EAC companies vs. total sample in three selected industries worldwide, 1994–97
Note: See app. D; the average pollution has been calculated without considering the values of two outlier companies.

and gas); and PowerGen (electric power generation). According to our database and the scoring system, such companies were among the first to adopt EAC in their industries and to produce the highest quality CERs. We look at their environmental performances considering their emissions reduction rates before and after the EAC adoption. We also relate their emissions to the quality of the environmental information produced to see whether information quality and quantity are related to emissions reduction.

The analysis of emissions at BP Chemicals shows that the introduction of environmental awards and compensation programs did influence the pollution growth rate, which has been diminishing faster since the implementation of a pioneering award scheme in 1994. The negative trend in emissions, after a slowdown in 1995, was strengthened by the introduction of a certification and compensation scheme in 1996 (see table 8.3 and fig. 8.6).

ELF is an equally interesting case. The company sequentially introduced an award scheme in 1995, and a certification-compensation scheme in 1996, constantly improving its environmental performance (see table 8.3 and fig. 8.7). PowerGen adopted, in sequence, an environment-related award scheme, a compensation mechanism, and finally a certification system. Its emissions constantly diminished at increasing rates, from -0.084 percent in 1993–94 to -0.136 percent in 1994–97 (see table 8.3 and fig. 8.8). In these case studies, we can also observe a negative relationship between the quality of corporate environmental information and the emissions index. Figures 8.6, 8.7, and 8.8 illustrate these findings, highlighting the year of adoption of the various management tools.

Table 8.3   **Average Pollution Growth Rates, Case Studies**

	Average Growth Rate, 1993–94	Average Growth Rate with EAC, 1993–94
BP Chemical	−0.1202683	−0.0204621
ELF	−0.1074995	−0.1729896
PowerGen	−0.0842359	−0.1358968

Source: See appendix D.

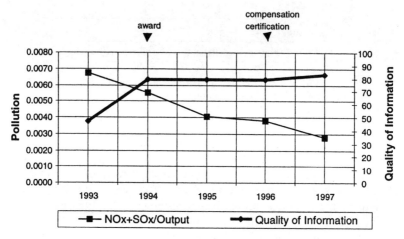

Fig. 8.6   BP Chemicals: the impact of information-based environmental policies, 1993–97

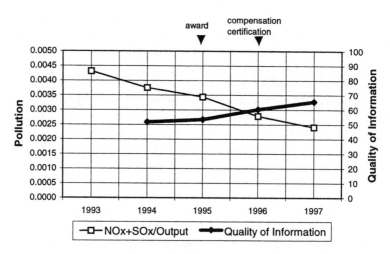

Fig. 8.7   ELF: the impact of information-based environmental policies, 1993–97

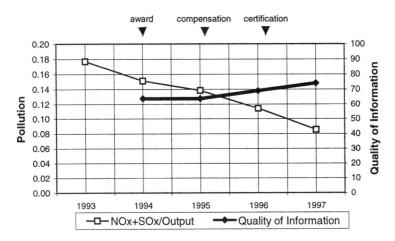

Fig. 8.8   PowerGen: the impact of information-based environmental policies, 1993–97

These results seem to be consistent with two ideas: (1) the adoption of information-based environmental tools improves company environmental performance, and (2) such management instruments are complementary with each other. Critics of information-based environmental strategies could object that our findings are possibly spurious because the analysis neglects standard environmental regulation, taxation, and several other variables that may influence emissions together with EAC. To overcome this objection, we carry out a statistical analysis that also includes other policy variables, together with some control variables. For this purpose, our panel is disturbingly small. But we believe that, at this stage, it is worthwhile to present, with many caveats, some tentative results.

8.6   A Statistical Analysis

We perform a cross-country analysis for 16 countries, assuming that governments adopt environmental policies based both on command and control and on economic instruments; and that companies comply and may also pursue tighter environmental strategies to avoid future risks and liabilities. In this setting, we check whether environmental policies (command and control instruments and energy taxation) affect the companies' economic and environmental performance. We also check whether the adoption of information-based environmental strategies (EAC) affects this relationship, influencing company behavior, given the energy tax burden and the severity of environmental legislation.

It is generally acknowledged that environmental policy reduces pollution, but harms economic performance. This trade-off, however, can be eased by information-based environmental policies. We want to test whether these policies, which affect company behavior, can make compliance more effective and less costly. The variables used in our estimates are listed in table 8.4.

The model we want to estimate is a random-effects model that can be written as

$$y_{it} = \overline{\beta}_1 + \sum_{k=2}^{K} \beta_k x_{kit} + \varepsilon_{it} + \mu_i.$$

We estimate the model using an instrumental variable (IV) procedure because we cannot include all the variables simultaneously in our estimation because of the endogeneity of the operating income with the dependent variable. At first we estimate the link between operating income (OPERATING INCOME) and the existence of environmental certification, award, and compensation schemes (EAC), and the energy taxation burden (TAX). We take into account the company dimension by using the number of employees (WORK) as a control variable. Second, we relate the environmental performance of companies (the logarithm of company pollution, LPOLL) to the quality of environmental information at time $t-1$ (INFO) to their economic performance (the instrumented operating income, IOPINC) and to the enforcement of legislation (ENFORCE). The results are shown in table 8.5. As we expected, OPERATING INCOME is positively related to EAC and to WORK (the company size). INFO (the quality of environmental information) is nonsignificant.

Table 8.6 shows the results of the regression of the logarithm of com-

Table 8.4   **Variables Used in Estimation**

Variable	Description
AWARD	Dummy, which is 1 when an environmentally based award program is implemented
CERTIFICATION	Dummy, which is 1 when the company environmental management system is certified for ISO 14000 and/or 1836/96 EMAS
COMPENSATION	Dummy, which is 1 when an environmentally based compensation program at company level is implemented
EAC	Sum of CERTIFICATION, COMPENSATION, and AWARD (index 0–3)
ENFORCE	Country index of environmental regulation enforcement
INFO	Index, which ranges from 0 to 100, assessing the accuracy of company environmental information
LPOLL	Logarithm of company pollution, computed as $SO_x + NO_x$ per TOE
OPERATING INCOME	Annual operating income in current U.S. dollars
SECTOR	Sectoral index, which is 1 for the electrical sector, 2 for oil and gas, and 3 for chemicals
TAX	Country index of burden energy taxes per GDP
WORK	Number of employees per firm

Table 8.5   **Operating Income**

Mean of dependent variable	1,889.92	R^2	.421142	
Standard deviation of dependent variable	2,489.03	Adjusted R^2	.404116	
		LM heterogeneity test	17.1105 [.000]	
		Durbin-Watson	.075581 [.000, .000]	
Sum of squared residuals	.381484E+09			
Variance of residuals	.374004E+07			
Standard error of regression	1,933.92			

Variable	Estimated Coefficient	Standard Error	t-statistic	p-value
EAC	413.964	81.1986	5.09818	** [.000]
WORK	.050996	.948368E−02	5.37726	** [.000]
INFO (1)	6.01047	9.13842	.657715	[.511]
C	−77.6047	766.808	−.101205	[.919]

Hausman test of H_0:RE vs. FE: CHISQ(3) = 29.548, p-value = [.0000]

Table 8.6   **Pollution (LPOLL)**

Mean of dependent variable	−8.95649	R^2	.276393	
Standard deviation of dependent variable	3.49287	Adjusted R^2	.234806	
		LM heterogeneity test	.015806 [.942]	
		Durbin-Watson	.022598 [.000, .000]	
Sum of squared residuals	812.667			
Variance of residuals	9.34100			
Standard error of regression	3.05630			

Variable	Estimated Coefficient	Standard Error	t-statistic	p-value
IOPINC	−.186477E−03	.108226E−03	−1.72303	* [.085]
INFO(1)	−.025010	.977345E−02	−2.55899	** [.010]
ENFORCE	−.110039	.155892	−.705866	[.480]
SECTOR	2.61997	.911784	2.87345	** [.004]
TAX	−.818044	.967608	−.845429	[.398]
C	−9.18296	3.70121	−2.48107	** [.013]

Hausman test of H_0:RE vs. FE: CHISQ(2) = 5.2067, p-value = [.0740]

pany SO_x plus NO_x emissions (LPOLL) at time t on the instrumented OPERATING INCOME at time t, on the quality of environmental information (INFO) at time $t − 1$, on the enforcement of legislation (ENFORCE), on the burden of energy taxation (TAX), and on the industry dummy (SECTOR). LPOLL is negatively related to IOPINC (the IV operating income), which is consistent with the idea that the adoption of EAC reduces emissions. Moreover, INFO (the quality of environmental information) is negatively related with LPOLL, suggesting that managers' and employees' efforts on environmental matters are significantly influenced not only by the presence of EAC but also by the accuracy of environmental information. SECTOR is positively related with LPOLL, simply reflecting the structural and technological characteristics of production in the three industries under review. Finally, the relation between ENFORCE and LPOLL is negative, but not significant ($p = 0.480$).

8.7   Concluding Remarks

Information-based environmental strategies play a significant role in our sample. Given environmental regulation, which is costly, they positively influence operating income and negatively influence pollution. Being primarily implementation tools, they cannot substitute for more traditional policies, but can play a useful role.

Our findings are consistent with a whole class of models on environmental information, incentives, and company behavior. In our panel data estimation, the accuracy of environmental information is negatively related with pollution and the relation is significant. That is, information quality is crucial for companies' environmental management and there are explanations for corporate noncompliance that are not related to the level of the penalties, but instead to the company's scarcity of internal information (Brehn and Hamilton 1996).

In contrast from our results we cannot infer the role of environmental information accuracy on financial performance. In our analysis, we used operating income as a proxy of companies' financial health since we wanted to investigate the existing relation between environmental management tools and company results in the short period. Existing literature on environmental information and corporate financial performances finds a significant relationship between these variables, but it refers to external environmental information (information provided to external stakeholders) and to long-term performances such as shareholder value or liabilities (Tietenberg 1997; Lanoi, Laplant, and Roy, 1997; Khanna and Damon 1999). These differences help in understanding the differences between our analysis and prior analyses.

However, generic pleas for better and wider "environmental information" or "eco-management" are too vague and may be misleading. In order to exert a positive influence, environmental information needs to be integrated with a set of incentives, as recommended by economic theory for any company objective. This explains the nature of many integrated environmental and management schemes (such as ISO 14000 or EMAS) adopted by firms and recommended by policymakers. Our empirical model confirms the positive role of self-regulated environmental audits and compensation programs on corporate environmental performance, and this is consistent with an emerging research field that explores the possible patterns for integrating environmental issues with concrete management systems (Sinclair-Desgagné and Gabel 1997; Pfaff and Sanchirico 1999).

These conclusions, of course, are just tentative, given the preliminary nature of our empirical analysis. In order to reach more robust conclusions, better data must be collected and better estimates must be carried out. But the preliminary results we have obtained so far seem to be consistent with economic theory and with common sense.

Appendix A

Table 8A.1   List of the Companies

Company	Sector[a]	Country[b]	First Year of CER Publication	First Year of EMAS ISO 14000 Implementation	First Year of Award Scheme Adoption	First Year of Compensation Program Adoption
Agip	OG	IT	1995	1998	1994	1996
Agip Petroli	OG	IT	1993	n.i.	1994	1996
APS	OG	US	1994	n.i.	1994	1996
Bayer Italia	C	IT	1994	1996	1994	1995
British Gas	OG	UK	before 1993	n.a.	n.a.	n.a.
British Petroleum (BP)	OG	UK	before 1993	1996	1994	1996
BP Chemical	C	UK	before 1993	1996	1994	1996
CIBA	C	CH	before 1993	1995	1994	1995
Conoco	OG	US	1993	1998	1994	1995
Dong	OG	DK	1994	n.i.	n.i.	n.i.
Edison	E	IT	1994	n.i.	1998	1998
Enel	E	IT	1995	1998	1994	1997
Eni	OG	IT	1995	1998	n.i.	n.i.
Exxon	OG	US	1995	1998	1994	n.a.
ELF	OG	FR	1994	1996	1995	1996
Eskom	OG	SA	1994	n.i.	1995	n.a.
Ivo	E	FL	1996	1997	n.a.	n.a.
Mobil	OG	US	1995	1996	1995	1996
National Power	E	UK	1994	1995	n.a.	n.a.

Company	Sector[a]	Country[b]				
Neste	OG	FL	before 1993	1997	1997	1996
Neste Chemicals	C	FL	before 1993	1997	1997	n.a.
Norsk Hydro	OG	FL	1995	1995	1995	n.a.
Novo Nordisk	C	DK	1993	1998	n.a.	n.a.
Ontario Hydro	E	US	1994	1997	1995	n.a.
Petrofina Downstream	OG	BE	1994	1997	1995	n.i.
Pacific Gas and Electric	OG	US	1995	1996	1994	n.a.
PowerGen	E	UK	1994	1996	1994	1995
Repsol	OG	SP	1996	n.i.	n.a.	n.a.
Royal Dutch/Shell Group Downstream	OG	UK	1996	1996	1996	n.i.
Shell International Exploration and Production B.V.	OG	NL	1996	1997	1996	n.a.
Shell Chemicals	C	UK	1994	1997	1996	n.a.
Shell UK	OG	UK	1993	1997	1996	n.a.
Snam Transport	OG	IT	1993	n.i.	n.i.	n.i.
Snam Gas	OG	IT	1993	n.i.	n.i.	1996
Solvay	C	BE	1994	1996	1995	n.a.
Statoil	OG	NO	1993	n.i.	1995	n.i.
Tepco	E	JA	1994	n.a.	n.a.	n.a.
Texaco	OG	US	1994	1996	1996	1997

Source: ERM (1999).

Note: n.a. = not available. n.i. = not implemented.

[a]C = chemicals; E = electrical power generation; OG = oil and gas.

[b]BE = Belgium; CH = China; DK = Denmark; FL = Finland; FR = France; IT = Italy; JA = Japan; NL = Netherlands; NO = Norway; SA = South Africa; SP = Spain; UK = United Kingdom; US = United States.

Appendix B

Forum on Environmental Reporting Guidelines

In order to guarantee a minimum standard of CERs as a voluntary document, FEEM organized in 1994 the Forum on Environmental Reporting (FER) by inviting some large companies emerging in the field of environmental management and reporting, and some interested target groups for environmental reports, environmental groups, and public administration, to work together to draw up guidelines. The aim of the FER is to set guidelines for companies seeking to produce an effective environmental report, providing stakeholders with the information needed from other similar initiatives for a consensus approach. Here follows the list of minimum and recommended requirements to be included in CERs. These requirements have been used as the basis for the ERM scoring system, aimed at evaluating the quality of environmental information.

Table 8B.1   Qualitative Information (Notes to the Environmental Balance Sheet)

1. Company description
    a. Company size and activities — Minimum requirement
    b. Number and location of production sites — Minimum requirement
    c. General description of production processes — Minimum requirement
    d. Description of the main environmental issues — Minimum requirement
       related to production and distribution
2. Environmental policy
    a. Year of introduction of environmental policy and — Minimum requirement
       content
    b. Expected achievements — Minimum requirement
    c. Achievements monitoring (comparison with — Minimum requirement
       prevous reported objectives)
3. Environmental management systems
    a. Organization structure (environmental department — Minimum requirement
       and relationships with other business units)
    b. Programs for environmental policy implementation — Minimum requirement
    c. Training activity — Recommended requirement
    d. Implementation level of environmental — Recommended requirement
       management system and certifications (EMAS,
       ISO, or UNI [Ente Italiano per l'Unificazione])
4. Risk management
    a. Audits, mesures taken, and achievements — Recommended requirement
       regarding risk management
    b. Description of cleanup operations carried out — Recommended requirement
    c. Description of major accidents — Recommended requirement
5. Compliance with environmental legislation
    a. Description of the way the company ensures — Recommended requirement
       compliance with environmental regulations (in
       relation to previous violations as well as to
       prevention measures)

b. Description of measures adopted to comply with new environmental regulations (EU, national, and local) that became operational during the period to which the report refers	Recommended requirement
6. Product policy	
a. Description of product's life cycle and of the related impacts and description of the most relevant measures to mitigate them	Recommended requirement
b. Product innovation	Recommended requirement
c. Product's energy efficiency (when relevant)	Recommended requirement
d. Company responsibility at the end of product use	Recommended requirement
e. Cooperation programs with consumers and clients	Recommended requirement
f. Ecolabel (where applicable)	Recommended requirement
7. Conservation of natural resources	
a. Energy-saving programs	Minimum requirement
b. Water-saving programs	Minimum requirement
c. Other programs for the protection of natural heritage	Recommended requirement
8. Stakeholders' relations	
a. Participation in voluntary agreement schemes	Recommended requirement
b. Relations with stakeholders (public administration, environmentalists, universities, etc.)	Recommended requirement
c. Department or name of the person to contact for further information	Minimum requirement
9. Certification	
a. External certification	Recommended requirement
b. Certification by EMAS-accredited verifiers	Recommended requirement

Table 8B.2   Quantitative Information (the Environmental Balance Sheet)

1. Environmental expenditures	
a. Data on environmental expenditures	Recommended requirement
b. Explanation of accounting criteria	Minimum requirements
2. Emissions and consumption of raw materials	
a. Site-by-site quantitative information (for main sites)	Minimum requirement
b. Raw materials	Recommended requirement
c. Energy as input	Minimum requirement
d. Wastes, air emissions, water discharges, soil pollution, and other pollutants relevant to company's activity	Minimum requirement
e. Quantity of products or a relevant figure to describe production level	Minimum requirement
f. Impacts (scientifically accounted) related to production activity	Recommended requirement
g. Reduction objects for raw materials, energy, pollutants, and impacts	Recommended requirement
3. Environmental performance indicators	
a. Environmental performance indicators compared with previous periods	Minimum requirement

Source: FER (1995).

Appendix C

Environmental Reporting Monitor (ERM)

Starting from the Forum on Environmental Reporting (FER) guidelines the FEEM has set up an Environmental Reporting Monitor (ERM) defining a three-section checklist as a scoring system. The first two sections represent of the two parts of the report: the first section checks for the qualitative information, the second one for the quantitative information, following the FER requirements (see FER 1995, app. 3); the third one is the comments section, explained here. The structure of the checklist is as follows:

- Qualitative section: It verifies that four minimum requirements and eleven recommended requirements are met. The score—the report can receive from 0 to 2 points for every minimum requirement met and from 0 to 1 for every recommended requirement met.
- Quantitative section: It verifies that nine minimum requirements and five recommended requirements are respected. The score—the report can receive from 0 to 2 points for every minimum requirement respected and from 0 to 1 for every recommended requirement respected.
- Comment: First, it checks that the CER structure complies with the FER guidelines. Then, it checks whether the report is complete. The score for data quantity—if it is exhaustive it receives 2 points, if medium 1 point, if it is not enough 0 points. The score for data quality— whether the CER refers to a sample, whether the report maker used a specific methodology for CER data collection, and whether an audit has been implemented to check the data from 0 to 2 points. Then it checks report legibility (from 0 to 2 points), and it verifies whether the report gives other information and whether there is a positive evolution in act from the last reports to the present one (if yes, 1 point).

Each CER can receive up to 19 points in the qualitative section, 23 points in the quantitative section, and 16 points in the comments section. The maximum score is 58 points. For this paper, each score has been normalized.

Appendix D

Table 8D.1   FEEM Company Database

Company	Sector[a]	Country[b]	Year	Quality of Information	Operating Income (millions of dollars)	NO$_x$+SO$_x$ per Unit Output[c]
Azienda Elettrica Milanese	E	ITA	1993	n.a.	76.081	0.0000868605
			1994	n.a.	75.013	0.0000782558
			1995	n.a.	79.863	0.0000519767
			1996	n.a.	103.373	0.0000534884
			1997	79.31	107.974	0.0000554651
Agip	OG	ITA	1993	n.a.	1,884.633	0.0000001634
			1994	n.a.	2,332.917	0.0000001490
			1995	58.62	2,487.967	0.0000001720
			1996	63.79	3,272.926	0.0000001220
			1997	82.76	2,944.338	0.0000001192
Agip Petroli	OG	ITA	1993	77.59	347.328	0.0049643298
			1994	81.03	377.858	0.0044940935
			1995	81.03	246.157	0.0041504887
			1996	84.48	563.851	0.0039790576
			1997	87.93	654.036	0.0041195417
APS	OG	USA	1993	n.a.	21.844	0.0000320440
			1994	60.34	32.941	0.0000315160
			1995	60.34	24.139	0.0000230695
			1996	67.24	3.839	0.0000216442
			1997	67.24	−46.566	0.0000222777
Bayer Italia	C	ITA	1993	n.a.	27.828	0.0001927980
			1994	51.72	55.841	0.0002004310
			1995	55.17	0.614	0.0001831091
			1996	58.62	15.554	0.0001929390
			1997	67.24	11.155	0.0001442700

(*continued*)

Table 8D.1 (continued)

Company	Sector[a]	Country[b]	Year	Quality of Information	Operating Income (millions of dollars)	NO_X+SO_X per Unit Output[c]
British Gas	OG	UK	1993	60.34	219.313	0.0264302326
			1994	65.52	1,134.206	0.0386511628
			1995	67.24	646.481	0.0011782100
			1996	67.24	1,288.052	0.0002417087
			1997	81.03	1,965.630	0.0000971710
British Petroleum (BP)	OG	UK	1993	81.03	3,653.153	0.0010081561
			1994	81.03	3,785.397	0.0011294054
			1995	82.76	4,034.091	0.0010782359
			1996	82.76	5,804.255	0.0010777591
			1997	82.76	6,203.075	0.0010614843
BP Chemical	C	UK	1993	46.55	−102.071	0.0067420907
			1994	79.31	385.734	0.0055033609
			1995	79.31	1,347.854	0.0040575132
			1996	79.31	743.494	0.0038010450
			1997	82.76	793.417	0.0027912249
CIBA	C	CH	1993	53.45	n.a.	0.0030669894
			1994	53.45	1,598.999	0.0035397271
			1995	51.72	2,000.381	0.0025118727
			1996	62.07	2,579.498	0.0016202434
			1997	62.07	1,650.181	0.0016204986
Conoco	OG	USA	1993	48.28	1,149.000	0.0015133333
			1994	51.72	1,096.000	0.0014460741
			1995	56.90	1,257.000	0.0013788148
			1996	56.90	1,887.000	0.0013451852
			1997	65.52	2,003.000	0.0012779259
Dong	OG	DK	1993	n.a.	147.220	0.0000000273
			1994	n.a.	145.673	0.0000000260
			1995	79.31	179.804	0.0000000275

Company	Sector	Country	Year			
Edison	E	ITA	1996	81.03	238.030	0.0000000251
			1997	81.03	318.421	0.0000000206
Enel	E	ITA	1993	n.a.	225.800	0.0000050576
			1994	60.34	267.100	0.0000035840
			1995	60.34	269.800	0.0000044763
			1996	60.34	403.121	0.0000015552
			1997	68.97	565.383	0.0000015597
Eni	OG	ITA	1993	n.a.	3,672.392	0.0000750148
			1994	n.a.	4,875.387	0.0000730349
			1995	53.45	5,130.626	0.0000687906
			1996	84.48	5,337.461	0.0000660019
			1997	86.21	5,101.308	0.0000591548
Exxon	OG	USA	1993	n.a.	3,631.924	0.0000124362
			1994	n.a.	4,611.855	0.0000113043
			1995	79.31	6,317.827	0.0000081900
			1996	70.69	6,224.392	0.0000076405
			1997	79.31	6,074.789	0.0000071072
ELF	OG	FRA	1993	n.a.	7,655.000	0.0002530000
			1994	n.a.	8,390.000	0.0001830000
			1995	70.69	7,897.000	0.0001715000
			1996	74.14	10,185.000	0.0001780000
			1997	n.a.	11,916.000	n.a.
Eskom	OG	SA	1993	n.a.	1,132.602	0.0043146603
			1994	51.72	210.512	0.0037527115
			1995	53.45	3,550.086	0.0034368737
			1996	60.34	4,733.260	0.0027870461
			1997	65.52	3,862.262	0.0023506366
			1993	n.a.	n.a.	0.0001293497
			1994	31.03	1,002.398	0.0001268752
			1995	63.79	1,031.836	0.0001270482
			1996	77.59	1,360.636	0.0001262553
			1997	79.31	1,396.682	0.0001282214

(continued)

Table 8D.1 (continued)

Company	Sector[a]	Country[b]	Year	Quality of Information	Operating Income (millions of dollars)	$NO_x + SO_x$ per Unit Output[c]
Ivo	E	FL	1993	n.a.	531.138	0.0000697500
			1994	n.a.	531.138	0.0000774000
			1995	n.a.	531.138	0.0000531000
			1996	74.14	531.138	0.0000123407
			1997	84.48	531.138	0.0000083056
Mobil	OG	USA	1993	n.a.	1,488.000	0.0014592593
			1994	n.a.	2,224.000	0.0013481481
			1995	51.72	2,231.000	0.0011777778
			1996	58.62	2,846.000	0.0011851852
			1997	67.24	3,097.000	0.0011481481
National Power	E	UK	1993	n.a.	899.129	0.0001782162
			1994	72.41	1,123.527	0.0001470563
			1995	75.86	1,213.699	0.0001056091
			1996	79.31	1,282.372	0.0000812075
			1997	79.31	1,242.582	0.0000727480
Neste	OG	FL	1993	74.14	94.394	0.0011792747
			1994	74.14	430.820	0.0010602409
			1995	82.76	491.011	0.0010733680
			1996	82.76	234.866	0.0008895546
			1997	86.21	311.516	0.0009296448
Neste Chemicals	C	FL	1993	74.14	n.a.	0.0008224736
			1994	74.14	n.a.	0.0000893939
			1995	82.76	n.a.	0.0000789130
			1996	82.76	n.a.	0.0000969231
			1997	86.21	n.a.	0.0000938043
Norsk Hydro	OG	FL	1993	n.a.	553.425	0.0003254156
			1994	n.a.	1,013.462	0.0003642039
			1995	58.62	1,689.127	0.0003937569

Company	Sector	Country	Year			
			1996	79.31	1,494.907	0.0003727004
			1997	82.76	1,511.138	0.0004111197
Novo Nordisk	C	DK	1993	82.76	269.824	0.0382871126
			1994	81.03	341.195	0.0370400000
			1995	81.03	372.808	0.0351500790
			1996	84.48	428.971	0.0367078825
			1997	86.21	460.228	0.0388978930
Ontario Hydro	E	USA	1993	n.a.	6,569.258	0.0000113953
			1994	56.90	6,586.616	0.0000111628
			1995	56.90	6,554.463	0.0000087209
			1996	65.52	6,516.096	0.0000102326
			1997	81.03	6,449.165	0.0000147674
Petrofina Downstream	OG	BEL	1993	n.a.	243.13	0.0017000000
			1994	53.45	134.49	0.0015500000
			1995	67.24	7.97	0.0014000000
			1996	67.24	167.69	0.0013300000
			1997	n.a.	n.a.	n.a.
Pacific Gas and Electric	OG	USA	1993	n.a.	2,560.000	0.0000093440
			1994	n.a.	2,424.000	0.0000089744
			1995	56.90	2,763.000	0.0000076547
			1996	56.90	1,896.000	0.0000073907
			1997	63.79	1,831.000	0.0000063349
PowerGen	E	UK	1993	n.a.	678.475	0.0000019325
			1994	63.79	742.385	0.0000017697
			1995	63.79	860.164	0.0000015128
			1996	68.97	1,082.440	0.0000013849
			1997	74.14	836.334	0.0000011418
Repsol	OG	SPA	1993	n.a.	1,320.223881	n.a.
			1994	n.a.	1,704.121893	n.a.
			1995	n.a.	1,567.561168	n.a.
			1996	56.90	1,116.120219	0.0020656441
			1997	70.69	36,189.80806	0.0020721761

(continued)

Table 8D.1   (continued)

Company	Sector[a]	Country[b]	Year	Quality of Information	Operating Income (millions of dollars)	$NO_x + SO_x$ per Unit Output[c]
Royal Dutch/Shell Group Downstream	OG	UK	1993	n.a.	8,925	n.a.
			1994	n.a.	9,600	n.a.
			1995	n.a.	12,476	n.a.
			1996	70.69	17,128	0.0000009012
			1997	79.31	15,965	0.0000008082
Shell International Exploration and Production B.V.	OG	NL	1993	n.a.	n.a.	n.a.
			1994	n.a.	n.a.	n.a.
			1995	n.a.	n.a.	n.a.
			1996	58.62	17,128	0.0000007803
			1997	81.03	15,965	0.0000009830
Shell Chemicals	C	UK	1993	n.a.	n.a.	0.0000128333
			1994	68.97	n.a.	0.0000103158
			1995	72.41	n.a.	0.0000091000
			1996	72.41	n.a.	0.0000071364
			1997	77.59	n.a.	0.0000065417
Shell UK	OG	UK	1993	44.83	n.a.	0.0028361582
			1994	51.72	n.a.	0.0024583333
			1995	75.86	n.a.	0.0022235294
			1996	77.59	n.a.	0.0021371429
			1997	77.59	n.a.	0.0020340909
Snam Transport	OG	ITA	1993	67.24	1,919.604	0.0009832
			1994	72.41	1,920.934	0.0008121
			1995	77.59	2,039.850	0.0007465
			1996	81.03	2,547.697	0.0007154
			1997	89.66	2,287.953	0.0005989

Snam Gas	OG	ITA	1993	67.24	1,919.604	0.0000000888
			1994	72.41	1,920.934	0.0000000838
			1995	77.59	2,039.850	0.0000001016
			1996	81.03	2,547.697	0.0000000925
			1997	89.66	2,287.953	0.0000000608
Solvay	C	BEL	1993	n.a.	35.948	0.0036727024
			1994	58.62	293.364	0.0027416178
			1995	58.62	426.870	0.0027011023
			1996	79.31	382.968	0.0026338028
			1997	79.31	466.578	0.0025826133
Statoil	OG	NOR	1993	79.31	1,752.044	0.0002457615
			1994	79.31	2,003.194	0.0001825224
			1995	79.31	2,145.811	0.0002188761
			1996	75.86	2,823.798	0.0001928398
			1997	75.86	2,256.648	0.0002083322
Tepco	OG	J	1993	n.a.	1,437.95	0.0000098837
			1994	72.41	2,047.87	0.0000102326
			1995	79.31	1,781.78	0.0000091860
			1996	79.31	1,309.73	0.0000077907
			1997	79.31	1,795.96	0.0000072093
Texaco	OG	USA	1993	n.a.	938.000	0.0005182373
			1994	60.34	758.000	0.0003634330
			1995	65.52	567.000	0.0002308802
			1996	68.97	1605	0.0002350000
			1997	70.69	1987	0.0002054886

Source: ERM (1999).

Note: n.a. = not available.

[a]C = chemicals; E = electrical power generation; OG = oil and gas.

[b]See note b to table 8A.1.

[c]Our calculation from data published in CERs.

References

Brehn, J., and J. T. Hamilton. 1996. Non compliance in environmental reporting: Are violators ignorant or evasive of the law? *American Journal of Political Science* 40 (2): 444–77.

Environmental Reporting Monitor (ERM). 1999. *Database and analysis from 1995.* Milan: Fondazione Eni Enrico Mattei.

Forum on Environmental Reporting (FER). 1995. *Corporate environmental reporting guidelines.* Milan: Fondazione Eni Enrico Mattei.

Khanna, M., and L. Damon. 1999. EPA's voluntary 33/50 program: Impact on toxic releases and economic performance of firms. *Journal of Environmental Economics and Management* 37 (1): 1–2.

Lanoi, P., B. Laplant, and M. Roy. 1997. Can capital markets create incentives for pollution control? Policy Research Working Paper no. 1753. Washington, D.C.: Environment, Infrastructure and Agriculture Division, World Bank.

McIntosh, M., D. Leipziger, K. Jones, and G. Coleman. 1998. *Corporate citizenship.* London: FT Pitman.

Musu, I., and D. Siniscalco. 1993. *Ambiente e contabilità nazionale.* Bologna: Il Mulino.

Pendergast, C. 1999. The provision of incentives in firms. *Journal of Economic Literature* 37:7–63.

Pfaff, A. S., and C. W. Sanchirico. 1999. Environmental self-auditing: Setting the proper incentives for discovery and correction of environmental harm. Columbia University School of International and Public Affairs. Working paper.

Sinclair-Desgagné, B., and L. Gabel. 1997. Environmental auditing in management systems and public policy. *Journal of Environmental Economics and Management* 33:331–46.

Tietenberg, T. 1997. Information strategies for pollution control. Paper presented at the eighth annual European Association of Environmental and Resource Economists conference, Tilburg University, The Netherlands, 26–28 June.

Comment   Kevin Hassett

It is all too often the case in economics that researchers spend far more time devising elaborate methods to tease answers from existing data sets than they do performing the heavy lifting required to develop new data sources. This paper is a refreshing contrast. The authors have built a fascinating database that will be an invaluable resource to future researchers, who will likely be able to shed new light on a number of interesting questions with these new data.

When I teach econometrics to graduate students, I always try to emphasize the potentially large benefits from developing new data: One can often learn a great deal with a simple inspection of sample means. If I have a criticism of this paper, it is that the authors have taken this point a little

Kevin Hassett is a resident scholar at the American Enterprise Institute.

bit too literally. After developing their new data, they seem to have run out of gas, providing only a cursory set of simple regressions that are very poorly documented. Clearly, much work is left to be done, and readers might better spend their time staring at the individual observations presented in the appendixes than reading the empirical section carefully.

Now for the details. Many economists have long feared that managers have little incentive to worry too much about how much pollution their firm produces. Cutting back pollution is costly, and since managers' compensation depends on near-term profits, there is little incentive to be too aggressive. Costs from pollution are often long term, and the manager will be floating on his yacht in the Mediterranean by the time the firm has to pay for the damages its pollution has caused.

Regulators, and to some extent firms, have recognized this problem, and a number of complementary approaches have been adopted to overcome it. Recognizing that shining a light on pollution as it occurs might increase incentives to internalize long-run costs, the European Commission adopted the EMAS (Eco Management Audit Scheme), which recommends a method for evaluating the environmental performance of a firm. In addition, many firms have begun to increase executive compensation when particular environmental targets are met by management. The question is, are these measures effective? Does pollution go down when the policies are adopted?

Ex ante, there is no reason to believe that they would be effective. Information concerning effluents that is supplied voluntarily might be very unreliable. Managers may enjoy receiving bonuses for green behavior, but the monetary rewards of high profits are significant, and one might expect them to dwarf the bonuses associated with environmental performance.

To address the question, the authors constructed a database from firm environmental reports published from 1993 to 1997. They selected a sample that includes all the companies belonging to three polluting industries: petrochemicals, oil and gas, and electric power generation. The final sample consists of 39 firms based in 16 countries. The authors first show that firms that adopt reward schemes have slightly better environmental performance, although the difference is not statistically significant. Pollution decreases over time faster for firms that have incentive programs as well, but again the evidence is fairly weak.

The authors then perform a statistical analysis that proceeds in two steps. They show that operating income is higher for firms that have environmental compensation programs, and then show that pollution is lower for firms that have better environmental reports.

It is at the estimation stage that the work starts to have problems. It is not clear to me what the authors are attempting to establish by running a regression with a limited number of variables to predict operating income. None of the variables is scaled or, as far as I can tell, deflated, so that strong trends in the data (or swings in exchange rates) could be determining the results. A sign that trouble is afoot is the Durbin-Watson, which is very close to zero. With very strong trends in the data, there is almost certainly a spurious regression problem, and the t-statistics are essentially meaningless. So do these programs affect environmental performance? Should governments everywhere start to require better environmental audits and green compensation packages? It is impossible to say given what has been done here because the empirical work is incomplete.

A more thorough empirical analysis of the data here will be quite promising. A good place to start would be to perform some simple difference-in-difference comparisons that build on the work presented in table 8.2. Clever use of this technique should overcome the biggest empirical problem here: Firms that adopt programs might have a strong taste for environmental reform, and this unobserved heterogeneity might make it look like the program is effective, when in fact the program is only a signal of the firm's underlying preferences toward pollution.

Despite these criticisms, I enjoyed the paper very much. The authors have provided the profession an invaluable service in constructing the data set (and carefully describing the programs), and they should be commended for printing the entire data set in the appendixes.

Environmental Policy and Firm Behavior
Abatement Investment and Location Decisions under Uncertainty and Irreversibility

Anastasios Xepapadeas

9.1 Introduction

A firm's response to changes in environmental policy is an issue that has drawn considerable attention in the environmental economics literature.[1] Questions that usually arise when environmental policy is introduced or changed are primarily associated with how firms react regarding their choices of investment in productive or abatement capital, the mix of relatively more or less polluting inputs, the choice of labor input or the decisions about research and development (R&D) expenses (process R&D or environmental R&D),[2] or what kind of decisions firms make regarding location choices.[3]

This paper focuses on these questions, and in particular it explores the behavior of polluting firms regarding expansion of abatement capital and location decisions in the presence of environmental policy. Environmental policy takes the form of emissions taxes or tradable emissions permits, and subsidies for the costs of expanding abatement capital. In this context, accumulated abatement capital can be interpreted as the stock of knowledge in pollution and abatement processes. This knowledge is useful in designing new "cleaner" products or better abatement processes.

One of the major factors affecting the responses of firms when a regulatory policy—in our case environmental policy—is introduced or changed

Anastasios Xepapadeas is professor of economics at the University of Crete and dean of the Faculty of Social Sciences.

1. See, e.g., Xepapadeas (1992, 1997a), Kort (1995), and Hartl and Kort (1996).

2. See, e.g., Xepapadeas (1992, 1997a), Kort (1995), Hartl and Kort (1996), and Carraro and Soubeyran (1996a, 1996b).

3. See, e.g., Markusen, Morey, and Olewiler (1993, 1995), Motta and Thisse (1994), Hoel (1994), Rauscher (1995), and Carraro and Soubeyran (1999).

is uncertainty regarding important parameters of the model. In particular for the case of firms responding to environmental policy, uncertainty could be associated with output price movements; that is, demand could be affected by stochastic shocks, or technological uncertainty could affect the efficiency of the abatement process. Another type of uncertainty that could be important is policy uncertainty, which in our case can be associated with stochastic movements of tradable emissions prices or unpredictable (from the firms' point of view) policy changes. In all these cases, the analysis of firms' behavior under uncertainty could be important not only in explaining the effects of environmental policy on abatement investment or relocation decisions, but also as a guide for exploring issues of optimal environmental policy design.

A second important factor affecting the same problem is the fact that firms' decisions regarding abatement investment and location have irreversibility characteristics. Thus abatement investment expenses are irreversible once they are incurred by the firms; movement to a new location when the costs of returning to the old location are sufficiently high is also an irreversible decision.

Since abatement investment is a dynamic process of accumulating abatement capital, and the type of uncertainty described undoubtedly embodies a time dimension since output or tradable emissions price evolves dynamically through stochastic processes, it follows that the analysis of firms' responses to environmental policy might be more realistically explored in a dynamic framework. In a dynamic setup, the interaction of uncertainty with the irreversibility characteristics of investment decisions or relocation decisions generates well-known option value issues.[4]

Thus, the purpose of the present paper is to explore abatement investment and location responses to environmental policy under uncertainty and irreversibility. The problem is analyzed in a dynamic setup, where uncertainty is modeled by Itô stochastic differential equations, by using optimal stopping methodologies. The idea is to define continuation intervals during which firms do not expand abatement capital or relocate, and intervals during which firms take the irreversible decision to undertake abatement investment expenses or relocate. The optimal stopping methodology will define a free boundary. When a state variable—which could be output price, the price of tradable emissions permits, or a technological coefficient—crosses the boundary, the irreversible decision to increase abatement capital or relocate is taken. The structure of the free boundary determines, therefore, the conditions under which the firm will invest in abatement capital or relocate.

Using this methodological approach, free boundaries are determined or

---

4. See Arrow and Fisher (1974), Fisher and Hanemann (1986, 1987), and Xepapadeas (1998) for related issues or Dixit and Pindyck (1994) for a more general treatment.

characterized for a variety of cases that include output price uncertainty, policy uncertainty expressed both in terms of continuous fluctuations of permit prices and unpredictable policy changes, and technological uncertainty. The advantage of this approach is that although a complex mathematical model is used, the emerging results regarding the structure of the free boundary are relatively simple and depend on estimable parameters. Thus it allows the analysis of firms' responses to environmental policy in a framework that combines uncertainty and irreversibility effects.

A second advantage of the approach is that when uncertainty is not associated with environmental policy parameters, but is either demand or technological uncertainty, the free boundaries are defined parametrically in terms of policy instruments, such as emissions taxes or abatement investment subsidies. This allows the meaningful performance of comparative statics regarding the irreversible decisions. Thus it is possible, given the parameters characterizing the stochastic processes associated with output price or technological uncertainty, to analyze how firms respond to changes in emissions taxes or abatement subsidies, regarding abatement investment or relocation decisions, by analyzing the shifts of the boundary.

Finally, the optimal stopping approach makes possible the design of optimal policies under uncertainty. The optimal policy is determined such that the firm's free boundary under the optimal policy is identical to a free boundary determined by maximizing the objective function of a regulator. In this case, the firm reaches decisions regarding abatement investment or relocation, given the stochastic movements of the state variables (output price or technological uncertainty), which are the same as the decisions that a regulator would have taken in the same stochastic environment. Thus this methodology introduces an approach to optimal policy design in which, under uncertainty, the target is not to choose the instrument such that the firms choose the same value of the variable of interest (e.g., abatement), but rather the target is to induce them to base their decision rule on the same rule that the regulator would have selected. In this case, the decision rule is determined by the free boundary.

9.2 Abatement Investment Decisions under Uncertainty

We assume an industry consisting of n identical firms producing in a small open economy. The firms behave competitively and sell their product in the world market where international competition prevails. We consider the representative firm producing at each instant of time output $q(t)$ at a cost determined by a cost function $c(q(t))$, with $c'(q) > 0$, $c''(q) > 0$. Output is sold in the world market at an exogenous world price $p(t)$.

The production of output generates emissions. Emissions per unit of output are determined by the function $E(t) = v(t) e(R(t))$, where $v(t) > 0$ is an efficiency parameter associated with the abatement process and $e(R(t))$

is a function of the accumulated abatement capital, up to time t.[5] A reduction in v indicates an improvement in the efficiency of the abatement process, while abatement capital, denoted by $R(t)$, is defined as

$$R(t) = \int_0^t r(s)ds,$$

where $r(s)$ is the abatement investment flow undertaken at instant of time s. This flow can, for example, represent resources devoted to the firm's lab in order to design cleaner processes at time s. It is assumed that $r(s) \geq 0$; thus, the abatement capital accumulation process is irreversible.[6] For the abatement process, we assume that

$$e(0) = 0, \quad e'(R) < 0, \quad e''(R) > 0,$$
$$\lim_{R \to 0} e'(R) < -\infty, \quad \lim_{R \to \infty} e'(R) = 0.$$

Therefore an increase in abatement capital reduces unit emissions at a decreasing rate, which means that diminishing returns in abatement capital are assumed. Thus, when the firm produces output $q(t)$, total emissions are defined as $v(t) e(R(t)) q(t)$.

The cost for increasing the stock of accumulated abatement capital by ΔR is defined as $(1 - s) h\Delta R$, where h is the exogenous unit-abatement investment cost and $s \in [0, 1)$ is a subsidy potentially given by the government to cover some of the expenses for expanding abatement capital. Assume that the firm pays an exogenously determined emission tax $\tau(t)$. Then the tax payments are defined as $\tau(t)[v(t) e(R(t)) q(t)]$.

Given this setup, the firm has to decide about output production and abatement investment. At each time the firm decides about the optimal output level given the stock of abatement capital. Thus output is regarded as an operating variable and output decisions can be regarded as short-run decisions, while abatement investment decisions are long-run decisions. The optimal choice of output for any given level of abatement capital determines a reduced-form instantaneous profit function, which can be defined as

(1) $\pi(p, v, \tau, R) = \max_q \{p(t)q(t) - c(q(t)) - \tau(t)[v(t)e(R(t))q(t)]\}.$

The first-order conditions for the optimal output choice, assuming interior solutions and dropping t to simplify notation, are given as

5. A more general formulation would be to define the $e(\cdot)$ function as $e(R, R^{AG})$, where $R^{AG} = nR$, is aggregate abatement knowledge. In this case there could be positive spillovers from aggregate abatement capital to the individual abatement function. When firms consider aggregate knowledge as fixed, there is a divergence between the private return of abatement capital and the social return of abatement capital (Xepapadeas 1997b).
6. To simplify things, we ignore depreciation issues.

$$p - c'(q) - \tau v e(R) = 0,$$

with optimal output determined as

$$q^* = q^*(p, \tau, v, R).$$

Using the first-order conditions for the optimal output choice, we obtain the following short-run comparative static results:

$$\frac{\partial q^*}{\partial p} > 0, \quad \frac{\partial q^*}{\partial \tau} < 0, \quad \frac{\partial q^*}{\partial v} < 0, \quad \frac{\partial q^*}{\partial R} > 0.$$

Thus an increase in the tax rate or a reduction in the abatement efficiency (increase in v) reduces optimal output, while an increase in the stock of abatement capital increases optimal output. From the short-run comparative statics and the envelope theorem, we obtain the derivatives of the profit function as[7]

$$\frac{\partial \pi}{\partial p} = q^*(p, v, \tau, R), \quad \frac{\partial^2 \pi}{\partial p^2} = \frac{\partial q^*}{\partial p} > 0,$$

$$\frac{\partial \pi}{\partial v} = -(\tau e(R)q^*) < 0, \quad \frac{\partial^2 \pi}{\partial v^2} = -\tau e(R)\frac{\partial q^*}{\partial v} > 0,$$

$$\frac{\partial \pi}{\partial \tau} = -(v e(R)q^*) < 0, \quad \frac{\partial^2 \pi}{\partial \tau^2} = -v e(R)\frac{\partial q^*}{\partial \tau} > 0,$$

$$\frac{\partial \pi}{\partial R} = -\tau v e'(R)q^* > 0, \quad \frac{\partial^2 \pi}{\partial R^2} = -\tau v\left(e''(R)q^* + e'(R)\frac{\partial q^*}{\partial R}\right).$$

Thus the profit function is convex in prices for fixed (τ, v, R), decreasing in (τ, v), and increasing in R.

Uncertainty can be introduced into this model in three ways. First, it can be assumed that the world demand is affected by stochastic shocks giving rise to a geometric Brownian motion price process. In this case, output price is the exogenous state variable,

$$(2) \qquad\qquad dp(t) = ap(t)dt + \sigma p(t)dz_p(t),$$

where $[z_p(t)]$ is a Wiener process,[8] and a and σ are constants. If the current price is a given constant $p(0) = p_o$, then the expected value of $p(t)$ is $E[p(t)] = p_o e^{at}$ and the variance of $p(t)$ is[9]

---

7. It is assumed that $[e''(R)q^* + e'(R)(q^*/R)] > 0$, so that the profit function is concave in R for fixed (p, τ, v). The concavity assumption requires sufficient curvature of the unit emissions function $e(R)$.

8. For definitions see Malliaris and Brock (1982).

9. See Dixit and Pindyck (1994).

$$V[p(t)] = p_\sigma^2 e^{2at}(e^{\sigma^2 t} - 1).$$

It should be noted that the Brownian motion assumption causes price to move away from its starting point. If, however, price is related to long-run marginal costs, then a better assumption about price movements could be a mean-reverting process. Under this assumption, price tends toward marginal costs in the long run and price movements can be modeled as

$$dp(t) = a(\tilde{p}(t) - p(t))p(t)dt + \sigma p(t)dz_p(t),$$

where $\tilde{p}(t)$ can be interpreted as long-run marginal costs.[10]

Second, it can be assumed that environmental efficiency evolves stochastically according to the geometric Brownian motion:

(3) $$dv(t) = \gamma \tau(t)dt + \delta v(t)dz_v(t).$$

The interpretation of this type of uncertainty can be associated with the stochastic operating conditions of abatement equipment. It can also be associated with stochastic effects of the general level of abatement knowledge in the economy that is external to the firm, but can affect the firm's abatement efficiency through spillover effects.[11]

Third, it can be assumed that environmental regulation takes place through a system of tradable permits, in which case $\tau(t)$ can be interpreted as the competitive market price for permits, which can evolve stochastically according to the geometric Brownian motion:[12]

(4) $$d\tau(t) = \eta \tau(t)dt + \omega \tau(t)dz_\tau(t).$$

Given the firm's instantaneous profit function (1), the next stage is to define the optimal abatement investment policy for the types of uncertainty described.

9.2.1   Abatement Investment Decisions under Price Uncertainty

Having optimally chosen the output level, the next step is to analyze the decision to undertake new abatement investment, denoted by ΔR, from the existing abatement capital level of R_0, under price uncertainty modeled by equation (2) and assuming that (τ, v, s, h) are fixed parameters. Consider the firm's decision to undertake new abatement investment by ΔR from the existing abatement capital level R_0; then the new abatement capital level becomes

---

10. If we consider entry and exit decisions in the world market, then an upper reflecting barrier $\bar{p}$ to the price movement can be considered. When price moves to the reflecting barrier, new entry is triggered, quantity increases, and price decreases.

11. Stochastic delays in the R&D processes can be modeled by assuming that v follows a Poisson process.

12. Then $v(t) e(R(t)) q(t)$ can be interpreted as the excess demand for permits. The expected values and the variances for $v(t)$ and of $\tau(t)$ are defined in a similar way as for $p(t)$.

$$R_{0+} = R_0 + \Delta R.$$

The cost of this change in abatement capital is defined as

(5) $$(1 - s)h(R_{0+} - R_0).$$

In the model developed here, the optimal abatement investment strategy takes the form of a free boundary, $p = p(R; \tau, v, s, h)$, relating price and accumulation of abatement capital. This boundary is parametrically defined for the vector of parameters (τ, v, s, h). When observed price p^{ob} is less than $p(R; \tau, v, s, h)$, no abatement investment is undertaken, while when p^{ob} is greater than $p(R; \tau, v, s, h)$ enough abatement investment is undertaken in the current period to restore equality on the boundary. Changes in the parameter vector (τ, v, s, h) shift the boundary and can accelerate or decelerate abatement investment accumulation for any given price. Thus we can determine, by using comparative statics associated with the free boundary, the effects of environmental policy on the firm's decision rule regarding abatement investment.

Assume that the initial price is p_0 and the firm's initial abatement capital stock is R_0. Given a discount rate ρ, the firm seeks the nondecreasing process $R(t)$, which will maximize the present value of profits less the cost of development. The value function[13] associated with this problem can be written as

(6) $$V(p, R) = \max_R \mathcal{E} \int_0^\infty e^{-\rho t} \pi(p(t), v(t), \tau(t), R(t)) dt,$$

subject to equation (2).

At each instant of time, the firm has two choices: to undertake the new abatement investment or not. The time interval when no new abatement investment is undertaken and the existing abatement stock is used to determine the unit emission coefficient, can be defined as the continuation interval. A stopping time is defined as a time $\mathcal{T}$ at which new abatement investment is undertaken.

Let $R^*(\mathcal{T})$ be the optimal development process at time $\mathcal{T}$. If $\mathcal{T}$ is a stopping time, then

(7) $$V(p, R; \tau, v)$$

$$= \max_R \mathcal{E} \left[\int_0^{\mathcal{T}} e^{-\rho u} \pi(p(u), v, \tau, R(u)) du + e^{-\rho \mathcal{T}} V(R^*(\mathcal{T}), p(\mathcal{T})) \right],$$

where $R^*(\mathcal{T})$ is the optimal process at time $\mathcal{T}$ (see Fleming and Soner 1993). Assume that in the time interval $[0, \theta]$, the firm undertakes no new

---

13. By the concavity of the profit function in R and the linear dynamics, the value function is also concave in R (Dixit and Pindyck 1994).

abatement investment, but keeps it constant at R_0. By the principle of dynamic programming, the value function should be no less than the continuation payoff in the interval $[0, \theta]$, plus the expected value after θ, or

(8) $V(p, R; \tau, v)$

$$\geq \mathcal{E} \left[\int_0^\theta e^{-\rho u} \pi(p(u), v(u), \tau(u), R(u)) du \ + \ e^{-\rho \theta} V(R(\theta), p(\theta)) \right],$$

with equality if R_0 is the optimal policy in $[0, \theta]$. Applying Itô's lemma to the value function on the right-hand side of equation (8), dividing by θ, and taking limits as $\theta \to 0$, we find that the value function should satisfy[14]

(9) $$\rho V \geq \frac{1}{2} \sigma^2 p^2 V_{pp} \ + \ ap V_p \ + \ \pi(p, v, \tau, R),$$

with equality if $R(t) = R_0$ in the interval $[0, \theta]$.

Consider now the decision to undertake abatement investment instantaneously by $\Delta R = R_{0+} - R_0$. Then from the definition of the optimal stopping time, we have

(10) $V(p, R; \tau, v) \geq \mathcal{E}[V(R_{0+}, p; \tau, v) - (1 - s)h(R_{0+} - R_0)]$.

Since the value function is concave in R, the optimal abatement investment flow can be obtained by maximizing the right-hand side of inequality (10). The necessary and sufficient condition for the optimal abatement investment choice is

(11) $V_R(R, p; \tau, v) - (1 - s)h \leq 0$,

with equality if $\Delta R > 0$.

Thus when no new abatement investment is optimal, inequality (9) is satisfied as equality, whereas when new abatement investment is optimal, inequality (11) is satisfied as equality. Combining inequalities (9) and (11), the Hamilton-Jacobi-Bellman (HJB) equation can be written as

(12) $\min \left\{ \left[\rho V - \frac{1}{2} \sigma^2 p^2 V_{pp} - ap V_p - \pi(p, v, \tau, R) \right] \right.$

$$\left. - [V_R - (1 - s)h] \right\} = 0.$$

The optimal free boundary will divide the (p, R) space into two regions: the "no new abatement investment" region, which we call region I, and the "new abatement investment" region, which we call region II.

In region I, the first term of the HJB equation (12) is 0, since $\Delta R = 0$,

---

14. Subscripts associated with the value function denote partial derivatives.

and the second term of the HJB equation is positive by inequality (11); while in region II, the second term of equation (12) is satisfied as 0 and $\Delta R > 0$. These conditions allow the determination of the value function and the free boundary as functions of the policy parameters.

PROPOSITION 1. *Given the structure of the model as defined above and a quadratic cost function $c(q) = \frac{1}{2}cq^2$, the value function and the free boundary are defined respectively as*

$$V(p,R) = A_1(R)p^{\beta_1} + \Pi(p,R;\tau,v),$$

$$V_R(p,R) = A_1'(R)p^{\beta_1} + \Pi_R(p,R;\tau,v) = (1-s)h, \quad A_1'(R) < 0,$$

$$p(R;\tau,v,s,h) = \frac{-\beta_1}{\beta_1 - 1} \frac{(\rho - a)}{\rho} \frac{[c\rho(1-s)h + \tau ve(R)e'(R)]}{\tau ve'(R)}.$$

PROOF. See the appendix.

The solution of the value function $V(p, R)$ indicates that the maximized expected value consists of the term $\Pi(p, R; \tau, v)$, which can be interpreted as the present value of net profits when abatement capital is kept constant, and the term $A_1(R)p^{\beta_1}$, which is the current value of the option to expand abatement capacity. When the firm increases abatement capital, it sacrifices the option value of the incremental abatement capacity; thus $A_1'(R) < 0$. Therefore an increase in abatement is desirable if its contribution to net profit $\Pi_R(p, R; \tau, v)$, realized through savings in emissions taxes less the cost of giving up the option to wait $A_1'(R)p^{\beta_1}$, equals the marginal expansion cost $(1 - s)h$. The free boundary $p(R; \tau, v, s, h)$ can be determined for estimated parameter values that characterize the price process, the cost structure, and the discount rate. Since $p(R) > 0$, the free boundary is defined for parameter values such that $c\rho(1-s) h > |\tau ve(R)e'(R)|$. In order to describe the free boundary we have, by the assumptions on the unit emissions function, $p(0) > 0$ and $\lim_{R \to \infty} p(R) = +\infty$. Furthermore,

$$\frac{\partial p}{\partial R} = \frac{-\beta_1}{\beta_1 - 1} \frac{(\rho - a)}{\rho} \frac{[(\tau v)(e')^3 - c\rho(1-s)he'']}{\tau v(e')^2} > 0.$$

The free boundary is shown in figure 9.1. For any given level of abatement capital, random price fluctuations move the point (R, p) vertically upward or downward. If the point goes above the boundary, then new abatement investment is immediately undertaken so that the point shifts on the boundary. Thus optimal abatement capital accumulation proceeds gradually. In the terminology of Dixit and Pindyck (1994), this is a "barrier control" policy.

By inverting the free-boundary function $p(R; \tau, v, s, h)$, we can obtain the optimal boundary function $R^* = p^{-1}(p; \mathbf{z})$, which determines the optimal abatement investment boundary as a function of the state variable p and

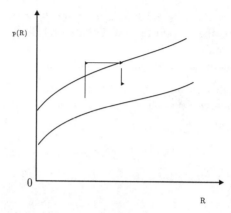

Fig. 9.1   Free boundary under price uncertainty

the vector **z** of the parameters of the problem. For price movements to the right of the boundary, new abatement investment is undertaken. If price stays on the left of the boundary, no new abatement investment is undertaken.

If price follows a mean reverting process, then the HJB equation for region I that corresponds to equation (12) becomes

$$(12')\qquad \rho V - \frac{1}{2}\sigma^2 p^2 V_{pp} - a(\tilde{p} - p)pV_p - \pi(p, R; \tau, v) = 0.$$

The steps for solving for the optimal boundary are the same as before; however, due to the more complicated structure of equation (12'), the analysis of the effects of mean reversion requires numerical solutions (see Dixit and Pindyck 1994).

9.2.2   The Impact of Changes in Policy Parameters

We can examine the shifts of the free boundary in response to changes in the tax parameter τ or the subsidy parameter s. These effects are determined as

$$\frac{\partial p}{\partial \tau} = \frac{\beta_1}{\beta_1 - 1}\frac{(\rho - a)}{\rho}\frac{c\rho(1 - s)h}{\tau^2 ve'} < 0,$$

$$\frac{\partial p}{\partial s} = \frac{\beta_1}{\beta_1 - 1}\frac{(\rho - a)}{\rho}\frac{c\rho h}{\tau ve'} < 0.$$

Thus an increase in the tax rate or the subsidy rate shifts the boundary downward and induces more abatement investment for any given price level, as is also shown in figure 9.1. An increase in the tax rate might not, however, increase abatement investment if there is a drop in prices below the boundary. This is because the reduction in equilibrium output and the consequent emissions reduction do not necessitate an increase in abatement. Abatement investment might increase at a future time when prices go up. This reveals that uncertainty affects the timing of the impact of environmental policy. At declining prices, a change in environmental policy might not induce any abatement investment. The impact of the policy might, however, be realized with a delay.

By performing the same type of comparative statics, we obtain

$$\frac{\partial p}{\partial v} < 0.$$

A reduction in abatement efficiency induces more abatement investment for any given price level.

9.2.3 Optimal Environmental Policy

In section 9.2.2 the tax and the subsidy parameters were treated as fixed. The analysis can, however, be extended to analyze the case of an environmental regulator who can choose the policy parameters optimally. Optimal policy choice is considered in the following way. From proposition 1, the free boundary that determines the profit-maximizing abatement investment depends on the tax and subsidy parameters. Consider the case of an environmental regulator that determines a socially optimal free boundary by explicitly taking into account environmental damages. An optimal environmental policy can then be defined by determining the values of the policy parameters such that the profit-maximizing free boundary will coincide with the socially optimal free boundary, as determined by the environmental regulator. Define a social profit function by

$$W(p,v,R) = \max_{q}[p(t)q(t) - c(q(t)) - D(v(t)e(R(t))q(t))],$$

where $D(v(t)e(R(t))q(t))$ is a strictly increasing and convex damage function. By following the steps in section 9.2.1, a free boundary that determines the socially optimal abatement investment under price uncertainty can be defined. Denote this free boundary by $p^{s}(R)$, and consider the free boundary defined in proposition 1 as a function of the policy parameters, or $p(R; \tau, s)$. An optimal environmental policy can be defined as the pair

$$(\tau^*, s^*) : p(R; \tau^*, s^*) = p^{s}(R).$$

A solution of the form $\tau^* = \zeta(s^*)$ will determine the trade-off between emissions taxes and abatement investment subsidies in the design of environmental policy.[15]

In the simplest possible case of constant marginal damages at the level d, the optimal trade-off is determined as $\tau^* = d(1 - s)$. The tax rule for

---

15. The relationship between the two policy instruments can be further elaborated to include budget-balancing schemes, where total tax revenues equal total subsidy expenses.

this case is very simple and, given the parameters of the model and output price observations, the regulator can determine the firms' responses regarding abatement investment.

It is interesting to note that under uncertainty and irreversibility, the optimal environmental policy equates the privately optimal and the socially optimal free boundaries and not the privately optimal and the socially optimal levels of the choice variables as in the case of optimal policy design under certainty.

9.3   Abatement Investment Decisions under Environmental Policy or Abatement Efficiency Uncertainty

When, under fixed prices, the environmental policy uncertainty is present in the form of stochastic evolution of prices for tradable emissions permits, or abatement efficiency is stochastic, then the mathematical treatment is similar, although the sources of uncertainty are different. Policy uncertainty can be regarded as uncertainty outside the firm, while abatement uncertainty can be regarded as internal to the firm. So although the mathematical results are the same, their interpretation and their policy implications are different.

In the case of policy uncertainty, the HJB equation can be written as

$$\min\left\{\left[\rho V - \frac{1}{2}\omega^2\tau^2 V_{\tau\tau} - \eta\tau V_\tau - \pi(\tau,R;p,v)\right], -[V_R - (1-s)h]\right\} = 0.$$

As before, the optimal free boundary will divide the (τ,R) space into two regions: the "no new abatement investment" region (region I) and the "new abatement investment" region (region II).

PROPOSITION 2. *For the quadratic cost function defined in proposition 1, the value function and the free boundary are determined as*

$$V(\tau,R) = B_1(R)\tau^{\xi_1} + \phi(p,R;\tau,v), \quad B_1'(R) < 0,$$

$$V_R(\tau,R) = B_1'(R)\tau^{\xi_1} + \phi_R(p,R;\tau,v) = (1-s)h,$$

$$\tau(R) = \frac{-\xi_1}{2(\xi_1 - 2)} \frac{\Delta_1'(R)(\xi_1 - 1/\xi_1) + \sqrt{\Delta}}{\Delta_2'(R)},$$

$$\Delta = \left[\Delta_1'(R)\left(\frac{\xi_1 - 1}{\xi_1}\right)\right]^2 + 4\Delta_2'(R)\left(\frac{\xi_1 - 2}{\xi_1}\right)(1-s)h,$$

$$\Delta_1'(R) = -\frac{\rho ve'(R)}{c(\eta - \rho)} > 0, \quad \Delta_2'(R) = \frac{2ve(R)e'(R)}{2c(\omega^2 + 2\eta - \rho)}.$$

PROOF. See the appendix.

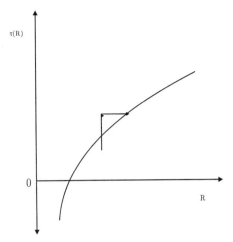

Fig. 9.2   Free boundary under policy uncertainty

The interpretations of the value function and the free boundary are similar to those under price uncertainty.

Using the assumptions about the unit-emissions function we have

$$\lim_{R \to 0} \tau(R) = -\infty, \quad \text{and} \quad \lim_{R \to \infty} \tau(R) = M > 0.$$

If the free boundary is monotonic, a property that can be checked by using specific functions, then its graph is shown in figure 9.2.

An increase of the policy parameter above the boundary induces more abatement investment. By inverting the $\tau(R)$ function, an optimal boundary function for abatement capital accumulation in terms of the policy parameter τ is defined as

$$R^* = \tau^{-1}(\tau, \mathbf{z}).$$

Policy uncertainty and abatement efficiency uncertainty can also be examined together by introducing the variable $z = \tau v$, with $\mathcal{E}(dz_\tau dz_v) = \rho_{\tau v} dt$. Using the fact that $\partial^2 z / \partial \tau^2 = \partial^2 z / \partial v^2 = 0$ and $\partial^2 z / \partial \tau v = 1$ we obtain

$$dz = (\gamma + \eta + \rho_{\tau v} \delta \omega) z dt + (\delta dz_v + \omega dz_\tau) z.$$

Thus changes in z have mean and variance

$$k_1 = \gamma + \eta + \rho_{\tau v} \delta \omega,$$

$$k_2 = \delta^2 + 2\rho_{\tau v} \delta \omega + \omega^2,$$

respectively. Following the same steps as before, the HJB equation is defined as

$$\min\left\{\left[\rho V - \frac{1}{2}k_2^2 z^2 V_{zz} - k_1 z V_z - \pi(z,R;p)\right], -[V_R - (1-s)h]\right\} = 0.$$

Then the free boundary can be defined in the context of correlated policy and technological uncertainty, as in the previous case of policy uncertainty.

9.3.1 Unpredictable Policy Changes

Policy uncertainty as analyzed previously is associated with continuous fluctuations of the tradable-emissions-permit price. It is possible, however, for a sudden change in policy due, for example, to an unexpected (from the firm's point of view) change in the supply of permits, to cause a discontinuous change in their price. In the context of our model, this unpredictable change introduces jump characteristics. Thus, while the usual fluctuations in prices are captured by the geometric Brownian motion, the sudden policy change should be captured by a Poisson process. Therefore, the price of permits is modeled by a mixed Brownian motion and jump process, or

$$d\tau = \eta\tau dt + \omega\tau dz_\tau + \tau dq^P,$$

where dq^P is the increment of a Poisson process with a mean arrival time of the change in the supply of permits λ. We further assume that the change in the supply of permits represents an increase and that this causes a fixed drop in the price[16] by a known percentage $\psi \in [0, 1]$ with probability 1, and that dz_τ and dq^P are independent.

To analyze this problem, the HJB equation is derived by using Itô's lemma for combined Brownian motion and jump process (Dixit and Pindyck 1994). Then the HJB equation can be written as

$$\min\left\{\left[\rho V - \frac{1}{2}\omega^2\tau^2 V_{\tau\tau} - \eta\tau V_\tau - \pi(\tau,R;p,v)\right]\right.$$

$$\left. + \lambda[V((1-\psi)\tau) - V], -[V_R - (1-s)h]\right\} = 0.$$

The solution to the value function is

$$V(\tau,R,\psi) = B_1^\psi(R)\tau^{\xi_1^\psi} + \phi^\psi(\tau,R;p,v,\psi),$$

where ξ_1^ψ is the positive solution of the nonlinear equation (see Dixit and Pindyck 1994),

---

16. A fixed increase in the price can be treated symmetrically.

$$\frac{1}{2}\omega^2\xi^\psi(\xi^\psi - 1) + \eta\xi^\psi - (\rho + \lambda) + \lambda(1 - \psi)^{\xi^\psi} = 0.$$

Once a solution for ξ_1^ψ is obtained, then the free boundary can be obtained as before.

It should be noted that special cases of the general mixed Brownian motion and jump process model can be used to analyze specific cases. For example, if $\eta = \omega = 0$ is set and τ is interpreted as an emissions tax, then the same model can be used to analyze the implications of unpredictable changes in the emissions tax rates.

The analysis of policy uncertainty provides a general way of analyzing firms' reactions to environmental policy. Given the structure of the free boundary, which can be determined for estimated parameter values, the regulator can obtain the firms' reactions to a wide range of policy changes using a unified model.

9.4   Location Decisions

When we examine location decisions, the problem can also be defined as an optimal stopping problem. In the waiting or continuation region, the firm stays in its present location, pays the emissions tax, and follows the optimal abatement capital-accumulation path, $R^*(t)$, given uncertainty described by the evolution of the state variable (price, policy parameter, or technology parameter).

Suppose that the firm examines the possibility of relocation to a new location (country) where there is no environmental policy. Assume that the setup costs are fixed, F, and are incurred once at the time of relocation, that the cost function remains the same, and that there are no transportation costs.[17] Suppose that relocation takes place at time t_d; then the profit function for the firm that chooses optimally operating output is defined as

$$\begin{cases} \pi_d(p(t)) - F, & \text{for } t = t_d, \\ \pi_d(p(t)), & \text{for } t > t_d, \end{cases}$$

where $\pi_d(p) = \max_q[pq - c(q)]$.

Assuming that price uncertainty exists, then at each period of time the firm faces a binary choice:

1. Relocate and take the termination payoff defined as $W(p(t_d),F)) = \mathcal{E} \int_{t_d}^\infty e^{-\rho t}\pi_d(p(t))dt - F.$
2. Continue operation at the initial location for one period, choosing output and abatement investment optimally; receive the operating profits; and then consider another binary choice in the next period.

17. See the proofs of propositions 1 and 2 in the appendix for these conditions.

The Bellman equation for this problem can be written as

$$(13) \quad V(p) = \max \left\{ \pi(p, \tau, v, R^*) + \frac{1}{1 + \rho dt} \mathcal{E}[V(p + dp | p)], W(p, F) \right\}.$$

In the continuation region, the first term on the right-hand side is the largest. Using Itô's lemma on this term, we obtain the usual differential equation,

$$(14) \qquad \rho V - \frac{1}{2}\sigma^2 p^2 V_{pp} - a p V_p - \pi(p; R^*, \tau, v) = 0,$$

with solution

$$(15) \qquad V(p(t); R^*, \tau, v) = K_1 p(t)^{\beta_1} + \Pi(p(t); R^*, \tau, v).$$

From the Bellman equation we have that at the critical relocation time t_d,

$$(16) \qquad V(p(t_d); R^*, \tau, v) = W(p(t_d), F).$$

This is the value-matching condition. The smooth-pasting condition requires that[18]

$$(17) \qquad V_p(p(t_d); R^*, \tau, v) = W_p(p(t_d), F).$$

Conditions (16) and (17) can be used to determine the constant K_1 and the free boundary $p = p^*(t_d)$. By inverting the boundary equation we obtain the optimal relocation time boundary function $t_d^* = p^{*-1}(p)$. This boundary determines the critical relocation time as a function of the observed price for given values of the parameters τ, v, and s.

Environmental policy uncertainty or abatement efficiency uncertainty can be treated in the same way. Suppose that policy uncertainty exists in the sense of stochastic permit prices. Then, following the same steps as before, the free boundary $\tau = \tau^*(t_d)$ is defined by the following conditions, using the quadratic cost function:

$$V(\tau(t); R^*, p, v) = \Lambda_1 \tau(t)^{\xi_1} + \Phi(\tau(t); R^*, p, v),$$

$$V(\tau(t_d); R^*, p, v) = W(F), \quad \text{value matching},$$

$$W(p, F) = \int_0^\infty e^{-\rho t} \frac{p^2}{2c} dt - F = \frac{p^2}{2c\rho} - F, \quad p \text{ fixed},$$

$$V_\tau(\tau(t_d); R^*, p, v) = W_\tau(F) = 0, \quad \text{smooth pasting}.$$

Using these conditions, the free boundary $\tau^*(t)$ is implicitly defined by

---

18. Alternative assumptions could include the existence of a different environmental policy abroad, for example command and control regulation, or differences in the political systems that affect the stringency of environmental policy.

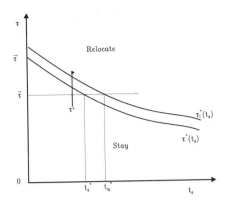

Fig. 9.3   Threshold policy parameter

$$(18) \quad \frac{-\Phi_\tau(\tau^*(t); R^*, p, v)}{\xi_1} \tau^*(t) + \Phi(\tau^*(t); R^*, p, v) = \frac{p^2}{2cp} - F.$$

By inverting the boundary function $\tau^*(t_d)$, we obtain the optimal relocation time boundary function $t_d^* = \tau^{*-1}(\tau)$ in terms of the environmental policy parameter and the rest of the parameters of the problem. Optimal relocation implies the existence of a threshold policy parameter such that when the actual policy parameter crosses this threshold, relocation takes place. This result is stated in the following proposition.

PROPOSITION 3. *Let* $\tau^o(t')$ *be the observed environmental policy parameter at time* t'. *If* $\tau^o(t') < \tau^*(t')$, *then it is optimal to remain at the initial site. If* $\tau^o(t') > \tau^*(t')$ *it is optimal to relocate at time* t'.

PROOF. See the appendix.

This proposition implies that for any time t a threshold environmental policy parameter exists such that when the policy parameter exceeds the threshold, the firm moves to the new location. The relation between the threshold policy parameter and the optimal relocation time is shown in figure 9.3, where $\bar{\tau}$ indicates the permit price that induces immediate relocation. The lower the permit price, the further away the optimal relocation time is. When the permit price crosses the boundary, it is optimal to take the irreversible relocation decision. Similar analyses, although with different interpretations, can be applied to the case where uncertainty relates to abatement efficiency, or to correlated policy and abatement uncertainty.

The analysis suggests that since firms are identical, they will relocate simultaneously when the critical time arrives. If firms are heterogeneous regarding characteristics of production cost or abatement technologies, then the optimal relocation time will be different across firms. In this case, there will more than one boundary such as the one depicted in figure 9.3.

Suppose that a second boundary, $\tau_j^*(t_d)$, exists and, as shown in figure 9.3, the two boundaries do not intersect. Then the same permit price $\bar{\tau}$ implies different optimal relocation times.

9.4.1   Relocation Time Policy under Uncertainty

The free boundaries and the optimal relocation policy functions derived here can be used to provide useful information regarding the effects of exogenous shocks on the critical relocation time and describe a framework for designing a policy that could affect relocation time. This can be obtained by performing comparative static analysis of the optimal boundary function.

Assuming, again, identical firms to simplify things, consider the boundary function $t_d^* = p^{*-1}(p; \tau, v, s)$ and take the derivative:

$$\frac{\partial t_d^*}{\partial \tau} = \frac{\partial p^{*-1}(p)}{\partial \tau}.$$

This derivative determines the effect on the critical relocation time of an exogenous change in the emissions tax for any given price level. In the same way, the effects from changes of other parameters of the model on relocation time can also be defined.

Consider now the total differential,

$$dt_d^*(p) = \frac{\partial t_d^*}{\partial \tau} dt_d + \frac{\partial t_d^*}{\partial s} ds.$$

This differential can express the rate of substitution between the emissions tax rate and the abatement investment subsidy rate in order to produce a given change in the critical relocation time at any given price level. For example by setting $dt_d^*(p) = 0$, the marginal rate,

$$\frac{ds}{d\tau} = \frac{\partial t_d / \partial \tau}{\partial t_d / \partial s},$$

expresses the necessary increase in the abatement investment subsidy in order to keep the critical relocation time constant after an increase in the emissions tax for any given price. The changes in the policy parameters shift the free boundary $p = p^*(t_d)$ and enlarge or shrink the stay or relocate regions as shown in figure 9.4. The particular forms of the policy functions, the comparative static derivatives, and the marginal rates of substitution can be explored under the quadratic cost function assumption.

9.5   Concluding Remarks

The responses of firms to environmental policy regarding their abatement investment and location decisions have been analyzed in an analyti-

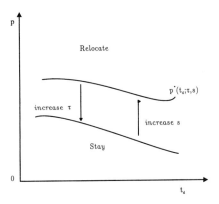

Fig. 9.4   Relocation time policy

cal framework characterized by uncertainty and irreversibilities. The optimal stopping time methodology adopted in this paper makes it possible to analyze firms' responses to environmental policy in terms of the impact that this environmental policy has on the barrier control policies followed by firms regarding their profit-maximizing decisions. The analysis of environmental policy impacts through their effects on barrier control policies makes it possible to view these impacts as shifts of a free boundary that determines firms' policies regarding abatement investment or relocation decisions. In this sense the approach developed in this paper can be regarded as another way of analyzing the effects of regulatory policy under conditions of uncertainty and irreversibility. Despite the mathematical complexity of the approach, the effects of regulatory policies on the free boundary are determined by parameters related to the stochastic process associated with uncertainty and the firms' production structure, which are in principle estimable.

It is also possible to use the optimal stopping time methodology in order to design optimal environmental policy under uncertainty and irreversibility, in the sense of choosing the policy parameters so that the free boundary, or equivalently the optimal policy function under profit maximization, coincides with the socially optimal free boundary (abatement-investment-policy function). The implication of this approach regarding optimal policy design under price uncertainty and irreversibility is that a regulator can in principle design a policy scheme consisting of two instruments: an emissions tax or a tradable permit system, and a subsidy on abatement investment. The policy scheme takes into account uncertainty through its dependence on the parameters of the price process and will induce individual firms to undertake the same output and abatement investment under uncertainty that a regulator would have undertaken. In this sense, the policy mix of emissions taxes (or emissions permits) and abatement investment subsidies will be welfare maximizing. It should be noticed that the function $\tau^* = \zeta(s^*)$, determining the optimal trade-off between taxes and subsidies, allows the regulator to determine the policy mix in order to obtain an optimal balance between the output-contracting pollution control by emissions taxes and pollution control through the subsidization of the accumulation of abatement capital.

A similar mix of emissions taxes (or emissions permits) and abatement investment subsidies can be used to affect location decisions. By linking location decisions to subsidies in emissions-reducing abatement investment, it was possible to derive rules relating to the amount of subsidy required in order not to accelerate relocation after the introduction of a stricter environmental policy. Given the function that determines optimal relocation time as a function of the observed price, an increase in emissions taxes may induce relocation of all or a subset of firms, depending on the heterogeneity assumptions, by bringing relocation time forward at the same price level. If relocation is not desirable, it may be prevented by an appropriate increase in the abatement subsidy. On the other hand, if price movements in the world market induce relocation, our results indicate that it could be prevented by an appropriate change of the policy mix, that is, changes in emissions taxes and/or abatement subsidies.

Further research could be directed toward the study of the relocation time when the country abroad follows a different environmental policy, or when the firms in the home country are heterogeneous. Further research could also be directed toward the study of the socially optimal relocation time. The optimal-stopping-time methodology could indicate the time at which it is socially desirable for a firm to relocate, and then help to design a policy scheme to prevent suboptimal relocation decisions.

Appendix

Proof of Proposition 1

From the HJB equation we obtain for region I

(A1) $$\rho V - \frac{1}{2}\sigma^2 p^2 V_{pp} - ap V_p - \pi(p, R; \tau, v) = 0.$$

The general solution of this second-order differential equation (A1) can be obtained as[19]

$$V(p, R) = A_1(R)p^{\beta_1} + A_2(R)p^{\beta_2} + \Pi(p, R; \tau, v),$$

where

19. The homogeneous part of this differential equation is an Euler equation.

$$\beta_1 = \frac{1}{2} - \frac{a}{\sigma^2} + \sqrt{\left(\frac{a}{\sigma^2} - \frac{1}{2}\right)^2 + \frac{2\rho}{\sigma^2}} > 1$$

is the positive root; β_2 is the corresponding negative root of the fundamental quadratic,

$$Q = \frac{1}{2}\sigma^2\beta(\beta - 1) + a\beta - \rho = 0;$$

and $\Pi(p, R; \tau, v)$ is the particular solution. We need to disregard the negative root in order to prevent the value from becoming infinitely large when the price tends to 0; thus we set $A_2(R) = 0$ (see Dixit and Pindyck 1994). So the solution becomes

(A2) $$V(p, R) = A_1(R)p^{\beta_1} + \Pi(p, R; \tau, v).$$

In order to obtain tractable results we need a better specification of the particular solution. To obtain such a specification, we consider a quadratic cost function $c(q) = \frac{1}{2}cq^2$; then the profit function becomes

$$\pi(p, \tau, v, R) = \frac{1}{2c}\{p^2 - 2\tau ve(R)p + [\tau ve(R)]^2\}.$$

Using the method of undetermined coefficients, we obtain the particular solution as

(A3) $$\Pi(p, R; \tau, v) = \Gamma_0 + \Gamma_1 p + \Gamma_2 p^2,$$

(A4) $$\Gamma_0 = -\frac{[\tau ve(R)]^2}{2c\rho}, \quad \Gamma_1 = -\frac{\tau ve(R)}{c(\rho - a)}, \quad \Gamma_2 = \frac{1}{2c(\rho - 2a - \sigma^2)},$$

$$\rho - a > 0, \quad \rho - 2a - \sigma^2 > 0.$$

In region II the second term of equation (12) is satisfied as 0 and $\Delta R > 0$, or

(A5) $$V_R(p, R) - (1 - s)h = 0.$$

Solving equation (A5) for p in terms of R, we can write the yet unspecified boundary equation as $p = p(R)$. From equations (A2) and (A5) we can determine the unknown functions $A_1(R)$ and $p = p(R)$ using the value-matching and the smooth-pasting conditions.[20] The value-matching condition means that on the boundary separating the two regions, the two value functions should be equal. Then we have, combining equations (A2) and (A5) and substituting for p,

---

20. For a presentation of these conditions, see Dixit and Pindyck (1994).

(A6) $V_R(p,R) = A'_1(R)p^{\beta_1} + \Gamma'_0(R) + \Gamma'_1(R)p = (1 - s)h, \quad p = p(R).$

The smooth-pasting condition means that the derivatives of the value functions with respect to p on the boundary are equal, or

(A7) $V_{Rp}(p,R) = \beta_1 A'_1(R)p^{\beta_1-1} + \Gamma'_1(R) = 0, \quad p = p(R).$

Combining equations (A6) and (A7), we can solve for the unknown functions $p(R)$ and $A'_1(R)$ to obtain

(A8) $p(R) = \dfrac{\beta_1}{\beta_1 - 1} \dfrac{(1 - s)h - \Gamma'_0(R)}{\Gamma'_1(R)},$

(A9) $A'_1(R) = -\left(\dfrac{\Gamma'_1(R)}{\beta_1}\right)[p(R)]^{1-\beta_1}.$

Relationship (A8) is the equation of the free boundary, which can be written, after substituting for $\Gamma'_0(R)$ and $\Gamma'_1(R)$, as

(A10) $p(R) = \dfrac{-\beta_1}{\beta_1 - 1} \dfrac{(a - \rho)}{\rho} \dfrac{[c\rho(1 - s)h + \tau ve(R)e'(R)]}{\tau ve'(R)}.$

Proof of Proposition 2

In region I, the first term of the HJB equation is 0, since $\Delta R = 0$, and the second term of the HJB equation is positive. Thus in region I,

$$\rho V - \frac{1}{2}\omega^2\tau^2 V_{\tau\tau} - \eta\tau V_\tau - \pi(\tau,R;p,v) = 0.$$

The general solution of this second-order differential equation can be obtained as before as

$$V(\tau,R) = B_1(R)\tau^{\xi_1} + B_2(R)\tau^{\xi_2} + \Phi(p,R;\tau,v),$$

where

$$\xi_1 = \frac{1}{2} - \frac{\eta}{\omega^2} + \sqrt{\left(\frac{\eta}{\omega^2} - \frac{1}{2}\right)^2 + \frac{2\rho}{\omega^2}} > 1$$

is the positive root; ξ_2 is the corresponding negative root of the fundamental quadratic,

$$Q = \frac{1}{2}\omega^2\xi(\xi - 1) + \eta\xi - \rho = 0;$$

and $\Phi(\tau, R; p, v)$ is the particular solution. As before, we set $B_2(R) = 0$, so the solution becomes

$$V(\tau, R) = B_1(R)\tau^{\xi_1} + \Phi(\tau, R; p, v).$$

Using again the quadratic cost function specification, $c(q) = \frac{1}{2}cq^2$, we obtain the particular solution as

$$\Phi(\tau, R; p, v) = \Delta_0 + \Delta_1\tau + \Delta_2\tau^2,$$

$$\Delta_0 = -\frac{p^2}{2c\rho}, \quad \Delta_1 = -\frac{pve(R)}{c(\eta - \rho)}, \quad \Delta_2 = \frac{[ve(R)]^2}{2c(\omega^2 + 2\eta - \rho)}.$$

In region II, the second term of the HJB equation is satisfied as 0 and $\Delta R > 0$, or

$$V_R(\tau, R) - (1 - s)h = 0.$$

The value-matching and smooth-pasting conditions imply that

(A11) $V_R(\tau, R) = B_1'(R)\tau^{\xi_1} + \Delta_1'(R)\tau + \Delta_2'(R)\tau^2 = (1 - s)h, \quad \tau = \tau(R),$

and

(A12) $V_{R\tau}(\tau, R) = \xi_1 B_1'(R)\tau^{\xi_1 - 1} + \Delta_1'(R) + 2\Delta_2'(R)\tau = 0, \quad \tau = \tau(R),$

respectively, where

$$\Delta_1'(R) = -\frac{pve'(R)}{c(\eta - \rho)} > 0, \quad \Delta_2'(R) = \frac{2ve(R)e'(R)}{2c(\omega^2 + 2\eta + \rho)}.$$

Combining equations (A11) and (A12), we obtain a quadratic expression that implicitly defines $\tau(R)$ as

(A13) $\Delta_2'(R)\left(\frac{\xi_1 - 2}{\xi_1}\right)\tau(R)^2 + \Delta_1'(R)\left(\frac{\xi_1 - 1}{\xi_1}\right)\tau(R) - (1 - s)h = 0.$

By taking the positive root of equation (A13), the free boundary is defined as

$$\tau(R) = \frac{-\xi_1}{2(\xi_1 - 2)} \frac{\Delta_1'(R)[(\xi_1 - 1)/\xi_1] + \sqrt{\Delta}}{\Delta_2'(R)},$$

$$\Delta = \left[\Delta_1'(R)\left(\frac{\xi_1 - 1}{\xi_1}\right)\right]^2 + 4\Delta_2'(R)\left(\frac{\xi_1 - 2}{\xi_1}\right)(1 - s)h,$$

with $\xi_1 > 2$ for $\Delta > 0$.

Proof of Proposition 3

The Bellman equation is

$$V(\tau) = \max\left\{\pi(\tau;p,v,R^*) + \frac{1}{1+\rho dt}\mathcal{E}[V(\tau + d\tau|\tau)], W(p,F)\right\}.$$

Following Dixit and Pindyck (1994), define $G(\tau) = V(\tau) - W$, $\tau + d\tau = \tau'$ and subtract W from both sides of the Bellman equation to obtain

$$(A14)\quad G(\tau) = \max\left\{0, \pi(\tau) - W + \frac{1}{1+\rho dt}\int V(\tau')d\Phi(\tau'|\tau)\right\}$$

$$= \max\left\{0, \pi(\tau) - W + \frac{1}{1+\rho dt}W + \frac{1}{1+\rho dt}\int G(\tau')d\Phi(\tau'|\tau)\right\}.$$

Since W does not depend on τ, we have that the expression,

$$m(\tau) = \pi(\tau) - W + \frac{1}{1+\rho dt}W,$$

is decreasing in τ, since $dm/d\tau = d\pi/d\tau < 0$. The function $m(\tau)$ reflects the difference between waiting for one period before relocating and relocating right away. Since $m(\tau)$ is decreasing in τ, continuation—that is, no relocation—should be optimal when τ is low.

Assume that the cumulative distribution $\Phi(\tau'|\tau)$ of the future values of the policy parameter shifts uniformly to the left as τ increases, so that the disadvantages of an increase in the current value of the environmental policy parameter in the original location are unlikely to be reversed in the future. This assumption, along with the decreasing $m(\tau)$ function, implies that $G'(\tau) < 0$ (see Dixit and Pindyck 1994, app. B).

Therefore, the second argument of equation (A14) is decreasing in τ. Thus, a unique critical time $\tau^*(t')$ exists such that the second argument of equation (A14) is negative if and only if $\tau^o(t') > \tau^*(t')$. Then it is optimal to relocate (optimal to stop) at time t'. If $\tau^o(t') < \tau^*(t')$, then it is optimal to remain at the initial site (continue).

References

Arrow, K. J., and A. Fisher. 1974. Environmental preservation, uncertainty and irreversibility. *Quarterly Journal of Economics* 88:312–19.

Carraro, C., and A. Soubeyran. 1996a. Environmental feedbacks and optimal taxation in oligopoly. In *Economic policy for the environment and natural resources*, ed. A. Xepapadeas, 30–58. Cheltenham, U.K.: Edward Elgar.

———. 1996b. Environmental policy and the choice of production technology. In *Environmental policy and market structure,* ed. C. Carraro, Y. Katsoulacos, and A. Xepapadeas, 151–80. Dordrecht: Kluwer Academic.

———. 1999. R&D cooperation, innovation spillovers and firm location in a model of environmental policy. In *Environmental regulation and market structure,* ed. E. Petrakis, E. Sartzetakis, and A. Xepapadeas, 195–209. Cheltenham, U.K.: Edward Elgar.

Dixit, A. K., and R. S. Pindyck. 1994. *Investment under uncertainty.* Princeton, N.J.: Princeton University Press.

Fisher, A. C., and M. Hanemann. 1986. Environmental damages and option values. *Natural Resource Modelling* 1:111–24.

———. 1987. Quasi-option value: Some misconceptions dispelled. *Journal of Environmental Economics and Management* 14:183–90.

Fleming, W., and H. M. Soner. 1993. *Controlled Markov process and viscosity solutions.* New York: Springer-Verlag.

Hartl, R. F., and P. M. Kort. 1996. Marketable permits in a stochastic dynamic model of the firm. *Journal of Optimization Theory and Applications* 89 (1): 129–55.

Hoel, M. 1994. Environmental policy as a game between governments when plant locations are endogenous. Paper presented at the 21st European Association for Research in Industrial Economics conference, Crete.

Kort, P. M. 1995. The effects of marketable pollution permits on the firm's optimal investment policies. *Central European Journal for Operations Research and Economics* 3:139–55.

Malliaris, A. G., and W. A. Brock. 1982. *Stochastic methods in economics and finance.* Amsterdam: North-Holland.

Markusen, J. R., E. R. Morey, and N. Olewiler. 1993. Environmental policy when market structure and plant locations are endogenous. *Journal of Environmental Economics and Management* 24:169–86.

———. 1995. Competition in regional environmental policies when plant locations are endogenous. *Journal of Public Economics* 56:55–77.

Motta, M., and J.-F. Thisse. 1994. Does environmental dumping lead to delocation? *European Economic Review* 38:563–76.

Rauscher, M. 1995. Environmental regulation and the location of polluting industries. *International Tax and Public Finance* 2:229–44.

Xepapadeas, A. 1992. Environmental policy, adjustment costs, and behavior of the firm. *Journal of Environmental Economics and Management* 23 (3): 258–75.

———. 1997a. *Advanced principles in environmental policy.* Aldershot, U.K.: Edward Elgar.

———. 1997b. Economic development and environmental pollution: Traps and growth. *Structural Change and Economic Dynamics* 8:327–50.

———. 1998. Optimal resource development and irreversibilities: Cooperative and noncooperative solutions. *Natural Resource Modelling* 11 (4): 357–77.

Comment   Charles D. Kolstad

This is a very interesting and impressive piece of work. Prof. Xepapadeas has tackled a very difficult question: In a dynamic context with irreversibilities in abatement investments and a pollution tax and abatement subsidy, as well as uncertainty, what is an optimal output and investment profile? The author examines three types of uncertainty: stochastic evolution of product price, stochastic evolution of the efficiency of abatement, and stochastic evolution of the pollution tax (or the price of pollution permits). Quite naturally, the author then asks what is the optimal environmental policy given this behavior of firms? Related to this issue, he closes the paper by asking what the optimal time is for firms to pull up stakes and move to a location with lighter environmental regulations.

While the paper provides an impressive use of stochastic calculus, the paper would be improved by refining or at least justifying some aspects of the model to provide a closer connection to realistic problems in pollution policy. For instance, of the three types of stochasticity, product-price evolution seems the most realistic and interesting. It is difficult to visualize the efficiency of abatement equipment following a random walk, sometimes increasing, sometimes decreasing. In the case of the level of a pollution tax, one would not expect it to wander around, since it is set by policymakers. On the other hand, it is plausible that the price of permits could follow a stochastic process, much as stocks do.

, It would appear that one of the primary purposes of this paper is to investigate the importance of uncertainty and irreversibilities. Because of this, it is somewhat disappointing that the paper focuses on only one type of irreversibility—abatement capital investment irreversibility. There are several interesting extensions that would push this issue further, for instance, a stock of pollution whose level is nondiminishing, corresponding approximately to the stock of carbon dioxide in the atmosphere or nuclear wastes in salt domes.

Another interesting extension would be to look at the stock of knowledge (as, in fact, the author hints at in the opening paragraph and examined in an earlier version of this paper). This endogenization of the R&D process would be an important contribution to this literature. The problem, which is very difficult to overcome, is that there are spillovers among firms. Without spillovers, the problem is less interesting (although still potentially of some interest).

The treatment of location is one of the more unique aspects of this pa-

Charles D. Kolstad is the Donald Bren Professor of Environmental Economics and Policy at the University of California, Santa Barbara, with appointments in the Department of Economics and the Bren School of Environmental Science and Management.

This work was supported in part by grant no. DE-FG03-96ER62277 from the U.S. Department of Energy.

per. The firm can choose to move to a regulation-free area, incurring a one-time fixed charge. The question is, when to do it? Basically, by paying a one-time charge, a firm can forever evade environmental regulation. This is really a shut-down and start-up question, which could be separated. When should a firm choose to shut down? When should a firm choose to restart in a pollution-free area, incurring a fixed start-up cost? This is an interesting problem, with many possible extensions.

One possible extension is to treat location as continuous, developing a Hotelling model of spatial competition. This would not be easy, but could be potentially very interesting. Where do firms locate in response to environmental regulations? How do regulations interact with spatial differences in pollution? Do environmental regulations provide an entry barrier? How do regulators compete with one another over space?

In summary, this is an ambitious paper that pushes our knowledge of firm behavior in a polluting environment subject to stochastic shocks. I would encourage the author to push this further, bringing more policy-relevant dimensions into the problem.

10

The Effects of Environmental Policy on the Performance of Environmental Research Joint Ventures

Yannis Katsoulacos, Alistair Ulph, and David Ulph

Over the last 10 years, a significant literature has developed on the effects of environmental policy on the incentives for firms to undertake research and development (R&D) that will lead to the development of new environmentally friendly products and/or processes. In what follows, we refer to this as environmental R&D.

The starting point of this literature is the recognition that market forces will produce very weak incentives for firms to undertake environmental R&D, and so government incentives are required to correct this market failure. There have been three main strands in this literature. The first[1] examines the effects of environmental policy (e.g., taxes and standards) on the incentives to undertake R&D. An important point that emerges from this literature is that while environmental policy does indeed give firms an incentive to undertake environmental R&D, a toughening of this policy will not necessarily increase the amount of R&D. This is because while a tougher environmental policy will have a direct effect of encouraging more environmental R&D, it can also have the indirect effect of raising costs and reducing output, and this will lower R&D incentives. Thus, contrary to the widely discussed Porter hypothesis, there is no theoretical presump-

Yannis Katsoulacos is professor of economics at the Athens University of Economics and Business. Alistair Ulph is deputy vice-chancellor and professor of economics at the University of Southampton. He is currently president of the European Association of Environmental and Resource Economists. David Ulph is professor of economics and executive director of the ESRC Centre for Economic Learning and Social Evolution (ELSE) at University College London.

The authors are grateful for the insightful comments of the discussant, Jerome Rothenberg, and of Gilbert Metcalf, Carlo Carraro, Sjak Smulders, John Whalley, and other conference participants.

1. See Ulph (1997) for a survey.

tion that tougher environmental policy will itself be sufficient to promote greater R&D incentives.

A second strand[2] looks at the combined effects of both technology policy and environmental policy on the levels of R&D, output, and emissions in an oligopolistic industry. Thus, in the context of a model where firms do only environmental R&D, Katsoulacos and Xepapadeas (1996) show that a combined R&D subsidy plus emissions tax can generate the first-best solution. Petrakis and Poyago-Theotoky (1997) explore the design of technology policy in the context where governments are constrained in the use of environmental policy.

The third strand[3] considers the setting of environmental policy in a multinational context, where governments are aware that the levels of environmental policies they set will affect the strategic competition between firms—particularly their choices of R&D. The issue here is whether trade concerns lead governments to set environmental policies that are too lax.

A feature of virtually all[4] this literature is that it assumes that firms undertake R&D in a noncooperative fashion. However, there are many potential benefits that are thought to flow from having firms undertake R&D cooperatively in a research joint venture (RJV): a reduction in risk, the achievement of economies of scale and scope, the elimination of wasteful duplication, and the greater appropriation of the returns to innovation. These benefits arise because RJVs are thought to promote greater information sharing and coordination of R&D decisions.[5]

There is now a considerable literature on the performance of RJVs. However, this literature focuses exclusively on the types of product and process R&D that firms undertake for conventional commercial benefit and, as such, ignores environmental innovation, which, as we have pointed out, is primarily undertaken in response to environmental policies.[6] Consequently, in this paper we wish to understand how environmental policy affects environmental innovation when we allow for the possibility that this innovation is undertaken cooperatively through the formation of what

2. See, e.g., Katsoulacos and Xepapadeas (1996) and Petrakis and Poyago-Theotoky (1997).

3. See e.g. A. Ulph (1996a, 1996b), Ulph and Ulph (1996), and D. Ulph (1994).

4. The exception is Petrakis and Poyago-Theotoky (1997). However, although they allow for R&D cooperation, they assume that governments are unable to implement environmental policies such as pollution taxes. Thus they are unable to address the central issue of this paper—the effects of environmental policies on RJV performance.

5. Of course, concern is also sometimes expressed that RJVs may use the ability to cooperate on R&D decisions to promote anticompetitive practices in the output market. Nevertheless, RJVs are widely thought to be beneficial on balance, and many governments promote their formation through reducing the ventures' antitrust liabilities, and, sometimes, by subsidizing R&D undertaken through an RJV. For an account of recent work on this topic, see, e.g., Poyago-Theotoky (1997) for a collection of recent papers.

6. These may be actual or anticipated environmental regulatory policies. Scott (1996) reports evidence that RJV formation takes place in response to both actual and anticipated regulation.

we will call environmental RJVs. In particular, we wish to understand how environmental policy affects the innovative performance of RJVs as compared to a noncooperative equilibrium.

The plan of the paper is as follows. In section 10.1, we set out some background discussion on the current understanding of RJV performance and sketch out the issues that need to be addressed in thinking about the interaction between environmental policy and the performance of environmental RJVs. In section 10.2 we set out a formal model that captures these issues. Section 10.3 uses the model to provide an analysis of the links between environmental policy and the performance of RJVs. Section 10.4 concludes.

Before we proceed, it is important to point out that in this paper we focus on the case where the central rationale for RJVs is the avoidance of duplication in R&D. This is captured by the assumption that the nature of research discoveries made by firms is duplicative. We fully recognize that an important alternative motivation of RJVs is to exploit complementarities, and in Katsoulacos and Ulph (1998a) we provide a positive analysis of this case. However, a full welfare analysis of this case would require a separate paper.

10.1 Preliminaries

In thinking about the interaction between environmental policy and the performance of environmental RJVs, it is important to recognize that there are a number of market failures in operation. One type of failure is *product market failures.* Taking as given the number of firms, the products they produce, and the technology[7] they employ, there are two market failures that can arise in relation to firms' output decisions. The first is the conventional pollution externality, which typically leads firms to overproduce. The second is imperfect competition arising from the oligopolistic nature of markets. This, in turn, may be attributable to entry barriers—in particular the scale economies generated by R&D. Imperfect competition typically leads to firms producing too little output. It is well known that in principle an emissions tax can be chosen to obtain the first-best level of output. Notice that the tax that achieves this first-best will depend on the technologies employed by the firms.

Another type of failure is *innovation market failure.* To facilitate discussion, here and throughout the rest of the paper we assume that (1) there are only two firms; (2) the products that firms produce are perfect substitutes; and (3) the research paths that firms are pursuing are perfect substitutes (or, more accurately, perfect duplicates) in the sense that if both firms make a discovery, they have discovered exactly the same thing and so can-

---

7. In particular, emissions technology.

not gain from any knowledge sharing. The first two assumptions are made in most of the RJV literature, but are by no means innocuous. The third captures one of the possible reasons why an RJV forms—to avoid duplication. However, it ignores another potential gain from RJV formation—exploiting research complementarities. We fully recognize the importance of this motive, and have recognized it in the positive analysis contained in Katsoulacos and Ulph (1998a). However, a full treatment of this case in the context of the welfare analysis we conduct here would warrant another paper.

It is well known in the industrial organization literature that there are significant market failures in the innovation process. These stem primarily from the public-good nature of knowledge, as something that is costly to produce, but virtually costless to reproduce. As is well known, this implies that the optimum allocation of resources involves having R&D undertaken in a relatively small number of labs, with the results being sold to others at a price equal to the value that society places on this knowledge. In this idealized market system, firms undertaking any R&D would perceive a return equal to the value that the entire industry places on it.

However, in the absence of any policy intervention, actual market mechanisms will not produce such an outcome. In particular, free riding on discoveries would so lower the rate of return to R&D that very little R&D would be undertaken. In the face of this market failure, virtually all governments institute a system of protection of intellectual property rights, of which the patent system is the major part.

While patents provide some correction of the fundamental market failures, they involve their own distortions, essentially because they reward firms for discovering information, but not for sharing it. To see what market failures still exist under a patent system, suppose for the moment that patents are fully effective (i.e., there are no involuntary information leakages) and that there are no mechanisms for sharing information.

To fully understand the nature of market failures, it is useful to follow Ulph and Katsoulacos (1998) and distinguish three stages in innovation decisions: (1) *research design,* in the case considered here, amounts to choosing the number of labs to operate; (2) *R&D* involves choosing the amount of R&D to do in each lab; and (3) *information sharing* entails choosing the amount of information to share with the other lab if a lab makes a discovery.

We then know that when firms act noncooperatively their decisions are subject to the following potential market failures.[8] Working backward: At the information sharing stage, firms will fail to share information when it is always socially desirable to do so. At the R&D stage there are three

---

8. The discussion that follows is based heavily on the analysis in Ulph and Katsoulacos (1998).

market failures: (1) the *undervaluation effect,* in which firms base decisions on profits rather than total surplus (profits plus consumer surplus), and so do too little R&D; (2) the *self-centeredness effect,* in which firms base decisions on the gains to themselves from having a new technology rather than the gain to the industry if everyone had the new technology, again resulting in too little R&D; and (3) the *"competition to be first" effect,* in which firms have incentives to try to be the first to introduce a new technology, leading to their overinvesting in R&D. At the research design stage, there may be excessive duplication because each firm will operate its own lab when it may be socially optimal to operate a single lab.

Suppose now that patents are not completely effective, and there are involuntary unpaid information leakages—spillovers. This will mitigate some of the welfare losses at the information sharing stage, but will introduce a fourth distortion at the R&D stage because firms acting independently will not internalize the externality arising from the spillover. We call this the *spillover effect.* It introduces a third reason why firms will underinvest in R&D.

Notice that it is not at all clear whether, on balance, firms are doing too little or too much R&D, either individually or collectively. What is clear is that a major reason why these market failures arise is that, by themselves, patents do not reward firms for sharing information. To resolve these difficulties, it is therefore necessary to introduce mechanisms that reward firms for sharing information. Two widely discussed methods of doing this are through licensing and through the creation of RJVs.

While licensing is certainly used in certain contexts, it is *not* a general solution to all these problems. In the first place, while it will often solve the information sharing problem that arises in the final stage of the decision-making process, it will not always do so. This is because even when there are only two firms—a buyer and a seller—a license will only be sold if the maximum amount the buyer is willing to pay exceeds the minimum amount the seller needs to receive, and this need not always be the case. More seriously, licensing will not solve the problems arising in the first two stages of innovation decision making. In particular, it does not eradicate the competition-to-be-first effect because, instead of competing to be the first to get exclusive use of a new technology, firms compete to be the first to be able to license it.

Given these problems with licensing, attention has focused recently on RJVs as a possible solution to these market failures. RJVs are arrangements under which firms are allowed to act cooperatively during all three stages of the innovation decision-making process, but are required to compete in the product market. Prima facie it would seem that, by making cooperative decisions about all aspects of innovation, RJVs could mitigate most of the market failures discussed here. By acting cooperatively, firms may be induced to share information and so mitigate the market failure at the information sharing stage. Second, acting cooperatively to maximize joint profits firms would internalize any spillovers, eliminate the competition-to-be-first effect, and remove the self-centeredness effect at the R&D stage. Finally, cooperative decision making means that firms may choose to operate a single lab rather than two independent labs, at the research design stage. These informal arguments suggest that the only market failure that RJVs may not potentially overcome is the undervaluation effect.

Given these potential benefits, it is not surprising that RJVs have received considerable attention. However, when one examines the theoretical literature on the subject, it turns out that, while it has provided some useful insights, nevertheless there are a number of weaknesses in the way it typically models RJVs and consequently the analysis fails to fully address many of these failure issues.[9] The major weaknesses of this literature are as follows.

First, spillovers are treated as exogenous. Either spillovers are the same in the cooperative equilibrium and the noncooperative equilibrium or else it is assumed that they are greater in the cooperative equilibrium. In neither case does the theory explain how cooperation might lead to greater information sharing.

Second, research discoveries by firms are assumed to be perfect complements. In both the cooperative and the noncooperative equilibria, it is assumed that information gained from other firms just adds to the progress that a firm makes on its own. This ignores the possibility of needless duplication of research and also that one of the benefits of an RJV is that it increases the degree of complementarity.

Third, a related problem is that equilibria are assumed to be symmetrical. In particular, in both the cooperative and the noncooperative equilibria, all firms are active in R&D and all do the same amount of R&D. This ignores the possible cost savings by concentrating R&D in a smaller number of labs.

Fourth, the noncooperative equilibrium is taken to be one in which no licensing is possible. Since many of the models assume that there are just two firms, it would seem sensible to allow the possibility of licensing, particularly if one wants to understand the full benefits of cooperation versus noncooperation.

To get a sense of just how limiting these assumptions are, it is worth noting the following result by Hinloopen (1997).

RESULT 1. *Suppose we have an industry comprising n identical firms. In addition, suppose R&D spillovers are the same in both the cooperative and*

---

9. Of course, not every paper suffers from every one of the weaknesses we identify. However, most of them arise in the paper by d'Aspremont and Jacquemin (1988), which has become a classic reference in the literature.

the noncooperative equilibria, the only policy instrument available to the government is an R&D subsidy, R&D discoveries are perfect complements in both the cooperative and noncooperative equilibria, both the cooperative and the noncooperative equilibria are symmetrical; and the R&D subsidy can be financed by nondistortionary taxation. Then the cooperative and noncooperative R&D equilibria achieve exactly the same level of welfare.

PROOF. The proof is simple. Given the assumptions, effectively the only variable that can be chosen by both firms and the social planner is the amount of R&D per firm. Work out the second-best[10] optimum level of R&D. R&D per firm will be monotonically increasing in the level of subsidy in both the cooperative and the noncooperative equilibria. So, whether firms act cooperatively or noncooperatively the subsidy can always be chosen to make these equilibria coincide with the second-best optimum.

The conclusion, then, is that as long as governments can subsidize R&D (which typically they do), then the promotion of RJVs is irrelevant. But this just emphasizes the point that the underlying model fails to capture virtually all of the factors that make RJVs interesting in the first place.

There is an immediate corollary.

COROLLARY 1. *Suppose now that there are environmental externalities; that firms undertake environmental R&D that lowers emissions per unit of output; and that, in addition to the R&D subsidy, the government can also impose an emissions tax. Suppose also that all the other assumptions of result 1 hold. Then, once again, the cooperative and noncooperative equilibria achieve exactly the same level of welfare.*

PROOF. The fact that the government has an emissions tax means that it can now control output and so can now achieve the first-best level. So choose output per firm and R&D per firm so as to achieve the first-best. Then, whether firms act cooperatively or noncooperatively, choose the tax rate and R&D subsidy per firm so as to achieve the first-best.

This corollary allows us to immediately generalize from the case of the noncooperative equilibrium analyzed in Katsoulacos and Xepapadeas (1996) to that of a cooperative equilibrium. This shows that in order to have an interesting theory of the interaction between environmental policies and RJV performance, one needs a more interesting model of R&D— one that gives scope for RJVs to achieve some of the objectives they are supposed to achieve. Recent papers by Katsoulacos and Ulph (1998a, 1998b) have gone some way toward correcting the weaknesses of the ex-

---

10. If governments have no output instrument, then we are confined to second-best optima.

isting RJV literature and so providing a better account of how RJVs perform. Their models have the following features.

1. Information sharing is endogenous in both the cooperative and noncooperative equilibria. In particular, the possibility of licensing is allowed in the noncooperative equilibrium.
2. They allow for the possibility of both complementarity and substitutability between research discoveries.
3. They determine the number of research labs that will be active in the cooperative equilibrium.[11]

For the case where there are just two firms, they obtain the following results.[12] First, firms may not share information within an RJV. When RJVs withhold information they do so for anticompetitive reasons. For example, this will happen whenever industry profits are higher and one firm has lower costs than the other and can exploit this to exercise some degree of monopoly power. However, the conditions under which information is shared in an RJV are exactly the same as those under which it is licensed in a noncooperative equilibrium. Hence, in this two-firm setting, RJVs perform no better or worse than the noncooperative equilibrium at the information sharing stage of innovation decisions. Second, RJVs may close a lab, but may do so for two reasons: to eliminate needless duplication and to avoid competition that arises when both firms discover the new technology. Third, while RJVs may give rise to a higher level of welfare than in the noncooperative equilibrium, there is a range of circumstances under which they do not.

In this paper we wish to explore the interaction between environmental policy and the performance of environmental RJVs within the framework for analyzing RJV performance proposed by Ulph and Katsoulacos (1998). To understand some of the issues involved in undertaking this exercise, we briefly set out the main features of a very simple version of the Ulph and Katsoulacos (1998) model adapted to the case of environmental innovation.

There are two firms producing a homogeneous product. Initially both firms use a technology whereby emissions per unit of output are $\bar{e} > 0$. Firms undertake R&D in order to discover a new technology for which emissions per unit of output are $\underline{e}$, $0 < \underline{e} < \bar{e}$. Each firm's probability of

11. Each firm operates its own lab in the noncooperative equilibrium.
12. While these papers do not allow for the possibility of an R&D subsidy, it is clear that in this framework such a subsidy will not, in general, enable governments to achieve a second-best outcome under either cooperation or noncooperation. This is particularly true when the social optimum involves closing a lab. An R&D subsidy may encourage an RJV to keep a lab open that it would otherwise have closed. Thus, the Hinloopen result will not generalize to this case.

discovery depends solely on the R&D that it does, and these probabilities are independent. There are three possible outcomes of the R&D process: (1) neither firm discovers the new technology; (2) both firms discover the new technology, but, since they have discovered the same thing, there is nothing to be gained from sharing the information;[13] and one firm alone discovers the new technology, in which case a decision has to be made whether to reveal the new technology to the firm that has not discovered it.

Assume for the moment that the only policy instruments open to the government are an emissions tax and the decision whether to allow firms to form RJVs. As in all the literature on RJVs, we assume that if firms are allowed to cooperate on decisions relating to innovation, they are forced to compete in the output market.

In modeling the impact of the emissions tax on the performance of RJVs, an important issue arises as to when in the decision-making process this decision is made. Notice that, leaving aside the tax-setting decision, the decisions made by firms and the government constitute a six-stage game.

1. *RJV policy:* The government decides whether or not to allow RJVs.

2. *RJV formation:* If RJVs are allowed, firms choose whether or not to form one.

3. *Research design:* Firms choose the number of labs to operate.

4. *R&D:* Firms choose the amount of R&D to do in each lab.

5. *Information sharing:* Depending on the outcome of the R&D decisions, firms choose whether or not to share information.

6. *Output:* Firms choose output in a noncooperative Cournot equilibrium.

To understand the impact of the environmental policy instrument (emissions tax), notice that there are many different assumptions one can make about when this instrument gets chosen and the information available to the government when the decision is made. This timing issue reflects the ability of the government to commit itself to the decision. To illustrate this point, consider two possible assumptions that we could make.

Assumption 1: Government Is Unable to Commit. Here the government sets the tax rate at the start of stage 6, after it has learned what technology each of the two firms is operating, but before firms have chosen output. The tax will be set to maximize welfare, conditional on the technologies employed by the two firms. This setup corresponds to the conventional

---

13. This reflects the assumption mentioned before that research discoveries are perfect substitutes. This is the opposite end of the continuum from most of the literature on RJVs, which focuses on the case of complementary research discoveries.

static analysis of environmental policy with given technologies. Notice that effectively the government sets three tax rates, t^{11}, t^{10}, and t^{00}, depending on whether both firms have the new technology, only one firm has the new technology, or neither firm has the new technology, respectively. The tax rate chosen by the government in each of these three states will involve balancing the two product-market failures already discussed. Thus, the government wants high taxes to discourage emissions, but low taxes to promote competition. Notice that since all the R&D and information-sharing decisions have already been made, the taxes chosen at this stage will be exactly the same whether or not firms have cooperated. So the taxes are conditioned only on technology, not on the previously made decision about whether or not to cooperate. Let $\hat{t}^{11}$, $\hat{t}^{10}$, and $\hat{t}^{00}$ be the optimum taxes set under this assumption.

Assumption 2: Government Has Some Ability to Commit. Here we could think of the tax being set at the start of stage 5, after firms have chosen R&D and after the outcome of the R&D race is known, but before the information-sharing decision is made. Notice that what the government announces are the taxes that it will set in stage 6 conditional on the technologies that each firm will have at that stage (i.e., conditional on the information-sharing decision that is about to be made at stage 5). Now there are three possible situations the government can be in at the start of stage 5: (1) If both firms have discovered the new technology, the only possible state that can arise at the start of stage 6 is state 11, and since there is no information-sharing decision to be made, it will simply announce $\hat{t}^{11}$. (2) If neither firm has discovered the new technology, by analogous reasoning the government will announce $\hat{t}^{00}$. (3) If one firm has discovered the new technology and the other firm has not, the economy will be in either state 11 or state 10 at the start of stage 6, but now the choice of tax rates in these two states can influence the information-sharing decision and obviously the optimal thing for the government to do is to announce a tax $\hat{t}^{10}$ that is so high that the firms will choose to fully share information. Thus, by being able to fully commit itself at the information sharing stage in the process, the government can induce the first-best level of information sharing in both the cooperative and noncooperative equilibria.

We could think of taxes being set at yet earlier stages in the decision-making process, and, in general, the taxes that are set will depend on the precise stage at which the decision is made. Rather than conduct an exhaustive analysis of all possible situations, we will confine our attention to the case where the government is unable to commit and explore the implications of this assumption for the desirability of permitting RJVs. We will contrast our conclusions with those obtained by Ulph and Katsoulacos (1998) for the case of nonenvironmental R&D.

Before proceeding to the detailed model, note the following two points. First, once we take account of the decision about environmental tax rates, we now effectively have the following seven-stage game:

1. *RJV policy:* The government decides whether or not to allow RJVs.
2. *RJV formation:* If RJVs are allowed, firms choose whether or not to form one.
3. *Research design:* Firms choose the number of labs to operate.
4. *R&D:* Firms choose the amount of R&D to do in each lab.
5. *Information sharing:* Depending on the outcome of the R&D decisions, firms choose whether or not to share information.
6. *Emissions tax:* The government sets environmental taxes t^{11}, t^{10}, or t^{00}, conditioning on the technology that each firm has as a result of the outcomes of stages 4 and 5.
7. *Output:* Firms choose output in a noncooperative Cournot equilibrium.

Second, given the stage at which they are set, the environmental taxes chosen by the government will be exactly the same irrespective of whether firms have made their decisions at stages 3, 4, and 5 in a cooperative or noncooperative fashion. It therefore follows that, if they are allowed to do so, firms will indeed choose to form an RJV at stage 2.

In section 10.2 we explore what decision the government should make at stage 1. From the discussion thus far it follows that this just reduces to the question of whether expected social welfare is greater if the decisions made at stages 3, 4, and 5 are made cooperatively or noncooperatively. It is important to appreciate that the issues arising in this comparison of the cooperative and noncooperative equilibria are exactly the same as in Ulph and Katsoulacos (1998). The crucial point is that the decisions made at stages 6 and 7 are very different, and the question is therefore how these differences affect our assessment of the balance of factors at stages 3, 4, and 5. There are three key differences: (1) The government has an instrument that can influence output decisions. (2) The presence of environmental damage means that there are two wedges between the profits that firms use to guide their decisions and the social welfare calculations that the government will use (consumer surplus and environmental damage). This will influence the magnitude of the undervaluation effect. (3) Since the tax rate will depend on decisions about information sharing, the relationship between profits and surplus will vary across states in a complex way. Taken together, these three factors will affect whether the cooperative or the noncooperative equilibrium comes closer to achieving the socially optimal innovation decisions.

10.2   The Model

Consider a closed economy in which there are just two goods: a "dirty" good whose production causes emissions of some pollutant and expenditure on all other goods. Let X denote aggregate consumption and production of the dirty good and Z the aggregate expenditure on all other goods.

There is a single consumer with utility function

$$u(X,Z) \equiv aX - \frac{1}{2} \cdot X^2 + Z.$$

There are two firms producing X. Denote the output of firm i by $x_i > 0$, $i = 1, 2$, so $X = x_1 + x_2$. There is a perfectly competitive sector producing Z under constant returns to scale. Z is numeraire.

There are two possible technologies for producing X: an old (i.e., existing) technology and a new one that has yet to be discovered. For each technology, unit costs of production are c, $0 < c < a$. Emissions per unit of output with the new technology are $\underline{e} = 1$, and with the old technology $\bar{e} = 1 + \theta, 0 < \theta < 1$. Thus technologies differ only in their environmental attributes; that is, we are dealing with purely environmental innovation. The parameter θ provides a measure of how much better the new technology is compared to the old one in terms of its emissions properties.

If firm i produces output x_i using a technology that generates emissions per unit of output $e_i \in \{1, 1 + \theta\}$, then the total emissions, E, produced by the two firms are $E = x_1 \cdot e_1 + x_2 \cdot e_2$. We assume that the damage done by these emissions is given by the function $D(E) = (d/2) \cdot E^2$.

In order to discover the new technology with lower emissions levels, firms undertake R&D. If the two firms act noncooperatively, they each operate their own independent lab. If they form an RJV, then at stage 3 they can choose either to continue to operate one lab each or to operate a single combined lab.

In stage 4, firms choose the amount of R&D to do in each lab. The amount of R&D a lab does determines the probability that it will discover the new technology. If both firms undertake R&D, then each has an independent probability of discovery that depends solely on the amount of R&D that it itself undertakes. Thus, as in Katsoulacos and Ulph (1998a), there are no R&D input spillovers, only R&D output spillovers.

As in Ulph and Katsoulacos (1998), we assume that the R&D expenditure that a lab needs to undertake in order to get a probability of discovery $p, 0 \leq p \leq 1$, is

$$\gamma(p) \equiv \frac{1}{1 - \beta}[1 - (1 - p)^{1-\beta}] - p, \quad 0 < \beta < 1.^{14}$$

----

14. It is worth noting that as $\beta \to 1$, $\gamma(p) \to -\log(1 - p) - p$.

Thus,

$$\gamma(0) = \gamma'(0) = 0; \quad \gamma''(0) = \beta > 0;$$

$$\forall p, \quad 0 < p < 1 \quad \gamma(p) > 0; \quad \gamma'(p) = (1 - p)^{-\beta} - 1 > 0;$$

$$\gamma''(p) = \beta(1 - p)^{-\beta-1} > 0;$$

$$\text{as } p \to 1, \quad \gamma(p) \to \frac{\beta}{1 - \beta}, \quad \gamma'(p) \to \infty.$$

The parameter β reflects the extent of decreasing returns to R&D in each lab. As discussed in Ulph and Katsoulacos (1998), this is an important determinant of whether RJVs will choose to operate one or two labs.

In stage 5, decisions are made about sharing the information resulting from the discoveries made in stage 4. There are three possible outcomes of stage 4. First, neither firm discovers the new technology. Since no information has been discovered, there is no information to be shared, and so both firms continue to operate with a technology in which emissions per unit of output are $\bar{e} = 1 + \theta$. Second, both firms discover the new technology. We assume that in this case, while each firm has obtained some information, it is exactly the same information, and so there is absolutely nothing to be gained by sharing it.[15] The reason for making this assumption is that, as we see later, it introduces the possibility of needless duplication of research effort—one of the market failures that RJVs are supposed to alleviate. Thus, in this case each firm now operates with a technology in which emissions per unit of output are $\underline{e} = 1$. Third, one firm alone discovers the new technology. The firm that has discovered the new technology will now have emissions per unit of output of $\underline{e} = 1$. Now there is some information to be shared, and something to be gained (at least by the recipient) from sharing it. We assume that there are just two possible decisions to be made about information sharing:[16] either no information is shared, in which case the emissions per unit of output of the firm that has not discovered the technology will be $\bar{e} = 1 + \theta$,[17] or full information sharing takes place, in which case the emissions per unit of output of the firm that has not discovered the technology will be $\underline{e} = 1$.[18]

Notice now that if one firm alone discovers the new technology, and if

15. In the terminology used by Katsoulacos and Ulph (1998a), the research discoveries are perfect substitutes.

16. We show in Katsoulacos and Uiph (1998a) that there is no loss of generality in reducing the options to just two.

17. This ignores the possibility that there may be some involuntary information leakage whereby the emissions of the firm that has not discovered the technology are $1 + \theta \cdot \underline{\delta}, 0 < \underline{\delta} < 1$.

18. This ignores the possibility that the firm that has not discovered the technology may have limited capacity to use the information it receives, so its costs of production are $1 + \theta \cdot \bar{\delta}, 0 < \bar{\delta} < 1$.

it fully shares the information with the other firm, then the outcome will be precisely the same as if both firms had discovered the technology—each firm will have emissions $e = 1$. Thus there is nothing to be gained by having both firms discover the new technology that could not be gained by having just one firm discover the new technology and sharing the results. In this sense, there has clearly been a duplication of research effort.

In stage 6, the government sets the environmental tax rate. There are three possible situations that can occur at the start of stage 6, depending on which technology each firm has. First, neither firm has the new technology. This situation can arise only if neither firm discovers the new technology. Variables relating to this situation are described by a superscript 00. Second, both firms have the new technology. This situation can arise if either both firms discover the new technology or if only one does and information is fully shared. Variables relating to this situation are described by the superscript 11. Third, only one firm has the new technology. This situation can arise only if just one firm discovers the new situation and information is not shared. Aggregate variables relating to this situation are described by the superscript 10. Variables relating to individual firms will carry the superscript 10 for the firm that has the new technology and 01 for the firm that does not. Thus the government effectively sets three tax rates: t^{11}, t^{10}, and t^{00}.

Finally, in stage 7 output decisions are made conditional on the marginal costs that each firm will have as a result of the decisions made in the previous three stages. We assume that collusion over output is forbidden. Consequently, the output of the two firms is determined as a noncooperative equilibrium. In this paper, we confine attention to the noncooperative Cournot equilibrium. However, the nature of this equilibrium depends on the taxes set by the government in stage 6.

10.3   The Solution and the Welfare Measures

As is conventional we solve the model backward.

10.3.1   Stage 7: The Cournot Equilibrium

If firm i has technology $e_i \in \{1, 1 + \theta\}$, and if the government has imposed an emissions tax t per unit of emissions, then, in an interior Cournot equilibrium, the output of firm i is

(1)
$$x_i = \frac{(a - c) + t(e_j - 2e_i)}{3}, \quad i = 1, 2; \quad j \neq i,$$

and the profits it makes are

$$\pi_i = (x_i)^2.$$

10.3.2   Stage 6: The Tax-Setting Decision

Since all the innovation decisions have already been made, all the government can influence at this stage is the output equilibrium in stage 7. All that matters to the government, then, is the flow of welfare, comprising consumer surplus, producer surplus (profits), and environmental damage.

If firm i produces output x_i using a technology that generates emissions per unit of output $e_i \in \{1, 1 + \theta\}$, then the total emissions, E, produced by the two firms are $E = x_1 \cdot e_1 + x_2 \cdot e_2$. As noted before, we assume that the damage done by these emissions is given by the function $D(E) = (d/2) \cdot E^2$. The flow of social welfare in this output market situation is, therefore,

$$(2) \qquad W(x_1, x_2; e_1, e_2) \equiv (a - c)(x_1 + x_2) - \frac{1}{2}(x_1 + x_2)^2$$
$$- \frac{d}{2}(x_1 \cdot e_1 + x_2 \cdot e_2)^2.$$

By substituting the equilibrium outputs in equation (1) into the social welfare function (2), we can determine social welfare as a function of the tax rate alone. Consequently, we can determine the optimum tax rate and the equilibrium to which it gives rise.

To understand this more fully, it is useful to consider in turn two separate cases. In case 1, both firms have the same emissions per unit of output, $e \in \{1, 1 + \theta\}$. Notice that in this case both firms will choose identical output, and so it follows from equation (1) that welfare depends solely on the aggregate level of output, X. Thus we can write

$$(3) \qquad W = (a - c)X - \frac{1 + d \cdot e^2}{2} \cdot X^2.$$

From equation (2) we know that equilibrium aggregate output is

$$(4) \qquad X = \frac{2}{3} \cdot (a - c - t \cdot e),$$

so it now follows that, by suitable choice of t, the government can achieve the first-best level of welfare.

From equation (3) it follows that the first-best level of aggregate output is

$$\hat{X} = \frac{a - c}{1 + d \cdot e^2}.$$

Combining equations (3) and (4), we see that the tax rate that achieves the optimum is

$$\hat{t} = \frac{(a - c) \cdot (2d \cdot e^2 - 1)}{2 \cdot e \cdot (1 + d \cdot e^2)}.$$

It is straightforward to check that welfare in the social optimum is

$$\hat{W}(e) = \frac{1}{2} \cdot \frac{(a - c)^2}{1 + d \cdot e^2},$$

and industry profits in the social optimum are

$$\Sigma(e) = \frac{1}{2} \cdot \left(\frac{a - c}{1 + d \cdot e^2}\right)^2.$$

Thus,

$$\Sigma(e) = \frac{\hat{W}(e)}{1 + d \cdot e^2}.$$

This shows that industry profits understate social welfare and that the gap is larger when both firms have the old technology than when they both have the new technology.

To ease notation later on, let t^{11}, W^{11}, Σ^{11} (t^{00}, W^{00}, Σ^{00}) denote the optimal tax rate and the flow levels of welfare and industry profits when both firms have the new (old) technology. We have

$$t^{11} = \frac{(a - c) \cdot (2d - 1)}{2(1 + d)}, \quad W^{11} = \frac{(a - c)^2}{2(1 + d)}, \quad \Sigma^{11} = \frac{1}{2} \cdot \left(\frac{a - c}{1 + d}\right)^2,$$

$$t^{00} = \frac{(a - c) \cdot [2d \cdot (1 + \theta)^2 - 1]}{2 \cdot (1 + \theta) \cdot [1 + d \cdot (1 + \theta)^2]}, \quad W^{00} = \frac{(a - c)^2}{2 \cdot [1 + d \cdot (1 + \theta)^2]},$$

$$\Sigma^{00} = \frac{1}{2} \cdot \left(\frac{a - c}{1 + d \cdot (1 + \theta)^2}\right)^2.$$

In case 2, firm 1 has the old technology and firm 2 the new technology. This case can only arise if one firm has discovered the new technology, but does not share the information. Notice first of all that, in general, the government can now no longer obtain the first-best output equilibrium. Since the two firms have different technologies, they will choose different output levels. The first-best level would require getting both of these output levels right, and with just a single instrument—the environmental tax rate—this typically is impossible to achieve. In particular, the first-best level would require that only firm 2 be active and so act as a monopolist. For this possibility to arise it would have to be the case that when the government sets the tax that would be optimal *if* firm 2 were a monopolist, then firm 1's costs would be so high that it would not be willing to enter.

It is straightforward to confirm that this can only arise if $\theta \geq d/(d - 1) > 1$. Since in this paper we assume that $\theta \leq 1$, we can rule this possibility out.

Indeed, throughout the paper we wish to confine our attention to the case where, in the optimum, both firms are active. To analyze this situation, we will begin by assuming that both firms are active, derive the optimum tax rate under this assumption, and then check that, given this tax rate, both firms are indeed active.

In an interior Cournot equilibrium, the individual and aggregate outputs of the two firms are as follows:

$$x_1 = \frac{a - c - t(1 + 2\theta)}{3},$$

$$x_2 = \frac{a - c - t(1 - \theta)}{3},$$

and

$$X = \frac{2(a - c) - t(2 + \theta)}{3}.$$

Total emissions are

$$E = X + \theta \cdot x_1 = \frac{(a - c) \cdot (2 + \theta) - 2t \cdot (1 + \theta + \theta^2)}{3}.$$

Social welfare is $W = (a - c) \cdot X - \frac{1}{2} \cdot X^2 - (d/2) \cdot E^2$, and so the optimal tax arises where

$$W' = [(a - c) - X] \cdot X' - d \cdot E \cdot E' = 0.$$

Carrying out the calculation, we find that

(5) $\qquad t^{10}(\theta, d) = \dfrac{(a - c) \cdot (2 + \theta) \cdot [2d(1 + \theta + \theta^2) - 1]}{d \cdot [2(1 + \theta + \theta^2)]^2 + (2 + \theta)^2}.$

Before discussing the properties of the optimum tax, we need to confirm that this tax rate is consistent with our assumption of an interior Cournot equilibrium (with a positive output for firm 1). It is easy to see that this condition requires that

$$t^{10}(\theta, d) < \frac{(a - c)}{1 + 2 \cdot \theta}.$$

We note below that t^{10} is a strictly increasing function of d, so this condition requires that

$$d < \bar{d}(\theta) \equiv \frac{2 + 3\theta + \theta^2}{2 \cdot \theta \cdot (1 + \theta + \theta^2)}.$$

Table 10.1   **Computed Values of d**

θ	$\bar{d}(\theta)$
0.1	10.405
0.3	3.585
0.5	2.142
0.7	1.497
1	1

It is straightforward to check that $\bar{d}(\theta)$ is strictly decreasing in θ. For later purposes, table 10.1 shows the values taken by this function for a range of values of $0 < \theta \leq 1$. We can also invert $\bar{d}(\theta)$ to obtain the function $\bar{\theta}(d)$, which gives, for any d, an upper bound on θ for which the optimal tax yields an interior Cournot equilibrium.

What are the properties of the optimal tax when firms do not share information? As noted previously, it is easy to check that t^{10} is a strictly increasing function of d. Thus, as we would expect, the more damaging are emissions, the higher is the optimal tax. The crucial question is how t^{10} varies with θ because later on we will want to know whether the tax rate is higher if firms do not share information than if they do share information. That is, we will want to know whether $t^{10}(\theta, d) \lesseqgtr t^{11}$. Now it follows by definition and from formula (5) that

$$t^{10}(0,d) = \frac{(a - c)(2d - 1)}{2(d + 1)} = t^{11},$$

so what we really want to know is whether $t^{10}(\theta, d) \lesseqgtr t^{10}(0, d)$.

To answer this question consider first two extreme cases. In one case, when $d = 0$,

$$t^{10}(\theta, 0) = -\frac{a - c}{2 + \theta} < 0, \quad \frac{\partial t^{10}}{\partial \theta} = \frac{a - c}{(2 + \theta)^2} > 0.$$

Thus, as we would expect, when there is no environmental damage the government imposes a subsidy to correct the loss arising from imperfect competition. However, because the subsidy is imposed on emissions rather than on output, the subsidy itself induces what is in this case an unwarranted asymmetry between the two firms. A larger value of θ means the subsidy has to be smaller (i.e., the negative tax larger) in order to reduce the unwarranted distortion.

In the second case, when $d \to \infty$,

$$t^{10}(\theta, \infty) = \frac{(a - c) \cdot (2 + \theta)}{2(1 + \theta + \theta^2)} > 0, \quad \frac{\partial t^{10}}{\partial \theta} = -\frac{(a - c) \cdot (1 + 4\theta + \theta^2)}{2(1 + \theta + \theta^2)^2} < 0.$$

Accordingly for the five values of d shown in the right-hand column of table 10.1, we have calculated the value of t^{10} as θ ranges over values in the interval $0 \leq \theta \leq \bar{\theta}(d)$. The results are presented in appendix tables 10A.1–10A.5. These show that when d is small the optimal tax first rises and then falls with θ, while when d is large the optimal tax is a strictly decreasing function of θ.

The intuition behind these results is as follows. In setting the tax, the government is trying to correct two distortions: (1) that caused by imperfect competition and (2) that caused by pollution. Consider these in turn. For the first distortion, the larger θ is, the lower is aggregate output, and so the greater the loss arising from imperfect competition. This suggests that the optimal tax should fall with θ. For the second distortion, when d is small, so too is the optimal tax, and, when $\theta \approx 0$, an increase in θ causes aggregate emissions to rise—which calls for an increase in the optimal tax to correct this distortion. On balance, the second factor outweighs the first when $\theta \approx 0$ causes the optimal tax to rise with θ. However, when d is large, so too is the optimal tax, and it is easy to check that in this case aggregate emissions are a strictly decreasing function of θ, so this second factor also calls for the optimal tax to fall with θ.

Having determined the optimal tax, we can substitute this back into the expressions for output, profits, and welfare, and so determine aggregate profits, Σ^{10}, and aggregate welfare, W^{10} as functions of the underlying parameters d and θ. These profit and welfare expressions can then be used to determine the equilibrium and optimum information-sharing and R&D decisions in stages 2 and 1, respectively. To these we now turn.

10.3.3   Stage 5: The Information-Sharing Decision

Suppose that just one firm has discovered the new technology. If it does not share information, then the resulting levels of aggregate welfare and aggregate profits will be W^{10} and Σ^{10}, respectively. If information is shared, then the resulting levels of aggregate welfare and aggregate profits will be W^{11} and Σ^{11}, respectively.

As discussed in Ulph and Katsoulacos (1998), information sharing will be socially desirable if and only if $W^{11} > W^{10}$, and will be privately profitable in both the cooperative and noncooperative equilibria (under licensing) if and only if $\Sigma^{11} > \Sigma^{10}$. Unfortunately, the expressions for aggregate profits, Σ^{10}, and aggregate welfare, W^{10}, that emerge from the analysis in stage 6 are sufficiently complex that it is extremely difficult to explore these inequalities analytically, so we have had to explore them numerically. Appendix tables 10A.1–10A.5 give the computed values for profits and welfare as θ ranges over values in the interval $0 \leq \theta \leq \bar{\theta}(d)$. In reading these tables, it is important to bear in mind that the first row, corresponding to the case $\theta = 0$, corresponds to the situation where both firms have a technology with emissions levels of 1, and so the levels of welfare and profits here are W^{11} and Σ^{11}, respectively.

An inspection of these appendix tables shows that in all cases welfare is lower when $\theta > 0$ than when $\theta = 0$; that is, $W^{11} > W^{10}$, and so, as we would expect, full information is socially desirable. This is because, when information is fully shared, both firms have the least-polluting technology, and the government can then use its tax powers to achieve the first-best levels of output.

Let us now consider the private information-sharing decision of the two firms. From Ulph and Katsoulacos (1998), we know that firms will choose not to share information whenever the firm with the new technology has a sufficient cost advantage. This enables it to exercise significant market power, and so the resulting industry profits are higher than would be the case if both firms had the new low-cost technology and the industry was consequently very competitive. The difference now is that the cost difference between firms depends on both the difference in technology, as reflected in the parameter θ, and on the tax rates t^{11}, t^{10}.

To understand what is going on here, consider first the case where the government sets some arbitrary tax rate, t, which is independent of the technologies actually used by the two firms, and thus is the same whether or not information is shared. Then it follows from the result in Ulph and Katsoulacos (1998) that information will definitely be shared if

$$(6) \qquad\qquad t(2 + 5\theta) < a - c,$$

and will definitely not be shared if inequality (6) is reversed. This shows that the information-sharing condition depends on both the tax rate, t, and on the technology gap parameter, θ, and that information is shared when both t and θ are small.

As a variant of this thought experiment, suppose now that instead of setting an arbitrary tax rate, the government sets the same tax rate t^{11} whether or not information is shared. Then inequality (6) becomes

$$(7) \qquad\qquad \theta < \tilde{\theta}(d) \equiv \frac{6}{5(2d - 1)}.$$

In the appendix, we present the values of $\tilde{\theta}(d)$ corresponding to each of the values of d in table 10.1. We see that when $d = 1$, then $\tilde{\theta}(1) > \bar{\theta}(1)$, and so, if the government sets the tax rate t^{11}, whether or not the information is shared, then, for this value of d information will always be shared— that is, will be shared for all values of $\theta \in [0, \bar{\theta}(1)]$. However, when $d > 1$, $\tilde{\theta}(d) < \bar{\theta}(d)$, and so information will not always be shared—that is, will not be shared for all values of $\theta \in [0, \bar{\theta}(d)]$. In particular, information will only be shared if emissions under the old technology are sufficiently close to those with the new technology.

An alternative way of seeing what is going on here is to let $\tilde{\lambda}(d) \equiv \min\{[\breve{\theta}(d)]/[\overline{\theta}(d)], 1\}$ measure the fraction of the range of feasible values of θ over which information is shared. Then we see from appendix tables 10A.1–10A.5 that $\tilde{\lambda}(1) = 1$, but that $\tilde{\lambda}(d)$ is a strictly decreasing function of d. In this sense, we conclude that information sharing becomes less likely the larger the value of d.

These results are in line with the work of Ulph and Katsoulacos (1998). Information sharing is likely when the cost differences are small. When the damage is small, then so too is the optimal tax rate, and so cost differences are small for all $\theta \in [0, \overline{\theta}(d)]$. However when damage is large, so too is the tax rate, and so information will only be shared when the underlying technology gap parameter θ is sufficiently small.

Finally, when we recognize that the government will in fact set a different tax rate t^{10} if information is not shared from the tax rate t^{11} that will be set if information is shared, then we need to know whether $\Sigma^{11} > \Sigma^{10}$. From the discussion so far we know that Σ^{11} is the value of industry profits when $\theta = 0$. So, in looking at the appendix tables, we can determine whether or not information is shared in any given situation by simply comparing the value of industry profits Σ^{10} with that given in the first row of appendix tables 10A.1–10A.5.

We see that when $d = 1$, for all positive values of $\theta \leq \overline{\theta}(1)$, industry profits are lower than in the case where $\theta = 0$—that is, information is always shared. However, when $d > 1$, then, by interpolation, there exists a $\breve{\theta}(d)$, $0 < \breve{\theta}(d) < \overline{\theta}(d)$ such that industry profits are the same when $\theta = \breve{\theta}(d)$ as when $\theta = 0$. This implies that information will be shared when $0 < \theta < \breve{\theta}(d)$ and will *not* be shared when $\breve{\theta}(d) < \theta \leq \overline{\theta}(d)$.

By analogy with what we did before we can let $\hat{\lambda}(d) \in \min\{[\breve{\theta}(d)]/[\overline{\theta}(d)], 1\}$ denote the fraction of the range of feasible values of over which information is shared when the government sets different tax rates t^{11}, t^{10}. From the appendix tables we see that when d is small, then $\hat{\lambda}(d) > \tilde{\lambda}(d)$; but that when d is large, then $\hat{\lambda}(d) < \tilde{\lambda}(d)$. Thus, when the government sets different taxes if firms share information than if they do not, then, compared to the situation where the same tax rate is set irrespective of the information-sharing decision, information sharing becomes more likely when the damage is small, but less likely when the damage is large. This latter result follows from the result noted before. When damage is small, taxes tend to increase with θ, which ceteris paribus lowers profits when information is not shared and so makes information sharing more attractive than in the case where taxes do not change with θ. However, when damage is large, taxes tend to fall with θ, which ceteris paribus raises profits when information is not shared and so makes information sharing less attractive than in the case where taxes do not change with θ. Having understood the information-sharing decision we can turn finally to the R&D decisions.

10.3.4   Stages 3 and 4: The R&D Decisions

We have to determine the amount of R&D done by each of the two labs in each of the two equilibria—cooperative and noncooperative. In particular, in considering the cooperative (RJV) equilibrium we have to allow for the possibility of an asymmetrical solution in which one of the labs does no R&D.

Before turning to the more detailed analysis, it is important to make three general points about the nature of the R&D decisions.

1. As noted in Katsoulacos and Ulph (1998a) and in Ulph and Katsoulacos (1998), although, ex ante, the two firms are identical, the R&D outcomes need not be. In particular, in both the social optimum and in the cooperative equilibrium, it may turn out to be optimal to have only one firm undertake R&D. Whether or not this is the case turns on a trade-off between diminishing returns and needless duplication. The greater the extent of diminishing returns (the larger the parameter β in the R&D cost function) the more likely it is that both labs will be kept open.

2. In the noncooperative equilibrium, both firms will undertake R&D (and the equilibrium will be symmetrical). This means that the noncooperative equilibrium may be prone to a welfare loss of needless duplication.

3. When firms share information in the noncooperative equilibrium, they do so through licensing. We assume that the license fee is determined by negotiation, and that the licenser and the licensee have equal bargaining power. So the license fee is just halfway between the maximum price that the licensee is willing to pay and the minimum price at which the licenser is willing to sell.

The detailed analysis of the R&D decisions is now very similar to that given in Ulph and Katsoulacos (1998), so in what follows we just briefly summarize the main points of their analysis.

Point 1 makes it difficult to undertake any general analysis of the R&D decisions in the social optimum and in each of the two equilibria. Most of the analytical conclusions are therefore obtained in the easier case where both firms undertake R&D. As noted in our earlier papers, there are then two incentives driving the R&D decisions—competitive threat[19] and profit incentive.[20]

Whether or not information is shared, the profit incentive is weakest in the cooperative equilibrium and strongest in the social optimum, with the profit incentive in the noncooperative equilibrium lying between the other two. The profit incentive is strongest in the social optimum because welfare

19. For each firm, this is defined as the difference between the payoff (in terms of profits or welfare) if both firms innovate and the payoff if the other firm alone innovates.

20. For each firm, this is defined as the difference between the payoff if the firm alone discovers the technology and the payoff if neither discovers it.

is greater than profits—essentially because of the wedge caused by environmental damage. The profit incentive is greater in the noncooperative equilibrium than in the cooperative equilibrium, because most of the gains from having a firm innovate go to the firm that innovates.

When information is shared, the competitive threat is zero in both the social optimum and in the cooperative equilibrium, since the aggregate outcome is exactly the same whether one firm or both discover the new technology. However the competitive threat is positive for the noncooperative equilibrium since an individual firm will always lose if it fails to discover technology when the other firm has done so. The only difference that arises when information is not shared is that the competitive threat in the cooperative equilibrium is then negative, since, by definition, industry profits are lower when both firms discover the technology than when only one does so.

The general conclusions are that when both firms undertake R&D, then compared to the social optimum, the cooperative equilibrium always underinvests; R&D spending will be higher in the noncooperative equilibrium than in the cooperative equilibrium; and R&D spending in the noncooperative equilibrium may be higher or lower than in the social optimum. It is therefore far from obvious that RJVs necessarily perform better than the noncooperative equilibrium in terms of getting the levels of R&D spending right. Where they may potentially prove beneficial is in eliminating the needless duplication of R&D effort.

10.3.5   Stage 2: RJV Membership

As noted previously, this is trivial. Since the tax rates set in stage 6 do not depend on the RJV-membership decision, and since in a given environment firms are always better off cooperating than not cooperating, the firms will always join an RJV if RJVs are allowed.

10.3.6   Stage 1: RJV Policy

The government now has to decide whether or not to allow firms to form an RJV. We know that the information-sharing decision in stage 5 will be exactly the same whether firms act cooperatively or noncooperatively. So, let

$$\tilde{W}^{10} \equiv \begin{cases} W^{11} & \text{if } \Sigma^{11} > \Sigma^{10} \\ W^{10} & \text{if } \Sigma^{10} > \Sigma^{11} \end{cases},$$

denote the level of welfare that will be achieved if, in stage 4, one lab discovers the new technology and the other does not, allowing for the information sharing decision that will subsequently be made in stage 5.

Let p_i^n, $i = 1, 2$, denote the noncooperative equilibrium probabilities of discovery by each of the two labs as determined in stages 3 and 4 of the game, and p_i^c, $i = 1, 2$, denote the corresponding probabilities in the co-operative (RJV) equilibrium. Let $\overline{W}^n$ and $\overline{W}^c$ denote the expected level of social welfare in the noncooperative and cooperative equilibria, respectively. Then the formula,

$$\overline{W}^t = p_1^t \cdot p_2^t \cdot W^{11} + p_1^t \cdot (1 - p_2^t) \cdot \tilde{W}^{10} + p_2^t \cdot (1 - p_1^t) \cdot \tilde{W}^{10}$$
$$+ (1 - p_1^t) \cdot (1 - p_2^t) \cdot W^{00} - \gamma(p_1^t) - \gamma(p_2^t),$$

gives the expected level of social welfare in the equilibrium of type t.

To complete our analysis of the model, we have undertaken a numerical comparison of both the cooperative and noncooperative equilibria and their associated levels of expected welfare. What we have done is as follows. We have set $a - c = 4$,[21] and we have chosen values for the parameter β and for a parameter λ, $0 < \lambda < 1$, which determines the magnitude of the parameter in relation to its upper bound $\overline{\theta}(d)$. Initially we have chosen values $\beta = 0.2$ and $\lambda = 0.5$. For each of the values of d given in table 10.1 and for the value of $\theta = \lambda \cdot \overline{\theta}(d)$, we have solved for the following: (1) the probability of discovery per firm in the social optimum, which we denote by $\hat{p}_i$, $i = 1, 2$ (in doing this calculation, we take into account the fact that in the social optimum there would be full information sharing); and (2) the equilibrium levels of R&D spending in both the cooperative and non-cooperative equilibria, which we denote by p_i^C, p_i^N, $i = 1, 2$, respectively (in doing these calculations, we take into account the fact that information will be shared if $\lambda \leq \check{\lambda}(d)$ and will not be shared otherwise).

We use the convention that if one of the firms does not undertake R&D it is always the second. We have then computed the expected levels of welfare in the social optimum and in both the cooperative and noncoop-erative equilibria. By definition, expected welfare in each of the two equilibria is less than the expected welfare in the social optimum. We can there-fore calculate the percentage welfare loss in each of the two equilibria. These are denoted by L^C and L^N for the cooperative and noncooperative equilibrium, respectively. For the case where $\lambda \leq \check{\lambda}(d)$, and so information is fully shared in both the cooperative and noncooperative equilibria, firms are making the socially optimal information-sharing decision, and so this welfare loss arises solely because firms are making the wrong R&D deci-sions. However, when $\lambda > \check{\lambda}(d)$ firms are, in addition, making the wrong information-sharing decision. We then calculate the hypothetical level of welfare that would have arisen for the given equilibrium levels of R&D spending, but assume now that information is fully shared. Using this, it is possible to decompose the overall welfare losses into welfare losses L_R^C

---

21. Higher values of this parameter resulted in outcomes where firms were innovating almost surely most of the time, while lower values of this parameter produced very low R&D probabilities. As we will see, this parameter produces quite a wide range of R&D probabili-ties as other parameters varied.

Table 10.2   **Comparison of Cooperative and Noncooperative Equilibria to Social Optimum**
$\beta = 0.2, \lambda = 0.5$

d	1	2.142	3.585
$\hat{p}_1$	0.99	0.659	0.532
$\hat{p}_2$	0	0.659	0.532
p_1^C	0.982	0.551	0.354
p_2^C	0	0.551	0.354
L^C	0.03	0.67	4.47
L_R^C	0.03	0.67	1.55
L_I^C	0	0	2.91
$p_1^N = p_1^N$	0.899	0.762	0.57
L^N	1.98	0.69	3.2
L_R^N	1.98	0.69	0.07
L_I^N	0	0	3.12

and L_R^N that arise in each of the two equilibria because of the wrong R&D decisions, and welfare losses L_I^C and L_I^N that arise because the wrong information-sharing decision has been made.

Table 10.2 presents the calculations for three values of $d = 1$, 2.142, 3.585 (no extra insights are obtained by including the calculations for the other two values of d). It should be borne in mind that the associated values of $\tilde{\lambda}$ are 1, 0.704, and 0.443, respectively. So, with $\lambda = 0.5$ information will be shared for the first two values of d, but not for the third.

A number of points emerge from this table. First, when damage is low, so too are taxes. The returns to R&D in terms of either welfare or profits are high, and this causes firms to undertake a lot of R&D with a high probability of discovery. But this also means that the probability of duplication is also high. For this reason, in both the social optimum and the cooperative equilibrium, it pays to shut a lab. However, in the noncooperative equilibrium both labs operate, giving rise to significant losses from excessive duplication. Since, when damage is low information is always shared, the only welfare losses arise from getting R&D decisions wrong, and the cooperative equilibrium performs significantly better than the noncooperative equilibrium.

Second, when damage is somewhat higher, so too are taxes and the returns to R&D in terms of both welfare and profits are lower. The risk of duplication is now lower, and it turns out that in both the social optimum and the cooperative equilibrium both labs are kept open. Compared to the social optimum, the cooperative equilibrium underinvests in R&D and the noncooperative equilibrium overinvests. The losses are of almost equal magnitude. Once again information is fully shared, so the R&D losses are the only ones that matter.

Third, when damage is higher still, so too are taxes and the returns to R&D fall further, so further reducing the risk of needless duplication. Both labs are therefore always used. The noncooperative equilibrium over-invests and the cooperative equilibrium underinvests, but now the loss from underinvestment is considerably greater than the loss from overin-vestment, and in terms of R&D decision making the noncooperative equi-librium now scores better than the cooperative equilibrium. However, with this level of damage, taxes fall with θ, which, ceteris paribus, raises the profits from not sharing information. Thus, in this case, information will not be shared, but, since the probability that just one firm will discover the technology is higher in the noncooperative equilibrium than in the cooperative equilibrium, the welfare loss from getting the information-sharing decision wrong is higher in the noncooperative equilibrium than in the cooperative equilibrium. Nevertheless, the difference is rather small and is dominated by the better performance of the noncooperative equilib-rium in terms of R&D decision making. Thus the overall conclusion from table 10.2 is that RJVs outperform noncooperative equilibria when dam-age is low, but that this is reversed when the damage is high.

Appendix tables 10A.6–10A.8 report the results of similar exercises for various parameter values. Thus in table 10A.6, we report the outcomes when there is more rapid diminishing to returns to R&D and $\beta = 0.5$. Now both labs are always kept open. Nevertheless, the conclusions of table 10.2 are broadly confirmed—RJVs do better when damage is small be-cause overinvestment in the noncooperative equilibrium dominates the un-derinvestment of the cooperative equilibrium. However, the positions are reversed as damage increases.

Appendix table 10A.7 reports the outcomes when $\beta = 0.2$, but now λ is raised to 0.75. Now information is not shared when damage takes its intermediate value, and in this case the information-sharing loss from the RJV exceeds that in the noncooperative equilibrium. Other than that, the results broadly confirm the findings of table 10.2.

Finally, appendix table 10A.8 reports the outcomes when $\beta = 0.2$, but now $\lambda = 0.25$. Now information is always shared and both labs are always kept open. The performance of the RJV gets steadily worse as damage increases and so too does the extent of underinvestment, whereas the per-formance of the noncooperative equilibrium steadily improves as the ex-tent of overinvestment is reduced.

10.4   Conclusion

We have examined the performance of environmental RJVs using a model in which information sharing is endogenous, firms can choose to license their technology, and the number of labs that are operated is endogenous. Welfare losses can arise through (1) the failure to make the right information-sharing decisions; (2) the failure to operate the right number of labs, leading to excessive duplication of effort; and (3) getting the R&D decisions wrong through under- or overinvestment. We have shown that an analysis of this issue requires a careful discussion of the informational and commitment powers of the government in setting its environmental policy.

In the context of a very simple model in which there is very limited commitment, we have shown the following. First, as in Ulph and Katsoulacos (1998), RJVs will share information under precisely the same circumstances as in the noncooperative equilibrium. Second, information sharing is more likely when damage and hence taxes are low. Third, when damage is low, RJVs perform better in R&D decision making than does the noncooperative equilibrium for two reasons: since the returns to R&D are high they are more likely to avoid the risk of needless duplication and the loss from underinvestment by the RJV is smaller than the loss from overinvestment in the noncooperative equilibrium. Fourth, as damage rises, the underinvestment by the RJV increases and the overinvestment in the noncooperative equilibrium falls. A tentative conclusion is that RJVs do better than noncooperative arrangements when environmental damage is low, but worse when environmental damage is high.

Appendix

Table 10A.1   **Computed Values of Tax Rate, Profits, and Welfare**
$d = 1, \bar{\theta} = 1, \tilde{\theta} = 1.2, \tilde{\lambda} = 1$

θ	t^{10}	Σ^{10}	W^{10}
0	25	1,250	2,500
0.1	27.4	1,130.1	2,384.9
0.2	29.6	1,027.4	2,291.5
0.3	31.4	950.0	2,220.8
0.4	32.8	902.2	2,172.7
0.5	33.8	884.3	2,145.3
0.6	34.3	893.8	2,136.1
0.7	34.5	926.3	2,142.1
0.8	34.3	976.7	2,160.3
0.9	34.0	1,039.8	2,187.9
1	33.3	1,111.1	2,222.2

Note: Information shared is $\forall \theta, 0 \leq \theta \leq \bar{\theta} \Rightarrow \tilde{\lambda} = 1$.

Table 10A.2   **Computed Values of Tax Rate, Profits, and Welfare**
$d = 1.497, \bar{\theta} = 0.7, \check{\theta} = 0.6, \tilde{\lambda} = 0.86$

θ	t^{10}	Σ^{10}	W^{10}
0	39.9	801.9	2,002.4
0.07	41.0	740.9	1,926.4
0.14	41.9	694.1	1,866.4
0.21	42.6	663.3	1,822.3
0.28	43.0	649.1	1,793.1
0.35	43.3	651.4	1,777.8
0.42	43.3	668.8	1,774.9
0.49	43.1	699.5	1,782.7
0.56	42.8	741.6	1,999.7
0.63	42.3	792.7	1,824.1
0.7	41.7	850.7	1,854.6

Note: By interpolation, $\check{\theta} = 0.641 \Rightarrow \check{\lambda} = 0.916$.

Table 10A.3   **Computed Values of Tax Rate, Profits, and Welfare**
$d = 2.142, \bar{\theta} = 0.5, \check{\theta} = 0.365, \tilde{\lambda} = 0.73$

θ	t^{10}	Σ^{10}	W^{10}
0	52.3	506.5	1,591.3
0.05	52.5	476.9	1,542.7
0.1	52.7	457.4	1,505.0
0.15	52.7	448.1	1,477.8
0.2	52.6	448.8	1,460.4
0.25	52.5	459.0	1,452.2
0.3	52.2	478.1	1,452.2
0.35	51.8	505.3	1,459.7
0.4	51.3	539.5	1,473.8
0.45	50.7	579.7	1,493.5
0.5	50.0	625.1	1,518.1

Note: By interpolation, $\check{\theta} = 0.352 \Rightarrow \check{\lambda} = 0.704$.

Table 10A.4   **Computed Values of Tax Rate, Profits, and Welfare**
$d = 3.585, \bar{\theta} = 0.3, \check{\theta} = 0.194, \tilde{\lambda} = 0.65$

θ	t^{10}	Σ^{10}	W^{10}
0	67.3	237.8	1,090.5
0.03	67.0	229.3	1,067.7
0.06	67.0	225.8	1,050.2
0.09	66.3	227.2	1,037.7
0.12	65.9	233.2	1,030.1
0.15	65.4	243.6	1,027.1
0.18	64.9	258.2	1,028.2
0.21	64.4	276.6	1,033.3
0.24	63.8	298.5	1,042.0
0.27	63.2	323.6	1,054.0
0.3	62.5	351.6	1,069.0

Note: By interpolation, $\check{\theta} = 0.133 \Rightarrow \check{\lambda} = 0.443$.

Table 10A.5   **Computed Values of Tax Rate, Profits, and Welfare**
$$d = 10.405, \bar{\theta} = 0.1, \check{\theta} = 0.06, \tilde{\lambda} = 0.6$$

θ	t^{10}	Σ^{10}	W^{10}
0	86.8	38.4	438.4
0.01	86.5	38.1	434.8
0.02	86.2	38.7	432.1
0.03	85.9	40.0	430.1
0.04	85.5	42.1	429.0
0.05	85.2	44.9	428.6
0.06	84.8	48.4	428.9
0.07	84.5	52.7	430.0
0.08	84.1	57.6	431.8
0.09	83.7	63.2	434.2
0.1	83.3	69.4	437.3

Note: By interpolation, $\check{\theta} = 0.015 \Rightarrow \check{\lambda} = 0.15$.

Table 10A.6   **Comparison of Cooperative and Noncooperative Equilibria to Social Optimum**
$$\beta = 0.5, \lambda = 0.5$$

d	1	2.142	3.585
$\hat{p}_1$	0.609	0.47	0.34
$\hat{p}_2$	0.609	0.47	0.34
p_1^C	0.573	0.358	0.197
p_2^C	0.573	0.358	0.197
L^C	0.13	1.07	3.73
L_R^C	0.13	1.07	1.61
L_I^C	0	0	2.11
$p_1^N = p_2^N$	0.665	0.48	0.31
L^N	0.33	0.0	2.93
L_R^N	0.33	0.0	0.07
L_I^N	0	0	2.85

Table 10A.7   **Comparison of Cooperative and Noncooperative Equilibria to Social Optimum**
$$\beta = 0.2, \lambda = 0.75$$

d	1	2.142	3.585
$\hat{p}_1$	0.996	0.708	0.595
$\hat{p}_2$	0	0.708	0.595
p_1^C	0.99	0.593	0.425
p_2^C	0	0.593	0.425
L^C	0.04	5.11	4.5
L_R^C	0.04	0.99	1.84
L_I^C	0	4.18	2.66
$p_1^N = p_2^N$	0.942	0.834	0.658
L^N	3.19	3.78	2.73
L_R^N	3.19	1.39	0.28
L_I^N	0	2.39	2.44

Table 10A.8   Comparison of Cooperative and Noncooperative Equilibria to Social Optimum
$\beta = 0.2, \lambda = 0.25$

d	1	2.142	3.585
$\hat{p}_1$	0.694	0.552	0.41
$\hat{p}_2$	0.694	0.552	0.41
p_1^C	0.676	0.448	0.248
p_2^C	0.676	0.448	0.248
L^C	0.01	0.39	0.85
L_R^C	0.01	0.39	0.85
L_I^C	0	0	0
$p_1^N = p_2^N$	0.779	0.599	0.395
L^N	0.34	0.08	0.00
L_R^N	0.34	0.08	0.00
L_I^N	0	0	0

References

d'Aspremont, C., and A. Jacquemin. 1988. Cooperative and non-cooperative R&D in a duopoly with spillovers. *American Economic Review* 78:1133–37.

Hinloopen, J. 1997. *Research and development, product differentiation and robust estimation.* Ph.D. diss., European University Institute, Florence.

Katsoulacos, Y., and D. Ulph. 1998a. Endogenous spillovers and the performance of research joint ventures. *Journal of Industrial Economics* 46:333–57.

———. 1998b. Innovation spillovers and technology policy. *Annales d'Economies et Statistiques* 49/50:589–607.

Katsoulacos, Y., and A. Xepapadeas. 1996. Environmental innovation, spillovers and optimal policy rules. In *Environmental policy and market structure,* ed. C. Carraro, Y. Katsoulacos, and A. Xepapadeas, 143–50. Dordrecht: Kluwer.

Petrakis, E., and J. Poyago-Theotoky. 1997. Environmental impact of technology policy: R&D subsidies versus R&D cooperation. University of Nottingham Discussion Paper no. 97/16.

Poyago-Theotoky, J., ed. 1997. *Competition, cooperation, research and development.* New York: Macmillan.

Scott, J. T. 1996. Environmental research joint ventures among manufacturers. *Review of Industrial Organisation* 11:655–79.

Ulph, A. 1996a. Environmental policy and international trade when governments and producers act strategically. *Journal of Environmental Economics and Management* 30:265–81.

———. 1996b. Strategic environmental policy. In *Environmental policy and market structure,* ed. C. Carraro, Y. Katsoulacos, and A. Xepapadeas, 97–127. Dordrecht: Kluwer.

Ulph, A., and D. Ulph. 1996. Trade, strategic innovation and strategic environmental policy—A general analysis. In *Environmental policy and market structure,* ed. C. Carraro, Y. Katsoulacos, and A. Xepapadeas, 181–208. Dordrecht: Kluwer.

Ulph, D. 1994. Strategic innovation and strategic environmental policy. In *Trade innovation, environment,* ed. C. Carraro, 205–28. Dordrecht: Kluwer.

———. 1997. Environmental policy and technological innovation. In *New direc-*

tions in the economic theory of the environment, ed. C. Carraro and D. Siniscalco, 43–68. Cambridge: Cambridge University Press.

Ulph, D., and Y. Katsoulacos. 1998. Endogenous spillovers and the welfare performance of research joint ventures. University College London. Mimeo.

Comment   Jerome Rothenberg

The authors examine the performance of environmental RJVs in an extremely rich analytical context, drawing together several important issues—environmental damage, government taxes, industry competitiveness, information sharing, and government policy toward cooperative research and development (R&D) behavior of environmental, growth, and regulatory economics. They do this with clarity and skill. Theirs is a very illuminating treatment.

They are rewarded with most interesting results. It is important that in their complex set of relationships there is no general answer to the question of whether environmental policy abets or hinders the performance of RJVs. What transpires is that for some combinations of parameter values the effort is positive, for others negative. But the results have intuitive thrust. Thus, we have an opportunity to see how the different portions of the model generate trade-offs and how these trade-offs are mediated by other portions of the model. The interconnectedness of parts becomes, in effect, the centerpiece of the exercise. And the reader's exercise of intuition in the system is most useful for understanding what such a system is really like. In particular, the disparate forces acting on social welfare become appreciated more fully.

The overall system is quite complex due to the variety of market failures that are embedded, and against which the performance of environmental policy and RJVs is evaluated. To render such a system reasonably tractable, a number of stringently simplifying assumptions are adopted. These make it possible to have enough transparency to understand intuitively the great variety of complex outcome scenarios. Greater complexity would risk this transparency, but I should like to venture to ruminate about some analytic avenues that have been precluded that would have permitted the inclusion of what I believe to be important dimensions of the real world. My remaining comments will sketch these dimensions and preliminary rudiments of how they might be incorporated in the present model.

The areas I will deal with are (1) the dimensionality of R&D, (2) the multiperiod gestation of R&D effects, and (3) various significant initial and achieved differences among firms.

Jerome Rothenberg is professor emeritus of economics at the Massachusetts Institute of Technology.

The Dimensionality of R&D

Level and "Angle"

The paper treats all R&D payoffs as improved emissions controls that are perfect substitutes for one another. Firms can engage in different levels of R&D spending, but there are only two kinds of payoffs: success or failure. The spending levels influence only the probability of success, but not its degree or character. All successes result from the same sort of new information and are perfect substitutes for one another.

An important fact about the real world is that in the face of the uncertainty represented by the problem of changing current technology so as to accomplish particular new goals, different researchers and research organizations will often try different approaches to the problem (the history of science is replete with this phenomenon). Moreover, especially where competitive market-oriented efforts are involved, these different routes will be deliberately kept secret from one another, notably in prepatent stages. So they are likely to have different kinds of failures and successes. The breakthroughs will generally be different. The resulting emissions improvements will therefore not usually be perfect substitutes for one another, either in amount or in kind. They will be variously imperfect substitutes. Over any one stretch of multifirm R&D behavior, there is likely to be a spectrum of emissions improvements—although all will be comparable in terms of the achieved degree of emissions decrease per unit of market product.

To facilitate the analysis, I propose to simplify this situation drastically. I suppose that at any point in time there exists a consensus about the identity of the most conventional breakthrough route. Let this route be the point of departure for classifying all other possible kinds of breakthrough. The others are less obvious, riskier, and more controversial, both with regard to the probability of any success and the probability distribution of extent of possible successes. Assume, then, that the most conventional, least controversial direction represents the R&D route with given mean and least variance of net gains (value of gross gains less cost of the R&D effort). Assume also that the other routes can be accorded at least an approximate net gain mean and variance, both mean and variance increasing with degree of unconventionality.

Let each R&D project be defined in the two dimensions of size—level of spending, r, and degree of unconventionality, which I call the research angle, α:

$$(1) \qquad\qquad R_i = (r_i, \alpha_i),$$

where (r_i, α_i) defines a probability distribution of emissions-decrease outcomes from zero upward.

At any given time there exists the scientific know-how for formulating

$n + 1$ different research directions, forming a boundary set.[1] Each (r_x, α_x) defines a different probability distribution of emissions decreases with the property that both mean, $M_{\Delta\mu}$, and variance, $\mu^2_{\Delta\mu}$, of outcome increase with research angle, the degree of unconventionality. These distributions are not simply multiples of r_x, since the different research angles are likely to have different scale economies.

Thus, associated with each choice of (r, α) is a stochastic outcome, Δe, reflecting the particular distribution relevant to that project—where Δe is the emissions decrease achieved by that project. The larger α_x, the greater is the probability that actual Δe will be zero, but also the greater is the expected value of gain. For the analysis of firm decision making, these physical outcomes can be used both where purely regulatory goals have to be met—in terms of minimizing research and penalty costs for achieving regulatory standards—and where emissions decreases have unit monetary value—in terms of maximizing net financial gains from research.

In the present context, we apply this two-dimensional R&D approach by supposing that there are several (m) firms in this industry and that they have different risk preferences in the relevant period. Then they will choose different (r_x, α_x), with both the level and angle different. As a result, these projects, (r_1, α_1), (r_2, α_x), . . . , (r_m, α_x), will result in different levels of achieved emissions decreases. Moreover, since R&D outcomes are all stochastic, even an identical set of projects by all firms would result, in the absence of information sharing, in a particular period of real time, in different achieved degrees of emissions decreases.

Substitutability of R&D Outcomes

Increasing the dimensionality of R&D has important implications for the nature of the substitutability of R&D outcomes. R&D always leads to stochastic outcomes. But these are influenced by the choice of both r and α—r as a scale factor and α as the particular direction and strategy by which it is hoped that research breakthroughs occur. Choice of α is the choice of the channel by which new knowledge is sought. In effect, it is the kinds of research issues that it is hoped will prove productive. Two firms selecting the same α, however different their research scales (r_1, r_2), will be on the same route, manipulating and discovering new knowledge of the same sort. Whatever research results transpire will be quite similar—even though the firm with smaller scale is not likely to have gone as far in discovery. Thus, the efforts of the smaller research project will carry little of interest to the larger project; the interest of the smaller in the larger will be primarily to find out how much further one can probably get to by continuing in the same direction. Thus, their results are highly substitu-

---

1. The set of directions that, for each variance, has the highest mean R&D outcome. Given the heterogeneous risk preferences, this represents the only set of inherently nondominated directions.

tive—each having little to learn from the other in that part of their quest with which both have experience.

Say, instead, that the two firms have chosen the same r and very different α, (α_1, α_2). Now they are setting out on very different tasks, dealing with different analytical issues and different strategies—although both aimed at the same final achievements, new technology for emissions reductions. Whatever results they stochastically achieve, they will have come by via different kinds of new knowledge. As when viewing statuary, you learn more about a statue by walking around it than by staring at it from a single perspective. Two R&D ventures with these different (α_1, α_2) will teach a lot more about the natural systems involved than two R&D ventures with the same α, $\alpha_1 = \alpha_2$. The former outcomes will not be substitutive, but complementary.

Generalizing, then, firms can influence the degree of substitutability or complementarity of their R&D results with those of other firms by their choice of similar or different relative to the others. Moreover, the degrees, as well as the sign, of outcome relatedness can vary substantially depending on the specific differences in chosen research direction.

We discuss later why degree and sign of relatedness are quite important for the issues taken up in this paper—notably, sharing of information and/or coordinating research efforts across firms.

R&D as a Multiperiod Process

The second major issue concerns the temporal dynamics of R&D. A firm's R&D commitment does not involve a single time-span effort after which the R&D is either attained or fails. The character of an R&D effort is often set by the nature of environmental regulation. Often the regulation specifies "ultimate" emissions control goals not presently attainable, along with a timetable of steps for gradually meeting them. Experience leads to the expectation that it will take a number of breakthroughs, not one, to achieve the goals. A particular R&D strategy (α) is a game plan for passing through a sequence of research efforts to reach the ultimate goals. Each R&D stage is expected to advance the search, but not end it.

Accordingly, R&D projects have intermediate effects—new understanding about technological processes and newly revised prospects for success. Future success depends on this new information and these successes so far. Choice of α at the outset is thus a strategy for generating an intermediate new understanding that will be productive for additional gains in the further future. Moreover, this further progress may depend on the willingness to combine this new information with that of intermediate new information generated by R&D in other firms. The potential research productivity of such combining depends on the degree of complementarity of such new intermediate understanding. R&D projects using nearly similar α will produce new information that is very much alike, and therefore not very productive for future progress. But projects with very dissimilar αs, exploring quite different aspects of the problem, are likely to generate intermediate results that are not only different but mutually enlightening, and thus with a high potential for productivity in combined efforts.

Thus, the impetus for either sharing research results along the way, or actually combining future efforts jointly, is influenced importantly by the choice of α at the outset of the R&D commitment. The key is complementarity of results—that is, nonsubstitutability. Indeed, degree of complementarity is appropriately defined in terms of extent of potential fruitfulness of combined, or shared, R&D efforts. Complementarity *now* makes *future* joint projects, or sharing of results (leasing), potentially attractive.

Therefore, the incentives for information transfer and/or RJV, and thus for an optimal new technology transfer, are directly related to the deliberate strategic choice of α, not merely of r. This relevance of α is expressly in the context of intertemporal, multiperiod innovation strategy.

The degree of interfirm informedness is important in this. Two extreme cases can be distinguished. In the first, no firm knows the α choice, and intermediate outcomes, of any other firm. Generally, each will know about the others only after new processes have been patented. This is plausible. The relevant information is highly proprietary, and firms strive to keep it secret. At the other extreme, every firm knows the α choices and intermediate results of all the others. The first case seems a much closer approximation to the truth. In such a situation, period-by-period R&D behavior by itself will have an important element of strategy about the attractions of future jointness in continuing efforts. As an example, the persistent choice of α near α_0 (a low-risk strategy) represents playing it safe. Such a firm will not be signaling a desire to initiate joint efforts with the other firms having very different α from theirs. It might subsequently be passively receptive to sharing or coordination with significantly nonconsensual efforts differing from its own markedly, but its low-risk propensities suggest that this would represent a gambling endeavor foreign to those propensities. On the other hand, a firm choosing a highly nonconsensual direction would probably be risk-preferring. It is likely to be signaling willingness both to share information—or even more, to collaborate with other firms selecting high but very different values of α—and also to accept such initiatives from other risk-preferring firms.

The impact of this strategic variable on leasing or RJV requires looking more richly at the R&D situation. In particular, we must know something about the initial similarities or differences between the two firms in their emissions rates per unit output.

Case 1. The two firms (F1 and F2) begin with very different emissions rates, but choose the same α. The firm with less prior R&D success (say F2) could gain here from a catch-up strategy by sharing F1's information. But F1 would gain little from such sharing, since F2 has little of interest for it. So only leasing by F1 to F2 is likely—and only on tough profit-maximizing terms.

Case 2. The initial difference in emissions rates for F1 and F2 is small, but the difference of the chosen α is large. The big strategy difference here suggests highly complementary R&D output. On the basis of pre-R&D technology, this complementarity would not make sharing or coordination attractive to either firm at the outset of R&D. But it makes future interchange probable. Since R&D is a multistage process, the large difference in α suggests that first-stage outcomes will be quite different for the two firms. In such a situation, their complementarity will be favorable for interchange.

The first two cases focus on firm differences in α and emissions rates. But the absolute strategy direction of each firm and the size of first-stage R&D outcomes matter as well.

Case 3. F1 has a high α, that is, is highly unconventional. It represents a high-risk strategy, carrying possibilities of large gains. Since much lower α strategies are likely to be adopted by other firms, F1 will have R&D outcomes that are highly complementary with those of many other firms. If its first-stage successes are meager, it will be anxious to share with more successful firms; but such interchange may well have little attraction for those firms. On the other hand, if it has important first-stage successes, its unique types of new information will be very attractive to other firms, successful or not. Moreover, the prospect of really major success from interchange with other complementary firms will make this firm seek sharing. So high risk taking may well signal a predisposition to share information.

Case 4. F1 and F2 adopt strategies that make their R&D results sufficiently complementary. If their R&D outcomes in the first stage are poor, they are likely to have little interest in sharing. Similarly, if both have nearly equal success in the first stage, their incentive to share is not likely to be great. But a large disparity in successful outcomes may well make the less-successful firm eager to share. If the strategy of F1, the less-successful firm, is one of high risk, the successes may well induce F2, the more-successful firm, to share, on the chance that their complementarity, fed by the moderate success of F1, could generate a large overall success due to the large stakes being pursued by F2.

These speculations suggest that in a multistage conception of the R&D process, initial and midcourse behavior depends on stochastic outcomes during the process, in the context of the strategic directions chosen by the various firms involved.

Asymmetries among Firms

Real-world firms have differences that are important to the issues dealt with in the paper. We have already suggested that differences in risk preferences will lead them to choose differently with respect to R&D, not only in whether or not they engage in R&D, individually or cooperatively, but also in how they invest in R&D, with regard to both level and angle. These differences create distinctive trajectories over time. Indeed, because the R&D process is inherently stochastic, they are likely to have different temporal trajectories even if they make identical choices at each time, although these differences will be smaller and less systematic than due to the former source.

But there are other systematic sources of difference as well: initial conditions. At any moment of time, firms in the same industry will generate different rates of emissions per unit of output due to historical differences in technology, in capital stock, in capital-output ratios, and in different vintages. Moreover, these emissions generate different unit damages. They have different locations, and these influence the size of damages because of differences in potential victim populations, density, and the assimilative capacity of the relevant local environments.

If environmental policy is used to maximize social welfare, then (1) firms with equal social-damage functions will be taxed (or otherwise regulated) to generate the same marginal trade-off between environmental impairment and net economic gain from market activities, and (2) firms with different social-damage functions will be taxed (or otherwise regulated) for different environmental and market trade-offs. In the presence of the two kinds of firm differences mentioned, both cases lead to different firms facing different required environmental performance for optimal performance compliance. If firms with similar damage functions are faced with regulation with the same final emissions-rate goal, then some will start with less far to go and others with farther to go to reach it. For some, adjustment to compliance will be easier, for others harder. Similarly, firms with different damage functions will face different degrees of difficulty in ultimate compliance.

These differences are pervasive in the real world and important. Indeed, the popularity of emissions trading in environmental policy is an explicit assertion of the importance of such interfirm differences; in their absence the policy would be meaningless.

The paper being reviewed minimizes such differences for purposes of tractability, but elaborations of such an analysis to bring the examination closer to the real world seem to warrant a compact treatment of these differences. Such a treatment is especially compatible with a multiperiod conception of R&D behavior.

In any stage of preparing for ultimate compliance, therefore, firms will confront one another with different requirements and opportunities, stemming from both their initial advantages and disadvantages, and the differential progress they have already made up to the present stage (and by different routes) via their R&D strategies and stochastic outcomes. This complex of differences certainly affects what decisions they will make about the next-stage strategies and actions. Symmetry will be a less relevant analytical predictor here, and bargaining power will differ from firm to firm. In such a context, leasing and trading schemes may become more important and RJVs less so, because the former have a more impersonal way of monetizing differences, while the latter must achieve monetization of these differences by personal negotiation and are embedded in the very structure of the more integral cooperation.

Competition and Competitive Advantage

The degree of competitiveness in an industry, as influence on and influenced by the various factors concerning R&D, is treated as having an influence on social welfare. It would be useful to bring together the impacts on the meaning and measure of competitive market advantage of the kinds of modifications I have been discussing here.

Given the multiperiod R&D framework, let us assume that full gestation to ultimate compliance from the outset involves a long time period. Then the regulator is likely to set stage-by-stage deadlines for achieving interim standards. If so, most R&D improvements would be implemented nearly as soon as they emerge, instead of being saved up until they could lend their insights and R&D experience to find ultimate R&D success on the most efficient path fed by trial and error. Likewise, new technology leased from others would be implemented steadily as it is acquired, with acquisition not awaiting the final perfection of outcomes in the particular direction of change.

Then the cost of acquiring these improvements would be annualized into firms' operating costs, and the resulting operating cost structures would affect the competitiveness of the various firms. Our discussion lends itself to the strong expectation that firms will differ substantially in these new technology costs, deriving from differences in firms' initial conditions, differential regulatory treatment, and differential R&D strategy and success. So environmental policy, in such a context, can lead to nontrivial changes in interfirm market power, with possibly further influence on market structure.

How these changes will play out against the outcomes in other dimensions of the complex system we have been discussing is far more likely to evade precise delineation than in our authors' worthy, far tidier model. But in what must be at least a highly speculative judgment, I must concur with the authors that the most useful attitude to adopt at the present about RJVs is skepticism.

Contributors

Edward B. Barbier
Department of Economics and
  Finance
PO Box 3985
University of Wyoming
Laramie, WY 82071

Randy A. Becker
Center for Economic Studies
U.S. Bureau of the Census
4700 Silver Hill Road, Rm 211-WPll
Washington, DC 20233

Stefania Borghini
Fondazione Eni Enrico Mattei
c.so Magenta 63
Milano 20123 Italy

A. Lans Bovenberg
Faculteit der Economische
  Wetenschappen
(Faculty of Economics and Business
  Administration)
P.O. Box 90153
Tilburg NL-5000 LE The Netherlands

David F. Bradford
Woodrow Wilson School
Princeton University
Princeton, NJ 08544

Dallas Burtraw
Resources for the Future
1616 P Street NW
Washington, DC 20036

Carlo Carraro
Department of Economics
University of Venice
San Giobbe, 873
Venezia 30121 Italy

Marcella Fantini
Fondazione Eni Enrico Mattei
c.so Magenta 63
Milano 20123 Italy

Don Fullerton
Department of Economics
University of Texas
Austin, TX 78712

Lawrence H. Goulder
Department of Economics
Landau Economics Bldg.
Stanford University
Stanford, CA 94305

Gilbert H. A. van Hagen
CPB Netherlands Bureau for
  Economic Policy Analysis
P.O. Box 80510
The Hague NL-2508 GM, The
  Netherlands

Kevin Hassett
American Enterprise Institute
1150 17th Street, NW
Washington, DC 20036

J. Vernon Henderson
Department of Economics
Box B
Brown University
Providence, RI 02912

Inkee Hong
Department of Economics
University of Texas
Austin, TX 78712

Raghbendra Jha
Indira Gandhi Institute of
  Development Research
General Vaidya Marg, Goregaon (E)
Bombay 400 065 India

Yannis Katsoulacos
Department of Economics
Athens University of Economics and
  Business
76 Patission
Athens 104-34 Greece

Charles D. Kolstad
Department of Economics
University of California,
  Santa Barbara
Santa Barbara, CA 93110

Arik Levinson
Department of Economics
Georgetown University
Washington, DC 20057

Gilbert E. Metcalf
Department of Economics
Tufts University
Medford, MA 02155

Ruud A. de Mooij
European Comparative Analysis Unit
CPB Netherlands Bureau for
  Economic Policy Analysis
POB 80510
The Hague 2508 GM The Netherlands

Federica Ranghieri
Fondazione Eni Enrico Mattei
c.so Magenta 63
Milano 20123 Italy

Michael Rauscher
Lehrstuhl Volkswertschaftslehre
  Außenwirtschaft
Department of Economics
Rostock University
Parkstr. 6
Rostock D-18051 Germany

Jerome Rothenberg
Department of Economics
Massachusetts Institute of Technology
Cambridge, MA 02142

Domenico Siniscalco
Executive Director
Fondazione Eni Enrico Mattei
c.so Magenta, 63
Milano 20123 Italy

Sjak Smulders
Center for Economic Research
University of Tilburg
POB 90153
Tilburg 5000 LE The Netherlands

Alistair Ulph
University of Southampton
Department of Economics
Highfield
Southampton S09 5NH England

David Ulph
Department of Economics
University College London
Gower Street
London WC1E 6BT England

Herman R. J. Vollebergh
Erasmus University Rotterdam
H 7-23 FEW, POB 1738
Rotterdam 3000 DR The Netherlands

John Whalley
Department of Economics
Social Science Centre
University of Western Ontario
London, Ontario N6A 5C2 Canada

Anastasios Xepapadeas
University of Crete
Department of Economics
Perivolia 74100 Rethymno
Crete, Greece

Aart de Zeeuw
Department of Economics CentER
Tilburg University
PO Box 90153
Tilburg 5000 LE The Netherlands

Author Index

Subject Index